QUATERNARY OF SOUTH AMERICA AND ANTARCTIC PENINSULA

Edited by
JORGE RABASSA & MÓNICA SALEMME

Centro Austral de Investigaciones Científicas and Universidad Nacional de la Patagonia, Ushuaia, Tierra del Fuego

VOLUME 11 (1995)

A.A.BALKEMA / ROTTERDAM / BROOKFIELD / 1998

Published by
A.A. Balkema, P.O. Box 1675, 3000 BR Rotterdam, Netherlands (Fax: +31.10.4135947)
A.A. Balkema Publishers, Old Post Road, Brookfield, VT 05036-9704 USA (Fax: 802.276.3837)

ISSN 0168-6305
ISBN 90 5410 453 8

QUATERNARY OF SOUTH AMERICA
AND ANTARCTIC PENINSULA

VOLUME 11

Morenelaphus sp. (after E. P. Tonni 1995, *Mamíferos extintos de la Provincia de Buenos Aires*. La Plata, CIC, 43 pp.)

Contents

The presence of *Neochoerus* Hay, 1926 (Rodentia, Hydrochoeridae) in Pleistocenic sediments of Southwestern Buenos Aires Province, Argentina

CECILIA M. DESCHAMPS
Comisión de Investigaciones Científicas (CIC) & Departamento Paleontología Vertebrados, Museo de La Plata, La Plata, Argentina.

ABSTRACT: An almost complete skull of *Neochoerus* sp. (Rodentia, Hydrochoeridae) from Bajo San José (lat. 38°34'S, southwestern Buenos Aires province, Argentina) is described. The bearing unit is considered of Ensenadan Land-Mammal Age (Early-Middle Pleistocene). This is the record of highest latitude of the genus, and of the Family Hydrochoeridae since the end of the Pliocene. Bajo San José is placed in the border between two zoogeographic subregions. The presence of *Neochoerus* at this latitude can be taken as an example of the historical dynamism of the Pampasic fauna during the Quaternary, which was signed by the strong climatic changes characteristic of the Pleistocene. The accompanying fauna is conformed by genera or species inhabiting today under either more humid and temperate conditions or more arid ones than those of the area of Bajo San José. The former may suggest more arid conditions than today for this area, and the latter, may either be relict forms of more humid and warmer prior conditions, or indicate environmental changes towards higher temperature and humidity which favoured their southern dispersion along rivers and streams.

RESUMEN: Se describe un cráneo casi completo de *Neochoerus* sp. (Rodentia, Hydrochoeridae), hallado en Bajo San José (lat. 38°34'S, sudoeste de la provincia de Buenos Aires, Argentina). La unidad portadora es considerada de Edad Ensenadense (Pleistoceno temprano-medio). Este es el registro más austral del género y de la Familia Hydrochoeridae desde el fin del Plioceno. Bajo San José está ubicado en el límite entre dos subregiones zoogeográficas. La presencia de *Neochoerus* en esta latitud puede tomarse como un ejemplo del dinamismo histórico de la fauna pampásica durante el Cuaternario, caracterizado por los grandes cambios climáticos del Pleistoceno. La fauna acompañante está compuesta por géneros o especies que habitan en la actualidad zonas más áridas o más húmedas y templadas que las del área de Bajo San José. Las primeras pueden sugerir condiciones más áridas que las actuales de esta zona, y las últimas, pueden ser tanto formas relictuales de condiciones más húmedas y cálidas previas, como indicar cambios ambientales hacia condiciones de mayor temperatura y humedad que habrían favorecido su dispersión hacia el sur a lo largo de ríos o arroyos.

1 INTRODUCTION

New remains of *Neochoerus* are described in this paper, coming from Bajo San José (Coronel Pringles county, Southwestern Buenos Aires province, Argentina). A preliminary list of the fossil fauna found in this locality was published by Deschamps & Borromei (1992); fish remains were studied by Cione & López Arbarello (1993), and the sigmodontine rodents by Pardiñas & Deschamps (1996). The fossil-bearing unit outcrops at both margins of the Sauce Grande river, being the best site a gravel quarry located near the bridge of the National Route 51 over this river (lat. 38°34' S, long. 61°41' W), 51 km northeast of the city of Bahía Blanca (Fig. 1). The geological setting has been discussed in Borromei

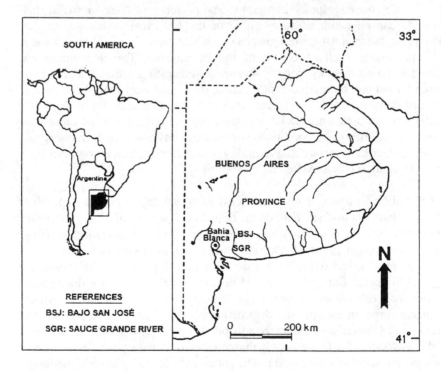

Figure 1. Location map.

(1989) and Deschamps & Borromei (1992), in whose papers the entire sequence was attributed to a Pleistocene deposit originated by a gravelly braided stream system. Deschamps (1995) assigned this deposit to the Ensenadan (Early-Middle Pleistocene) Land-Mammal Age (Pascual et al. 1965, Cione & Tonni 1995).

2 MATERIAL AND METHODS

The fossil material herein described has been compared with skulls of *Hydrochoerus hydrochaeris* ('carpincho'): MLP 650, MLP 1406, MLP 14-IX-55-1 and MLP 18-VIII-92-16 from the Departamento de Zoología Vertebrados, 43 and 427 from the Departamento de Paleontología Vertebrados, Museo de La Plata, La Plata, Argentina, and with the following fossil specimens from Museo Argentino de Ciencias Naturales 'Bernardino Rivadavia' (MACN), Buenos Aires, of related genera: MACN: 16688, 5302, 5313, 587 and 588.

The ontogenetic stage of this specimen was estimated following Mones (1991: 96, Fig. 12) who proposed five classes of ages (Ages 0-4) taking into account the ossifying sequence of the basicranium.

3 SYSTEMATICS

Family HYDROCHOERIDAE Gray, 1825
Subfamily HYDROCHOERINAE Gray, 1825
Genus *Neochoerus* Hay, 1926
Neochoerus sp. (Fig. 2 A, B, C)
Material: UNSGH (Universidad Nacional del Sur, Geología Histórica) 645: part of the skull with right P4-M3 and left M3.
Description: It is a very well preserved skull, though without the rostrum and part of the zygomatic arches. It is a young specimen of age 2 since its

A

Figure 2. Skull of *Neochoerus* sp. UNSGH 645. a) Upper view.

3

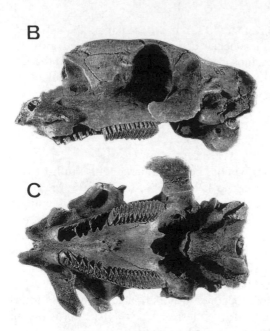

B

C

Figure 2. Continued. b) Lateral view, c) Lower view. Scale: 5 cm.

Table 1. Measurements (in mm) and distribution of *Neochoerus* sp. and related species.

	Neochoerus sp. UNSGH 645	*H. crespoi* MACN-16688	*H. dasseni* MACN-5302	*N. tarijensis* MACN 587-588	*H. hydrochaeris* (six skulls) mean	range
1	92.00	112.50	100.40	98.00	74.92	74.3-76.2
2	16.80	21.50	17.20	18.20	13.07	12.2-13.9
3	13.90	14.00	15.30		11.62	11.55-11.7
4	13.60	16.00	14.30		11.15	11.1-11.7
5	48.00	58.50	52.50	55.00	38.44	37.2-39.98
6	11.30	15.00	15.20	~13.00	9.60	9.45-10.05
7	11.60	15.60	15.00		9.88	9.48-10.04
8	12.70	13.50	15.60		10.34	10.0-11.0
9	18.00	21.00	20.60	19.00	14.88	14.0-16.15
10	14	15	13-14	14	12.6	12-13
11	33.40				31.9	29.92-35.6
12	15.40				14.61	13.08-16.5
13	77.30				61.06	52.3-64.6
14	49.40		54.00		41.3	37.9-43.7
15	74.30	67.70			48.3	46.4-51.6
16	89.00	110.00	92.00		61.48	57.8-63.2
17	115.00				86.98	82.0-90.0
18	102.50				82.54	79.8-85.8
19	IV to VI	No	I to IV	III and IV	No	
20	97.00	120.00	110.00		76.51	71.8-82.5
21	53.00				38.84	37.8-39.8
22	48.30	58.50	62.00		41.36	32.1-45.1

4

Table 1. Continued.

	Neochoerus sp. UNSGH 645	*H. crespoi* MACN-16688	*H. dasseni* MACN-5302	*N. tarijensis* MACN 587-588	*H. hydrochaeris* (six skulls) mean	range
23	156.50		180.50		126.0	122.6-128.5
24	76.50		84.60		62.38	58.4-68.1
25	14.70				10.46	8.7-11.7
26	170.00				132.58	120.2-138.0
27	72.30				63.5	62.5-65.1
28	41.00	42.00	57.30		37.23	32.4-39.7
29	72.90		76.00		57.75	52.5-61.3
30	106.50	132.00			67.51	63.8-72.6
31	Early-Mid. Pleistocene	L. Pliocene-L. Pleistoc.	L. Pliocene	Pleistocene	Recent	
Locality	Bajo San José (Argentina)	Chaco (Argentina)	Uquía (Argentina)	Tarija Valley (Bolivia)	B. Aires and Chaco (Argentina)	

References: 1. length of premolar-molar series, 2. AP (antero-posterior) diameter of P4, 3. AP diameter of M1, 4. AP diameter of M2, 5. AP diameter of M3, 6. T (transverse) diameter of P4, 7. T diameter of M1, 8. T diameter of M2, 9. maximal T diameter of M3, 10. number of M3 transverse plates, 11. intraorbital length, 12. postpalatine fissure width, 13. basion-palation length, 14. maximal bicondilar width, 15. maximal posterior width of nasals, 16. minimal interorbital width, 17. maximal width of anteorbitary apophysis, 18. width over anteorbital fissure, 19. external fissures in transverse plates of M3, 20. frontals length, 21. parietals width, 22. parietals length, 23. length from P4-P4 to basion, 24. width M3-M3 (outside the alveoli of plate XII), 25. minimal width of anteorbitary ramus, 26. frontals-occiput length, 27. basicranium-parietals height, 28. height of temporal crest-base of zygomatic arc, 29. height from alveoli of M3 (plate VII) to roof of the orbit, 30. width of the fronto-parietal suture, 31. stratigraphic distribution, L.: Late.

exoccipital-basioccipital and basisphenoid-presphenoid (Ages 1 and 2) sutures are ossified, but not its exoccipital-supraoccipital and basioccipital-basisphenoid ones (Ages 3 and 4 respectively). Its size is approximately 20% larger than that of the single living species and genus *Hydrochoerus hydrochaeris* (see Table 1). Consequently, it can be accepted that when adult, it could have reached the size of the species of the genus *Neochoerus*, that is, a third to half larger than the living species.

3.1 *Upper view (Figs 2A, 3A)*

The skull is stout, with a flat roof descending smoothly towards the occiput (see also Figs 2B, 3B). The frontals do not contract level with the orbits, consequently their width is almost equal from front to back. They are fused and their medial anterior part is forwardly projected between the

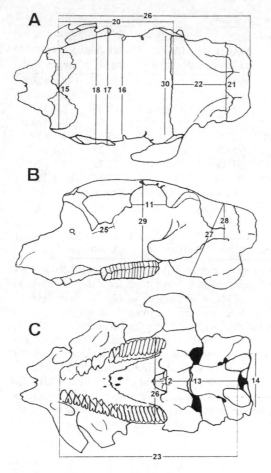

Figure 3. Diagrams of the skull of *Neochoerus* sp. UNSGH 645 showing the measurements of Table 1. Scale 5 cm.

nasals, forming an undulate nasal-frontal suture. This feature is also observed in *Hydrochoeropsis dasseni* MACN 5302 and in the type of '*Hydrochoeropsis crespoi*' (=*Neochoerus aesopi*) MACN 16.688, but not so conspicuously. In *Hydrochoerus hydrochaeris* instead, this suture is straight. The nasals extend backwards up to plate II of M2. In *Hydrochoerus hydrochaeris* and in *Hydrochoeropsis dasseni*, they extend only up to plate II of M1. The base of the rostrum is narrow as the representatives of the subfamily Hydrochoerinae (e.g. *Neochoerus tarijensis* MACN 587 and 588, *Neochoerus aesopi* MACN 16.688, or *Hydrochoerus hydrochaeris*), but unlike *Hydrochoeropsis dasseni* which is now placed within the subfamily Anatochoerinae Mones & Vucetich, 1991. The temporal crests are quite pronounced. A plane surface (both anteroposteriorly and transversely) extends at each side of them. The fronto-parietal suture is straight,

6

not slightly curved as in *H. hydrochaeris*. The parietal-occipital suture is not straight all through the plane surface between both temporal crests; instead, it deviates backwards recovering its transverse direction after crossing the temporal crests. These features resemble those of the type of '*Hydrochoeropsis crespoi*', but are quite different from *Hydrochoerus hydrochaeris*.

3.2 *Lateral view (Figs 2B, 3B)*

The lacrimal is very stout. The vertical zygomatic ramus is placed more anteriorly than in *H. hydrochaeris*, so, the lacrimal foramen is displaced backwards and can be seen clearly. The dorsal edge of the lacrimal is horizontal, forming an equilateral triangle, not descending backwards as in *H. hydrochaeris*. The orbital edge of the lacrimal is more concave and runs descending forwardly. The frontal goes down lateraly to contact with the dorsal edge of the lacrimal, more than what it does in *H. hydrochaeris*, especially in its anterior part. The lower part of the orbital cavity is forwardly extended, being its antero-posterior diameter larger than the vertical one. As in the type of '*Hydrochoeropsis crespoi*', but unlike *Hydrochoerus*, the post-orbital apophysis formed by the frontal and the squamosal is large.

3.3 *Lower view (Figs 2C, 3C)*

The mesopterygoid fossa is suboval. The pterygoid fossa is forwardly extended. It has two pairs of palatine foramina, one on the maxillaries and the other on the palatines, by the maxilo-palatal suture, which is levelled with plate I of M3. The zygomatic ramus of the maxillary is similar to that of *Hydrochoerus* and *Neochoerus*. The fossette of the superficial masseteric tendon is elliptic and occupies the whole anteroposterior extension of the base of the zygomatic ramus. This ramus merges with the rostrum between P4 and M1. In *Hydrochoeropsis dasseni* MACN 5302 and 5313 this part of the zygomatic ramus is displaced forwardly, the fossettes are subtriangular and divergent backwardly. The ventral side of the lacrimal is quite excavated. The auditory bullae are suboval, without the concavity observed in the anterior face in *H. hydrochaeris*.

3.4 *Molars (Fig. 4, Table 1)*

Both plates of P4 are separated and the posterior lobule of plate I has an extension of its labial edge towards plate II. This extension is also observed in *Neochoerus tarijensis* and *Neochoerus sulcidens*. In *Hydrochoeropsis dasseni* MACN 5302 both plates are joined, whereas in the type of

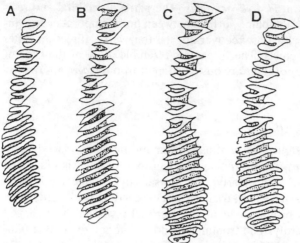

Figure 4. Upper toothrow of: A. *Neochoerus* sp. UNSGH 645 (right), B. *Neochoerus sulcidens* (right),C. '*Hydrochoeropsis crespoi*' (left, inverted), D. *Hydrochoeropsis dasseni* (right). Scale: 5 cm. B, from Mones 1991: 99; C from Kraglievich & Reig 1945: 271 and D, from Kraglievich 1930: 247.

'*Hydrochoeropsis crespoi*' and in *Hydrochoerus*, they are completely independent and lack such extension.

M1 and M2 are similar in shape with two Y shaped plates formed by the deep external fissures.

M3 has 14 plates, the last two are joined labially. Plate I is Y shaped, being the following, laminar. Plates IV to VI have slight external extraordinary fissures. The plates are closer to each other because the amount of cement between them is less than in other species (see Fig. 4).

Hydrochoeropsis dasseni has 13 plates in M3, from II to VI with more conspicuous external fissures. The genus *Neochoerus* has 13 to 17 plates in M3, exceptionally with external fissures (v. gr. *N. tarijensis*). The type of '*Hydrochoeropsis crespoi*' has 15 plates with no external fissures; *Hydrochoerus* has 10-11 to 14 plates and no external fissures either.

4 CONCLUSIONS

The skull UNSGH 645 is placed within the subfamily Hydrochoerinae because of the following cranial diagnostic characters of this subfamily: middle to large sized capybaras; M3 with a variable number of transverse plates (10 to 17), exceptionally with external fissures (Mones 1991).

Within this subfamily, the skull UNSGH 645 is assigned to the genus *Neochoerus* as it shares the following diagnostic cranial characters: size larger than *Hydrochoerus hydrochaeris*, anteorbitary bar and lacrimal more stout, cranial roof broad, flat and stout, M3 with 13 to 17 transverse plates, exceptionally with external fissures.

Mones (1991) in the revision of the family Hydrochoeridae, accepted the genus *Neochoerus* as independent from *Hydrochoerus*. He recognized five valid species: *Neochoerus dichroplax* Ahearn & Lance 1980 (Blancan Land-Mammal Age, Late Pliocene. United States: Arizona, Florida; Mexico: Guanajuato); *Neochoerus tarijensis* (Ameghino) (Ensenadan Land-Mammal Age, Early-Middle Pleistocene. Bolivia: Tarija valley); *Neochoerus sulcidens* (Lund) (Lujanian Land-Mammal Age, Late Pleistocene. Brazil: Lagoa Santa, Minas Gerais; Uruguay: Río Negro, Yapeyú creek and Colonia, Chileno creek); *Neochoerus fontanai* (Rusconi) ('Puelchense', 'Uquian' Land-Mammal Age, Late Pliocene-Early Pleistocene. Argentina: Buenos Aires, Villa Ballester) and *Neochoerus aesopi* (Leidy) (Rancholabrean Land-Mammal Age, Lujanian Land-Mammal Age, Late Pleistocene. North America: Florida, Georgia, South Carolina. Mexico: Mexico State, Tlapacoya. Guatemala. North of Nicaragua. Venezuela: Muaco, Estado Falcón. Ecuador: Santa Elena peninsula, La Carolina. Perú: Ucayali river. Argentina: La Sabana, Chaco; Carcarañá river, Santa Fé; Buenos Aires) (see Fig. 5).

This is the most complete skull of the genus. The types of the species with which it was compared are not so complete. Consequently, several characters could not be compared. Among the species of *Neochoerus*, it is more close to *N. tarijensis* (Ameghino) because of the following characters: P4 with a labially backward extension of the plate II; M3 with 14 plates, the IV, V and VI of which with external fissures; large postorbital apophyses; the nasal-frontal suture is undulate because of the interposition of an acuminate portion of the frontal on the median line; the curved parietal-occipital suture.

The fossil-bearing unit was considered as Ensenadan (conventionally regarded as Early to Middle Pleistocene, Cione & Tonni 1995) Land-Mammal age, because of the presence of *Macraucheniopsis ensenadensis* (Litopterna, Macraucheniidae), *Megatherium* cf. *M. gallardoi* (Tardigrada, Megatheriidae) (Deschamps & Borromei 1992) and *Ctenomys kraglievichi* (Rodentia, Octodontidae) (Verzi 1994, Verzi & Lezcano 1996) all of them found solely in Ensenadan-age sediments of Buenos Aires province.

Hydrochoerids arose in South America during the Late Miocene (Chasicoan Land-Mammal Age, Chasicó, southwestern Buenos Aires province) and had a relatively short dispersion towards Central America and southern North America by the end of the Pliocene and the Pleistocene. Their greatest abundance was recorded in the latest Miocene (Huayquerian Land-Mammal Age), as 50% of the known species come from the 'Mesopotamian' beds of the Paraná river banks (Mones 1991). This span of time

9

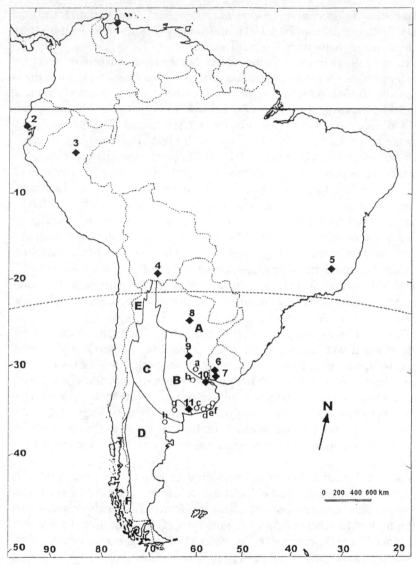

Figure 5. Distribution of the genus *Neochoerus* Hay in South America (♦): 1. Venezuela, Falcón State, 2. Ecuador, Santa Elena peninsula, 3. Perú, Ucayali river, 4. Bolivia, Tarija valley, 5. Brazil, Lagoa Santa, 6. Uruguay, Río Negro, 7. Uruguay, Colonia, 8, 9, 10, 11. Argentina. 8. Chaco, La Sábana, 9. Santa Fé, Carcarañá river, 10. Buenos Aires, 11. Bajo San José. Other localities mentioned in the text (O): a. Paraná river banks, b. Junín, c. Tres Arroyos, d. Necochea, e. Centinela del Mar, f. Mar del Plata, g. Chasicó, h. Río Negro province, Argentine Zoogeographic Dominions (sensu Ringuelet 1961). A. Subtropical, B. Pampasic, C. Central, D. Patagonian, E. Andean, F. Austral-cordilleran, A and B. Subtropical Subregion, C-F. Andean-patagonian Subregion.

10

is known as 'Edad de las planicies australes' ('Age of the Southern Plains', Pascual & Bondesio 1985) because of the development of wide plains with varying but always high hydric values. Hydrochoerids are recorded in sediments of these ages (Late Miocene) up to lat. 42° S (Rio Negro province, Northern Patagonia, Fig. 5). During the Pliocene, they also reached the southeastern Buenos Aires province, but from then on, the records are restricted only to lower latitudes. The skull of *Neochoerus* found at Bajo San José, is the record of highest latitude of the genus and of the Family Hydrochoeridae since the end of the Pliocene (Fig. 5). The only living genus of the Family, *Hydrochoerus*, inhabits tropical to temperate regions from Panamá to eastern Argentina. They are semiaquatic, confined to areas with permanent standing or running water and can occur in marshes or estuaries and along rivers and streams (Redford & Eisenberg 1992).

The location of Bajo San José is very important for the understanding of its faunal composition. According to Ringuelet (1961), Bajo San José is placed in the border between two zoogeographic Subregions (Fig. 5): the Guayanan-Brazilian Subregion with the Pampasic Dominion and the Andean-Patagonian Subregion with the Central Dominion, being very close to the Patagonian Dominion. This means that this area has been strongly affected by climatic variability, as the border marginal localities react first to environmental changes related to glacial-interglacial or stadial-interstadial cycles, and their faunistic composition show the influences of the neighbouring dominions. The Pampasic Dominion '... se puede considerar como un gigantesco ecotono entre fauna hílica o brasílica (guayano-brasileña) y fauna erémica y mesófila (andino-patagónica)' ('...may be considered as a gigantic ecotone between a hylian or Brazilian – Guayanian-Brazilian – fauna and an eremic and mesophyllous – Andean Patagonian – fauna'; Ringuelet 1961:161). It has been scarcely differentiated since the end of the Pleistocene due to the retraction of the subtropical fauna. This was caused by the strong climatic changes characteristic of the Pleistocene that can be detected throughout the palaeontologic record. More arid and/or colder climatic pulses are recognized by a northward expansion of the Patagonian fauna. Examples of this case are (see localities in Fig. 5) the records in Late Pleistocene sediments of *Lestodelphys halli* (Marsupialia, Didelphidae, 'comadrejita patagónica', Patagonian opossum) at Junín, northwestern Buenos Aires province (Odreman Rivas & Zetti 1969), *Pediolagus salinicola* (Rodentia, Caviidae, 'conejo del palo'), in Centinela del Mar, southeastern Buenos Aires province (Tonni 1981), *Lyncodon patagonicus* (Carnivora, Mustelidae, 'huroncito patagónico', Patagonian weasel), *Zaedyus pichiy* (Cingulata, Dasypodidae, 'piche') and *Dolichotis patagonum* (Rodentia, Caviidae, 'mara', Patagonian cavy) in Tres Arroyos, Buenos Aires (Fidalgo & Tonni 1981, Tonni et al. 1992). On the other hand, the record in southeastern Buenos Aires province of

11

several genera or species which inhabit at present warmer areas farther north, are indicators of a warmer and more humid pulse. Examples of this case (see localities in Figure 5) are the records in Ensenadan-age sediments of *Clyomys* (Rodentia, Echimyidae, spiny rat) in Necochea (Verzi & Vucetich 1994, Vucetich et al. 1997), *Akodon* cf. *A. cursor* (grass mouse) and *Nectomys squamipes* (Rodentia, Muridae, 'rata nadadora', water rat) in Mar del Plata, *Tapirus* (Perissodactyla, Tapiridae, 'tapir', anta) in Mar del Plata and Buenos Aires city and *Dolichotis patagonum* (Rodentia, Caviidae) in Tres Arroyos, Buenos Aires (Tonni & Cione 1994).

The accompanying fauna found at Bajo San José is conformed by genera or species inhabiting today either more humid and temperate conditions or more arid ones (Deschamps & Borromei 1992, Pardiñas & Deschamps 1996). *Lundomys molitor* (Rodentia, Muridae, marsh rat), *Hydromedusa tectifera* (Testudines, Chelidae, river turtle) and a Tayassuidae are among the former; *Akodon iniscatus* (Patagonian mouse), *Phyllotis* (Rodentia, Muridae, 'pericote', leaf-eared mouse), *Lestodelphys* (Marsupialia, Didelphidae), and *Tolypeutes* (Cingulata, Dasypodidae, 'quirquincho bola', three-banded armadillo) are among the latter. Considering the fossil fishes that have been found, *Pimelodella* (Siluriformes, Pimelodidae, catfish), *Callichthys*, and *Corydoras* (Siluriformes, Callychtyidae) live today in the Guayanian-Brazilian Subregion, but *Percichthys* (Perciformes, Percichthyidae, 'perca') is a Patagonian genus (Deschamps & Borromei 1992, Cione & López Arbarello 1995). Those taxa related to lentic or lotic enclosures, as the fishes, *Hydromedusa tectifera*, *Lundomys* and *Neochoerus* may either be relict forms of more humid and warmer conditions of a previous period, or they may indicate environmental changes towards higher temperature and humidity which favored their southern dispersion along rivers or streams.

ACKNOWLEDGEMENTS

To Dr. M.G. Vucetich for her critical review; Dr. Diego Verzi for his valuable comments; Dr. R. Pascual, whose suggestions as referee improved the manuscript; Dr. M. Quattrocchio for her constant support; Santiago Jara (Universidad Nacional del Sur) who found this material and L. Zampatti for her help with the figures. To the Departamento de Geología of the Universidad Nacional del Sur and the Comisión de Investigaciones Científicas de la Provincia de Buenos Aires (C.I.C.), for their financial support.

REFERENCES

Borromei, A.M. 1989. A braided fluvial system in pleistocenic sediments in southern Buenos Aires province, Argentina. *Quaternary of South America & Antarctic Peninsula* 6: 221-233. Rotterdam: Balkema Publishers.

Cione, A.L. & A. López Arbarello 1994. La ictiofauna de Bajo San José (Provincia de Buenos Aires; Ensenadense, Pleistoceno temprano). Su significación paleoambiental y paleobiogeográfica. *VI Congreso Argentino de Paleontología y Bioestratigrafía*, Resúmenes: 22-23. Trelew, Chubut, April 3-8, 1994.

Cione, A.L. & E.P. Tonni 1995. Chronostratigraphy and 'land-mammal ages' in the Cenozoic of southern South America: principles, practices, and the 'Uquian' problem. *Journal of Paleontology* 69(1): 135-159.

Deschamps, C.M. 1995. El registro de *Neochoerus* Hay (Rodentia, Hydrochoeridae) en Bajo San José, provincia de Buenos Aires. Reconsideración de la antigüedad de los sedimentos portadores. *XI Jornadas Argentinas de Paleontología de Vertebrados*, Resúmenes: 21. Tucumán.

Deschamps, C.M. & A.M. Borromei 1992. La fauna de vertebrados pleistocénicos de Bajo San José (Provincia de Buenos Aires, Argentina). Aspectos paleoambientales. *Ameghiniana* 29(2): 177-183. Buenos Aires.

Fidalgo, F. & E.P. Tonni 1981. Sedimentos eólicos del Pleistoceno tardío y Reciente en el área Interserrana Bonaerense. *VIII Congreso Geológico Argentino*, Actas III: 33-39. San Luis.

Kraglievich, L.J. 1930. Los más grandes carpinchos actuales y fósiles de la subfamilia Hydrochoerinae. *Anales de la Sociedad Científica Argentina* CX: 233-264. Buenos Aires.

Kraglievich, L.J. & O. Reig 1945. Un nuevo roedor extinguido de la subfamilia Hydrochoerinae. *Notas Museo de La Plata* X, Paleontología 85: 266-275. La Plata.

Mones, A. 1991. Monografía de la familia Hydrochoeridae (Mammalia, Rodentia). Sistemática, Paleontología, Filogenia, Bibliografía. *Courier Forschungsinstitut Senckenberg*, 134: 1-235. Frankfurt am Mein.

Odreman Rivas, O.E. & J. Zetti 1969. Apéndice paleontológico. In: O. De Salvo et al., Caracteres geológicos de los depósitos eólicos del Pleistoceno superior de Junín (provincia de Buenos Aires). *Actas IV Jornadas Geológicas Argentinas* I: 269-292. Mendoza.

Pardiñas, U.F.J & C.M. Deschamps 1996. Sigmodontinos (Mammalia, Rodentia) pleistocénicos del sudoeste de la provincia de Buenos Aires (Argentina): aspectos sistemáticos, paleozoogeográficos y paleoambientales. *Revista de Estudios Geológicos* 52(5-6): 367-379. Madrid.

Pascual, R. & P. Bondesio 1985. Mamíferos terrestres del Mioceno Medio-Tardío de las Cuencas de los Ríos Colorado y Negro (Argentina). Evolución ambiental. *Ameghiniana* 22(1-2): 133-145. Buenos Aires.

Pascual, R., E.J. Hinojosa, D. Gondar & E.P. Tonni 1965. Las edades del Cenozoico mamalífero de la Argentina, con especial atención a aquellas del territorio bonaerense. *Anales de la Comisión de Investigaciones Científicas de Buenos Aires* VI: 165-193. La Plata.

Redford, K.H. & J.F. Eisenberg 1992. *Mammals of the Neotropics. The Southern Cone*. Volume 2. The University of Chicago Press.

Ringuelet, R.A. 1961. Rasgos fundamentales de la zoogeografía de la Argentina. *Physis* 22(63): 151-170. Buenos Aires.

Tonni, E.P. 1981. *Pediolagus salinicola* (Rodentia, Caviidae) en el Pleistoceno tardío de la Provincia de Buenos Aires. *Ameghiniana* XVIII (3-4): 123-126. Buenos Aires.

Tonni, E.P., M.T. Alberdi, J.L. Prado, M.S. Bargo & A.L. Cione 1992. Changes of mammal assemblages in the Pampean Region (Argentina) and their relation with the Plio-Pleistocene boundary. *Palaeogeography, Palaeoclimatology, Palaeoecology* 95: 179-194. Elsevier.

Tonni, E.P. & A.L. Cione 1994. Los mamíferos y el clima en el Pleistoceno y Holoceno de la provincia de Buenos Aires. *Jornadas de Arqueología e Interdisciplinas:* 127-142. CONICET, Programa de Estudios Prehistóricos. Buenos Aires.

Verzi, D.H. 1994. Origen y evolución de los Ctenomyinae (Rodentia, Octodontidae): Un análisis de anatomía craneo-dentaria. Ph.D. Dissertation, Facultad de Ciencias Naturales y Museo, Universidad Nacional de La Plata. 1-227. La Plata. Unpublished.

Verzi, D.H. & M. Lezcano 1996. Un nuevo resto de tuco-tuco del Ensenadense (Pleistoceno inferior-medio) de Necochea y el *status* de '*Megactenomys' kraglievichi* (Rodentia, Octodontidae). *Revista del Museo de La Plata,* Paleontología, N.S. IX(60): 239-246. La Plata.

Verzi, D.H. & M.G. Vucetich 1994. Los roedores Echimyidae del Cenozoico superior de Argentina. *VI Congreso Argentino de Paleontología y Bioestratigrafía,* Resúmenes: 45. Trelew.

Vucetich, M.G., D.H. Verzi & E.P. Tonni 1997. The presence of *Clyomys* (Rodentia, Echimyidae) in the Pleistocene of central Argentina. Paleoclimatic implications. *Palaeogeography, Palaeoclimatology, Palaeoecology* 128: 207-214. Elsevier.

Radiocarbon Chronology of Paso Otero 1 in the Pampean Region of Argentina

2

EILEEN JOHNSON
Museum of Texas Tech University, Lubbock,Texas

GUSTAVO POLITIS & GUSTAVO MARTINEZ
Facultad de Ciencias Sociales, Universidad Nacional del Centro de la Provincia de Buenos Aires, Olavarría, Argentina

WILLIAM T. HARTWELL
Desert Research Institute, University of Nevada, Las Vegas, Nevada

MARIA GUTIERREZ
Facultad de Ciencias Sociales, Universidad Nacional del Centro de la Provincia de Buenos Aires, Olavarría, Argentina

HERBERT HAAS
Desert Research Institute, University of Nevada, Las Vegas, Nevada

ABSTRACT: Paso Otero 1, in the Pampas of Argentina, is a stratified, riverine site containing alluvium with a sequence of buried A-horizons and two guanaco bone beds. The site is the first-recorded guanaco kill/butchering locale for the Pampas and represents at least two events during different times. Bulk organic sediment samples were taken from the three buried A-horizons and the humates assayed to produce a chronology for the site and date the cultural activities. The resultant ages indicate stable land surfaces in the Early, Middle, and Late Holocene with the kill/butchering events taking place during the Middle and Late Holocene.

RESUMEN: El sitio Paso Otero 1, en la Región Pampeana de Argentina, se encuentra en sedimentos fluviales que contienen dos horizontes A enterrados en los cuales se han registrado dos niveles con hueso de guanaco. Este es el primer sitio de caza y despostamiento de guanaco hallado en la Región Pampeana y representa por lo menos dos eventos diacrónicos. Se tomaron muestras de sedimento orgánico de tres horizontes A enterrados con el objeto de obtener una cronología para el sitio y datar la actividad antrópica. Las edades resultantes indican superficies de estabilización en el Holoceno temprano, medio y tardío con los eventos de caza de guanaco en el Holoceno medio y tardío.

1 INTRODUCTION

Paso Otero 1 is the first-recorded guanaco kill/butchering site for the Pampas of Argentina, and is located in the banks of the middle Río Quequén Grande. The site is part of a complex of stratified sites occurring in the neighboring banks, which also contain guanaco bone beds (Fig. 1). Some of these sites, such as Paso Otero 3 and 5, are now in the process of being excavated (Martínez 1994).

The radiocarbon ages reported for Paso Otero 1 are the first set of radiometric dates obtained from bulk sediment samples in the Río Quequén Grande basin. The first radiometric ages from archaeological sites for the continental part of the Pampean Region were obtained less than two decades ago (Carbonari et al. 1982), and chronological data for Quaternary sediments in the region still are quite limited. Those ages now available are primarily bone dates (Barrientos et al. 1996, Fidalgo et al. 1986, Figini 1987, Politis 1984, Politis & Beukens 1991). Several radiocarbon dates also have been obtained from charcoal scattered in anthropogenic levels at several sites in the Tandilia mountain range (Flegenheimer 1987, Zárate & Flegenheimer 1997) as well as from an isolated organic sediment date from a rockshelter in the same range

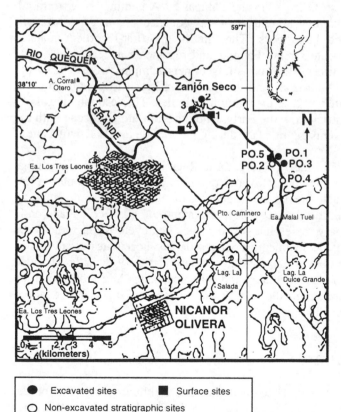

Figure 1. Middle basin of the Río Quequén Grande with archaeological sites noted including Paso Otero 1.

16

(Orquera et al. 1980). The radiocarbon ages presented here from Paso Otero 1 provide an expanded radiometric database and more detailed understanding of local and regional events during the Holocene.

Paso Otero 1 (Martínez 1994, 1995, Politis et al. 1991) is located on the left bank of the Río Quequén Grande in the Interserrana Bonaerense Area (Necochea District) in the province of Buenos Aires, Argentina. This area, the humid Pampa sub-region of the Pampean Region, is a grassland plain broken by two mountain ranges. In 1980, erosion caused by regional flooding resulted in the exposure of guanaco (*Lama guanicoe*) bones within the cut-bank (Fig. 2). Test excavations between 1989 and 1991 yielded seven lithic artifacts and 3781 bone elements from a 22 m^2 area, representing an MNI of 36 (Fig. 3).

The guanaco bones occurred in four discrete concentrations (piles) and stratigraphic positioning indicated that the piles represented two separate events. Taphonomic analysis demonstrates that the bones were not laid down by water and were modified by people (cut marks confirmed through SEM analysis; helicoidal fractures), rodents (gnaw marks of different sizes), and a carnivore (gnaw marks and tooth punctures) (Gutiérrez et al. 1994, Johnson et al. 1996). This situation where only one prey species is represented is highly unusual for archaeological sites in the Pampean Region, despite guanaco being the main game animal hunted by prehistoric peoples on the Pampas (Politis & Salemme 1989). One exception

Figure 2. Guanaco bones outcropping from the cut-bank at Paso Otero 1.

Figure 3. Two guanaco bone piles from the middle buried A-horizon exposed through archaeological excavations. Samples DRI-2829 and DRI-2830 were taken from this horizon.

is the site of Zanjón Seco 3, located approximately 6 km north of Paso Otero, where guanaco was the only species represented in a pile of intentionally discarded bone (Politis 1984, Politis & Tonni 1985).

2 STRATIGRAPHY AND RADIOCARBON AGES

The general stratigraphic sequence for Paso Otero 1 is similar to that of other fluvial valleys on the Pampean plains (Fidalgo & Tonni 1978, 1981). The Pleistocene sediments of the 'Pampiano Formation' are located at the base of the sequence. Overlying this Formation the fluvial sediments of the Luján Formation are found, composed of two members. One is the Late Pleistocene Guerrero Member (previously referred to as 'Lujanense' by Frenguelli 1950) and the other, the Holocene Río Salado Member (previously referred to as 'Platense' by Frenguelli 1950). The Late Holocene aeolian sediments of La Postrera Formation overlie the Luján Formation (Fig. 4).

The Río Salado Member at Paso Otero 1 is a stratified fluvial deposit that records episodes of landscape stability in the river valley. These stable land surfaces are represented by buried A-horizons of soils that developed in the sediments. The stratigraphy records an alternating pattern of allu-

18

Figure 4. Stratigraphy at Paso Otero 1 with the schematic location of the bulk organic sediment samples (marked with the Lab. N°).

viation-stability-alluviation. These A-horizons developed within a very moist setting that was producing large amounts of organic matter under poorly-drained conditions. This moist setting is indicated by thick, dark (10YR 3/1 and 2/1) A-horizons relatively high in organic matter (1.0-1.5%) and subhorizons with dull gray colors (10YR 7/2 and 6/2) that are indicative of gleying and that contain mottles of oxidized iron (V.T. Holliday, pers.comm. 1995). This setting is interpreted as a wet meadow immediately adjacent to the river, such as the context of a 'vega'. The modern soil has developed in the La Postrera Formation.

One of the bone piles is associated with the upper buried A-horizon, and the other three piles with the middle buried A-horizon. The lower buried A-horizon, at the base of the Río Salado Member, currently has no known archaeological materials associated with it at Paso Otero 1. Bulk organic sediment samples were taken from the three buried A-horizons and processed at the Radiocarbon Laboratory of the Desert Research Institute in Las Vegas, Nevada.

Previous attempts to radiocarbon date guanaco bones at Paso Otero 1, using both standard and AMS methods, failed due to a lack of sufficient collagen. Therefore, bulk sediment samples were collected from the buried A-horizons to date the soil humate (NaOH soluble) fractions. Some problems have been identified in results obtained from the dating of organic-rich A-horizons (Hammond et al. 1991, Martin & Johnson 1995,

19

Matthews 1985, Scharpenseel 1979, Yang et al. 1996). However, within non-leaching environments, they can provide reliable age control (Haas et al. 1986, Martin & Johnson 1995).

Samples were collected from exposed cut-bank profiles (Fig. 4) following the methodology of Haas et al. (1986). Profiles were cleaned and the samples removed beginning from a position extending 50 cm or more into the wall. Samples from the upper two buried A-horizons were recovered from a column adjacent to the main excavation area (column 1) whereas the sample for the lower A-horizon was taken from a column facing the river (Fig. 2).

The samples produced four radiocarbon ages (Table 1) for the three buried A-horizons. A pair of ages was run on humates from the sample from the middle buried A-horizon. All three samples were obtained by digestion of the soil samples with NaOH. From each sample, a moderately strong humate solution was decanted into storage bottles. More humate solutions were obtained by adding distilled water to the soils and by letting all particulate matter settle before decanting the progressively weaker solutions. This process was repeated five times with each soil sample. Each of these batches was processed individually into a solid humate crystal sample. This process involved acidification, filtration, and drying. Chemically and isotopically, the batches were similar. The date DRI-2830 (Table 1) was from the first humate batch of the middle A-horizon. It was heavier in ^{14}C and lighter in ^{13}C than the companion date DRI-2829 that was derived from the second humate batch. These two dates overlapped at the one sigma level. The older age of 4820 ± 105 ^{14}C yr BP, however, is likely the more accurate of the two, since radiocarbon assays of buried soils tend to produce ages that are younger than the true ages of the soil due to the continuous input of organic matter into soils (Yang et al. 1996). The dating of the two batches confirmed that no significant fractionation occurs during partial extraction of humates. Furthermore, the resultant ages indicated stable land surfaces in the Early, Middle and Late Holocene.

Table 1. Results of Radiocarbon Assays on Humates from Bulk Organic Samples from Paso Otero 1.

Lab. N°	$\delta^{13/12}$C	Location	Uncorrected date (yr BP)	Corrected date (yr BP)
DRI-2831	-20.63	lower buried A-horizon	9880 ± 65	9950 ± 65
DRI-2829	-22.92	middle buried A-horizon	4820 ± 105	4855 ± 105
DRI-2830	-23.22	middle buried A-horizon	4720 ± 60	4750 ± 60
DRI-2837	-22.82	upper buried A-horizon	2690 ± 40	2720 ± 40

3 DISCUSSION AND CONCLUSIONS

The new data obtained from the organic sediment samples from Paso Otero 1 contribute to the discussion of the environmental evolution, chronology, and culture history of the Holocene period for the Pampas. Isotope data have been used as an independent means to explore changes in the composition of vegetation through time and across grassland biomes, C_3 versus C_4 grasses (Bowen 1991, Cerling 1984, Tieszen 1994). The isotope data (Table 1) from Paso Otero 1 indicate a dominance of C_3 grasses (cool season grasses, aquatic, or moist grasses) throughout the Holocene with some variation in the composition or ratios. This dominance probably reflects the riparian setting of the wet meadows more than the extant grassland biome in general.

Dating of the Paso Otero 1 buried A-horizons using bulk sediment samples represents only the second time this method has been applied to sites in the Pampas. Previously, one organic-sediment age was available for the Pampas from a rockshelter in the Sierras de Tandil range. A bulk sample was taken from a buried soil within the rockshelter and processed by Beta Analytic; the sample yielded an age of ca. 7000 ^{14}C yr BP (Orquera et al. 1980). While significant as the first such date for the Pampas, it has been difficult to correlate the rockshelter's stratigraphy and occupations with the open-air sites of the rest of the Pampas.

On the other hand, the new radiocarbon ages from Paso Otero 1 fill a chronological gap between existing Late Pleistocene and Late Holocene radiometric data for the Río Quequén Grande basin based on bone collagen. Late Pleistocene radiocarbon ages have been obtained from a palaeontological site in the lower member of La Postrera Formation of ca. 10,700 yr BP (Figini 1987). In addition, a recent radiocarbon assay has been obtained on the bone of an unidentified extinct megamammal recovered from Paso Otero 5. This site is located in the bank of the Río Quequén Grande opposite Paso Otero 1 and displays a similar stratigraphic history. Megamammal remains, possibly butchered at the site, and associated lithic artifacts were recovered between 2.45 and 2.75 m below the surface. The samples returned an age of 10,190 ± 120 ^{14}C yr BP (Martínez 1997). The archaeological materials and sample were located in the transition between the two members of the Luján Formation in a stratigraphic position comparable to sample DRI 2831, and the ages confirm that the lower boundary of the Rio Salado Member represents the beginning of Holocene deposition in the area.

Other dates from the area correlate well with previously obtained results. A Late Holocene age of ca. 1450 yr BP comes from the archaeological site of Zanjón Seco 3 (Politis & Beukens 1991). This age was obtained from the upper part of the La Postrera Formation, which overlies the Río Salado member of the Luján Formation. Recent isotope research conducted in the vicinity of the Paso Otero locality has used terrestrial gastropod shell for radiometric dating (Bonadonna et al. 1995). The shell, taken from a level ca. 10-20 cm below the middle buried A-horizon, has yielded an age of ca. 5730 yr BP (J.L. Prado, pers. comm. 1995). However, this age should be considered only as a maximum age, due to the reservoir effect found in terrestrial gastropod shells of the Pampas (Figini et al. 1989 and other later papers). Therefore, the new radiometric data refine the local chronological sequence and fix the sedimentary time span of the Río Salado Member in the Río Quequén Grande basin from ca. 10,000 to 2700 yr BP. These limits are in agreement with the results obtained in a chronological sequence from the lower Río Quequén Grande where dates of 9340 ± 110, 9820 ± 100 and 10,270 ± 70 ^{14}C yr BP where obtained for the lowest section of the Río Salado Member (Zárate et al. 1996). These chronological limits relate well with other ages from the Río Salado Member in other stratigraphic sequences of the Pampean Region. In the Río Quequén Salado, the bottom of the Río Salado Member is dated to ca. 10,800 ^{14}C yr BP while in the Punta Hermengo sequence, the upper limit is dated at ca. 3400 ^{14}C yr BP (Bonadonna et al. 1995).

On a broader basis, few Middle Holocene radiocarbon ages are available for the continental part of the southeastern Pampean Region. The majority of these are from archaeological sites such as Fortín Necochea (González & Weiler 1987/1988) and Cueva Tixi (Mazzanti 1997). Most Middle Holocene chronological data for the Pampean Region are related to marine sequences resulting from sea-level fluctuations and were obtained from shell within ancient shoreline deposits (Fidalgo 1979, Figini 1992, González & Weiler 1983, Isla et al. 1986). This marine-based chronological information represents the time span of the locally-known Las Escobas Formation (a marine unit defined by Fidalgo 1979). Extending a chronological sequence inland through stratigraphic correlation has proven difficult. The Paso Otero 1 ages and stratigraphy, therefore, provide continent-based data in which to build an inland chronological framework for the Holocene. Recent dates obtained for the Lower Río Quequén Grande are also of great interest for the same purpose (Zárate et al. 1996).

In conclusion, aggradation of the Río Salado Member in the Río Quequén Grande basin took place between ca. 10,000 and 2700 ^{14}C yr BP. During this time, landscape stability within the floodplain occurred at least three times. For two of these times, at ca. 4800 and 2700 ^{14}C yr BP, the floodplain was used by ancient hunter-gatherers who hunted, killed, and butchered guanaco on the stable 'vegas'. Kill/butchering sites in general are rare in Argentina and guanaco kill/butchering sites unknown for the Southern Cone until now at Paso Otero 1. Moreover, Paso Otero 1 documents repeated use of the same landscape features through time for the same activity.

ACKNOWLEDGEMENTS

Support for the Paso Otero 1 research was provided by the National Science Foundation International Programs (INT-9218457) and Museum of Texas Tech University (EJ) and represents part of the ongoing Lubbock Lake research into grasslands hunter-gatherers; and CONICET (PID Nr. 3-065100/88) and the Universidad Nacional del Centro de la Provincia de Buenos Aires (GP) for ongoing work in the eastern Pampas. The friendly interaction of several colleagues and their constructive criticisms are greatly appreciated. In particular, thanks are due to Dr. Vance T. Holliday (University of Wisconsin) for sharing information and ideas, initial analysis of Paso Otero 1 sediments, and critical reading of this manuscript; and to Sr. Gesué Noseda, Director of the Museo de Ciencias Naturales de Lobería, and José Prado (Universidad Nacional del Centro de la Provincia de Buenos Aires, Olavarría) for sharing information and ideas, assistance, and great interest. Interpretations and any errors, however, are those of the authors.

REFERENCES

Bonadonna, F., G. Leone & G. Zanchetta 1995. Composición isotópica de los fósiles de gasterópodos continentales de la Provincia de Buenos Aires. Indicaciones paleoclimáticas. In: M.T. Alberdi, F.P. Bonadonna & E.P. Tonni (eds), *Registro Continental de la Evolución Climática y Biológica de los Últimos 5 Ma por Correlación entre el Hemisferio Norte (SO de Europa) y el Hemisferio Sur (Argentina)*: 75-104. Monografías Museo Nacional de Ciencias Naturales, Madrid, Spain.

Barrientos, G., A. Gil, J. Moirano & M. Saghessi 1996. Bibliografía Arqueológica de la Provincia de Buenos Aires. *Boletín del Centro de la Provincia de Buenos Aires*, 4: 27-78. Centro de Registro Arqueológico y Paleontológico de la Provincia de Buenos Aires. La Plata, Argentina.

Bowen, R. 1991. *Isotopes and Climate*. Elsevier, New York.

Carbonari, J., G. Gómez, R. Huarte & A. Figini 1982. Capítulo radiocarbónico. Determinación de la edad radiocarbónica de un resto óseo de *Megatherium* cf. *americanum* (localidad Arroyo Seco, Pdo. de Tres Arroyos, Pcia. de Buenos Aires, Rep. Argentina). *Actas del VII Congreso Nacional de Arqueología*: 151-152. Colonia Del Sacramento, Uruguay.

Cerling, T.E. 1984. Stable isotope composition of modern soil carbonate and its relationship to climate. *Earth and Planetary Science Letters*, 71:229-240.

Fidalgo, F. 1979. Upper Pleistocene-Recent marine deposits in northeastern Buenos Aires Province (Argentina). *Proceedings of the International Symposium on Coastal Evolution in the Quaternary*: 384-404. IGCP, Sao Paulo.

Fidalgo, F. & E. Tonni 1978. Aspectos paleoclimáticos del Pleistoceno Tardío-

Reciente de la Provincia de Buenos Aires. *Segunda Reunión Informativa del Cuaternario Bonaerense* 21-27. CIC, Provincia de Buenos Aires, La Plata.

Fidalgo, F. & E. Tonni 1981. Sedimentos eólicos del Pleistoceno Tardío y Reciente en el Area Interserrana Bonaerense. *VIII Congreso Geológico Argentino Actas,* III: 33-39. San Luis.

Fidalgo, F., L.M. Guzmán, G.G. Politis, M.C. Salemme, E.P. Tonni, J.E. Carbonari, G.J. Gómez, R.A. Huarte & A.J. Figini 1986. Investigaciones arqueológicas en el sitio 2 de Arroyo Seco (Pdo. de Tres Arroyos, Provincia de Buenos Aires, República Argentina). In: A.L. Bryan (ed.), *New Evidence for the Pleistocene Peopling of the Americas*: 221-269. Center for the Study of Early Man, University of Maine, Orono.

Figini, A. 1987. Datación radiocarbónica de restos óseos de la Formación La Postrera en el Partido de Lobería, Provincia de Buenos Aires, Argentina. *X Congreso Geológico Argentino Actas*: 185-188. San Miguel de Tucumán.

Figini, A. 1992. Edades C-14 de sedimentos marinos holocénicos de la Provincia de Buenos Aires. *III Jornadas Geológicas Bonaerenses*, 3: 245-247. La Plata.

Figini, A., J. Rabassa, E.P. Tonni, R. Huarte, G. Gómez, J. Carbonari & A. Zubiaga. 1989. Datación radiocarbónica de gasterópodos terrestres en sedimentos del Pleistoceno Superior y Holoceno del valle del río Sauce Grande, Pcia. de Buenos Aires. *Iras. Jornadas Geológicas Bonaerenses*, 1985, Tandil. Actas 809-824. Bahía Blanca.

Flegenheimer, N. 1987. Recent research at localities Cerro La China and Cerro El Sombrero, Argentina. *Current Research in the Pleistocene*, 4:148-149.

Frenguelli, J. 1950. Rasgos generales de la morfología y geología de la Provincia de Buenos Aires. *LEMIT*, 2(33): 1-72. La Plata.

González, M. & N. Weiler 1983. Ciclicidad de niveles marinos holocénicos en Bahía Blanca y en el delta del Río Colorado (Provincia de Buenos Aires), en base a edades carbono-14. *Simposio oscilaciones del nivel del mar durante el último hemiciclo deglacial en la Argentina*, IGCP-61: 69-90. Mar del Plata.

González, M. & N. Weiler 1987/1988. Sitio arqueológico Fortín Necochea (Pdo. de General Lamadrid, Provincia de Buenos Aires). Informe geológico preliminar. *Paletnológica*, 4: 56-63. Buenos Aires.

Gutiérrez, M., G. Martínez, G. Politis, E. Johnson & W. Hartwell 1994. Nuevos análisis óseos en el sitio Paso Otero 1 (Pdo. de Necochea, Pcia. de Buenos Aires). Actas y Memorias del XI Congreso Nacional de Arqueología Argentina, (Resúmenes). *Revista del Museo de Historia Natural de San Rafael* 14(1/4): 222-224, San Rafael, Mendoza.

Haas, H., V.T. Holliday & R. Stuckenrath 1986. Dating of Holocene stratigraphy with soluble and insoluble organic fractions at the Lubbock Lake archaeological site, Texas: an ideal case study. *Radiocarbon* 28(2A):473-485.

Hammond, A.P., K.M. Goh, P.J. Tonkin & M.R. Manning 1991. Chemical pretreatments for improving the radiocarbon dates of peats and organic silts in a gley podzol environment: Grahams Terrace, North Westland. *New Zealand Journal of Geology and Geophysics* 34:191-194.

Isla, F., L. Ferrero, J. Fasano, M. Espinosa & E. Schnack 1986. Late Quaternary marine-estuarine sequences of the southeastern coast of Buenos Aires Province, Argentina. *Quaternary of South America and Antarctic Peninsula* 4:137-157. Rotterdam: Balkema Publishers.

Johnson, E., M. Gutiérrez, G. Politis, G. Martínez & W.T. Hartwell 1996. Holocene Taphonomy at Paso Otero 1 on the Eastern Pampas of Argentina. In: *Proceedings of the 1993 Bone Modification Conference, Occasional Publication* 1. Hot Springs, Archeology Laboratory, Augustana College. South Dakota.

Martin, C.W. & W.C. Johnson 1995. Variation in radiocarbon ages of soil organic matter fractions from Late Quaternary buried soils. *Quaternary Research* 43(2): 232-237.

Martínez, G. 1994. Diversidad de ocupaciones arqueológicas en el curso medio del Río Quequén Grande (Pdos. de Necochea y Lobería, Pcia. de Puenos Aires). Actas y Memorias del XI Congreso Nacional de Arqueología Argentina (Resúmenes). *Revista del Museo de Historia Natural de San Rafael* 14(1/4): 228-229. San Rafael, Mendoza.

Martínez, G. 1995. Archaeological sites in fluvial and aeolian settings in the Quequén Grande River (Interserrana Area, Buenos Aires Province, Argentina). *World Archaeological Congress*, Abstracts, New Delhi, India.

Martínez, G. 1997. A Preliminary Report on Paso Otero 5, a late Pleistocene Site in the Pampean Region of Argentina. *Current Research in the Pleistocene* 14: 53-55.

Matthews, J.A. 1985. Radiocarbon dating of surface and buried soils: principles, problems and prospects. In: K.S. Richards, R.R. Arnett & S. Ellis (eds), *Geomorphology and Soils*: 269-288. London: Allen and Unwin.

Mazzanti, G.L. 1997. Excavaciones arqueológicas en el sitio Cueva Tixi, Buenos Aires, Argentina. *Latin American Antiquity* 8(1): 55-62.

Orquera, L., E. Piana & A. Sala 1980. La antigüedad de la ocupación humana de la Gruta del Oro (Partido de Juarez, Provincia de Buenos Aires): Un problema resuelto. *Relaciones de la Sociedad Argentina de Antropología* 14(1): 83-101. Buenos Aires.

Politis, G.G. 1984. Arqueología del Area Interserrana Bonaerense. Ph.D. Dissertation, Facultad de Ciencias Naturales y Museo, Universidad Nacional de La Plata, Argentina. 392 pages. Unpublished.

Politis, G.G. & R. Beukens 1991. Cronología radiocarbónica de la ocupación humana del Area Interserrana Bonaerense (Argentina). *Actas del X Congreso Nacional de Arquelogía Argentina, Revista Shincal* 3:151-157. Catamarca.

Politis, G.G. & E. Tonni 1985. Investigaciones arqueológicas en el Sitio 3 de Zanjón Seco (Partido de Necochea, provincia de Buenos Aires). *Sapiens*, 5: 14-30. Chivilcoy. Argentina.

Politis, G.G. & M.C. Salemme 1989. Pre-Hispanic mammal exploitation and hunting strategies in the Eastern Pampas Subregion of Argentina. In: L.B. Davis & B.O.K. Reeves (eds), *Hunters of the Recent Past*: 352-372. London: Unwin.

Politis, G., M. Gutiérrez & G. Martínez 1991. Informe preliminar de las investigaciones en el sitio Paso Otero 1 (Partido de Necochea, Provincia de Buenos Aires). *Boletín del Centro de la Provincia de Buenos Aires* 3: 80-90. Centro de Registro Arqueológico y Paleontológico de la Provincia de Buenos Aires. La Plata, Argentina.

Scharpenseel, H.W. 1979. Soil fraction dating. In: R. Berger & H. Suess (eds) *Radiocarbon Dating, Proceedings of the 9th International Radiocarbon Conference*: 277-283. Berkeley: University of California Press.

Tieszen, L.L. 1994. Stable isotopes in the Great Plains: vegetation analyses and diet determinations. In: D. Owsley & R. Jantz (eds), *Skeletal Biology in the Great Plains*: 261-282. Washington, D.C.: Smithsonian Institution Press.

Yang, W., R. Amundson & S. Trumbore 1996. Radiocarbon Dating of Soil Organic Matter. *Quaternary Research* 45: 282-288.

Zárate, M., M. Espinosa & L. Ferrero 1996. La Horqueta II, Río Quequén Grande: Ambientes Sedimentarios de la Transición Pleistoceno-Holoceno. *IV Jornadas Geológicas Bonaerenses,* Actas I: 195-204. Junín. Buenos Aires.

Zárate, M. & N. Flegenheimer 1997. Considerations on radiocarbon dates from Cerro La China and Cerro El Sombrero, Argentina. *Current Research in the Pleistocene* 14: 27-28.

The Late Quaternary of the Upper Juruá River, Southwestern Amazonia, Brazil: Geology and Vertebrate Palaeontology

EDGARDO M. LATRUBESSE
Universidade Federal do Acre, DEGEO, Rio Branco, AC, Brazil.

ALCEU RANCY
Universidade Federal de Santa Catarina,Geociências, Florianópolis, SC, Brazil.

ABSTRACT: A review of existing data and new data on the Juruá River, the most representative and typical locality for stratigraphical and palaeoecological studies in southwestern Amazonia, is presented. Three units have been recognized: Older Pleistocene sediments, Late Pleistocene sediments, and Holocene sediments. The Late Pleistocene sediments and the Holocene sediments are forming the Lower Terrace.

In the bone-bearing conglomerate of the Late Pleistocene sediments a rich Pleistocene Lujanan mammal fauna was recorded. The mammal fauna and the conglomeratic sediments indicate climate aridity in the area, probably during the Middle Pleniglacial of the Last Glaciation. Another phase of sedimentation, Holocene in age, has been recognized. The presence of fine sediments, lateral accretion deposits, and abundant remains of trunks, leaves and steams in the sediments indicate that the rain forest was present in the area during the Holocene. At this time the river had a meandering pattern.

RESUMEN: Se presenta una revisión de datos existentes y nuevos acerca del río Juruá, la localidad más típica y representativa para estudios estratigráficos y paleoecológicos en Amazonia Sud-Occidental. Se han reconocido tres unidades: sedimentos del Pleistoceno antiguo, sedimentos del Pleistoceno tardío y sedimentos del Holoceno. Los sedimentos del Pleistoceno tardío y Holoceno forman la Terraza Inferior.

Una rica fauna pleistocénica Lujanense se ha recuperado en un conglomerado de sedimentos del Pleistoceno tardío que contiene un nivel de restos óseos. La fauna de mamíferos y los sedimentos conglomerádicos indican clima árido en el área, probablemente durante el Pleniglacial medio de la Última Glaciación. Otra fase de sedimentación, de edad Holoceno, se ha reconocido. La presencia de sedimentos finos, depósitos de acreción lateral y abundantes restos de troncos, hojas y ramas en los sedimentos indican que hubo condiciones de selva pluvial en el área durante el Holoceno. Para este momento, el río mostraba un patrón meandroso.

1 INTRODUCTION AND BACKGROUND

The first references to fossil vertebrates from southwestern Amazonia refer to Chandless (1866). The same fossils collected by W. Chandless were erroneously referred to as late Cretaceous by Louis Agassiz (1868). Sporadic references to fossils from western Amazonia continued to be published through the turn of the century from material collected during the rubber boom that sent businessmen from Europe and North America to the area (Barbosa Rodrigues 1892, Gervais 1876, 1877, Gurich 1912, Mook 1921). At the same time the Museu Nacional in Rio de Janeiro and the Museo Paraense Emilio Goeldi in Belém also began to receive material sent from the western Amazonia by rubber barons and government officers. Some papers were published by Goeldi (1906), Kraglievich (1930, 1931), Miranda Ribeiro (1938) and Roxo (1921, 1937). Fawcett (1909), explorer of the Brazilian and Peruvian Amazon during the years 1906-1907, worked along the Acre river and referred to fossils in the area.

The first stratigraphical correlations between the western Amazon and Tertiary of Argentina using vertebrate fossils was made by Patterson (1942) correlating Eocene (Mustersan) and Miocene (Huayquerian) ages. Paula Couto (1944) produced the first of a series of papers about fossil mammals of the state of Acre. Spillmann (1949) studied mammalian fossils from the Ucayali River region (Peru). A fundamental advance in the study of stratigraphy and palaeontology of the western Amazonia resulted from the 1956 expedition of G.G. Simpson and L.I. Price. They spent months travelling along the Juruá River and its affluents within the state of Acre. The large collection that they produced was shared between the Departamento Nacional da Produção Mineral (DNPM) in Rio de Janeiro and the American Museum of Natural History (AMNH) in New York (Price 1957). Carlos de Paula Couto, following up on his earlier effort, used these collections to publish a brillant series of papers on the fossil mammals from the Cenozoic of Acre (Paula Couto 1956, 1976, 1978, 1981, 1982a, 1982b, 1983a, 1983b, Simpson & Paula Couto 1981). The researchers of Radambrasil (1976, 1977) collected vertebrate fossils in Acre that were deposited at the DNPM in Rio de Janeiro and were not subjected to detailed study or publication. Data on Tertiary fauna were published by Frailey (1986).

In 1983, the Universidade Federal do Acre organized the Laboratório de Pesquisas Paleontológicas (LPP/UFAC) under the coordination of one of the present authors (A. Rancy). During the past fourteen years the Laboratory gathered an impressive collection of Cretaceous, Tertiary and

28

Quaternary palaeovertebrates (fishes, birds, reptiles and mammals) of the area. With more than 4000 specimens catalogued, this represents a significant contribution to the palaeobiology of the Amazonia. At present, the group of the LPP/UFAC is working with the Laboratorio de Geomorfología e Sedimentologia of the Universidade Federal of Acre to carry out an integrated geological/palaeontological research project.

Here we review the existing data on Vertebrate Palaeontology and Geology in the Juruá River, the most representative and typical locality for stratigraphical and palaeoecological studies in Southwestern Amazonia.

Field work was carried out by the authors, along the Juruá River by boat during expeditions in 1977, 1982, 1992 and 1994 (Fig. 1).

The grain size analyses of some sedimentary samples was determined by sieved method for sand and gravel fractions and by the Appiani Cylinder Hydrometer Test for fine grain sizes (clay and silt). Mineralogical studies were performed by X-ray diffraction in a Phillips PW 1729 diffractometer at the Department of Geology, University of Modena, Italy. Samples were powdered to a homogeneous grain size and the dust was

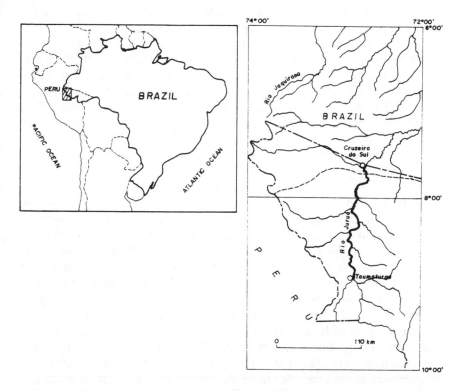

Figure 1. Location map. Field work was carried out from Cruzeiro do Sul to Taumaturgo.

29

analysed. Radiocarbon data analyses were performed at LATYR (La Plata, Argentina). The first radiocarbon datings for the Upper Juruá river region are presented in this paper.

2 GEOLOGY

Tertiary sediments cover almost the entire region of the western Brazilian Amazonia, west of Manaus. These sediments have been assigned to the Solimões Formation. During the last twenty years the Solimões Formation was considered to be Pliocene-Late Pleistocene in age (Radambrasil 1976,1977) or Late Pleistocene in the uppermost levels (Frailey et al. 1988, Kronberg et al. 1990). However, based on new geological and palaeontological data, we consider the Solimões Formation to be Late Miocene-Pliocene in the uppermost levels that outcrops in Acre (Latrubesse et al. 1997). The Solimões Formation was interpreted to be deposited in a continental, fluvio-lacustrine environment. A very rich vertebrate fauna is found in this unit. The palaeovertebrate fauna of the Solimões Formation at the Juruá and Purus basins are indicative of Huayquerian-Montehermosan Land Mammal Ages, as defined by Marshall et al. (1984).

The Quaternary sediments are of fluvial origin and are found in the alluvial belts of the Southwestern Amazonia rivers. In these belts three terrace levels were recognized. The Upper Terrace (UT) reaches 34 to 38 m, the Intermediate Terrace (IT) 15 to 20 m and the Lower Terrace (LT) 8 to 12 m (relative to the lowest dry season water level; Latrubesse 1992, Latrubesse & Ramonell 1993). The lower terrace is the more exposed unit along the banks of the Purus, Acre, Iaco and Juruá rivers. In the Moa River, the terraces only attain heights of 20 to 25 m, 9 to 12 m and 2 to 4 m, respectively.

Some detailed geological descriptions were carried out by G.G. Simpson (in: Simpson & Paula Couto 1981), who states about the sediments of the Upper Juruá River: 'The strata along the Upper Juruá are visible at lower water in isolated exposures and cannot be traced continuosly from one exposure to the next. They are not quite identical in any two exposures, and may change laterally from one end to the other of a somewhat extensive single exposure. Thus we could not establish clear correlations or dates for these beds and could not define formations in the usual geological sense of the term. Our ideas about the stratigraphy changed as we became familiar with more exposures. By the end of our work in 1956 we had distinguished lithological types that followed a definite sequence although not all were present at any locality. I gave these field designations for purposes of my notes, but these designations were placed in quotation marks and were not considered definitive of either age correlation. It is further probable that this sequence is not always internally consistent, and

30

it applies only to the Juruá as it was in 1956 between Cruzeiro do Sul and Fos do Breu, with slight additions from short excursions up tributaries, especially the Río Amonea.'

Simpson (in: Simpson & Paula Couto 1981) identified the following types of sediments:

– 'Recent' or 'R-Type' (possibly including some Late Pleistocene sediments) exposed almost everywhere along the banks and flooded at the highest river water level. The lithologies are buff, brown to yellowish, unconsolidated or little consolidated sands and silts, beach and shoal sands and gravels, completely unconsolidated (present day fluvial sediments), and laminated lake deposits (meander filling).

– 'Pleistocene' or 'Pl-type'. According to Simpson (in: Simpson & Paula Couto 1981) these sediments are at least in part Pleistocene although not necessarily all so. He explains that four lithologies can often, but not invariably, be distinguished: Pl Phase 2, mostly clay, light in color, frequently with a depositional dip. PL Phase 1 formed by a Sandy facies, a Conglomerate facies and a 'Pseudopuca' Facies.

– 'Puca type'. At the outcrops, it is always the lowest distinguishable lithology, presenting a clear erosional unconformity on top.

2.1 *The Solimões Formation*

The 'Puca Type' sediments make up the Solimões Formation. They are composed of clastic sediments such as sandstones, siltstones and claystones. The sediments can be differentiated into two groups of facies (Latrubesse 1992, Latrubesse et al. 1997): a low floodplain/lacustrine facies and a channel fluvial facies. The channel facies is composed of sands, silty and clayey sands and some intraformational small conglomerates. This facies is red to brown and shows cross-bedding stratification. The floodplain/lacustrine facies is formed by green to gray-green silty clays and clayey silts. These sediments have in general a massive structure and secondarily planar lamination. In the two facies Late Miocene-Pliocene (Huayquerian-Montehermosan) vertebrate fossils are found (Latrubesse 1992, Latrubesse et al. 1997).

3 QUATERNARY SEDIMENTS

The best exposures of Quaternary fossiliferous sediments outcrop upstream of the locality of Triunfo. In this area, both G.G. Simpson and ourselves have found the totality of the Pleistocene vertebrates (Fig. 2). The Quaternary sediments can be differentiated in three units: 'Older' Sediments, Late Pleistocene Sediments and Holocene Sediments (Table 1).

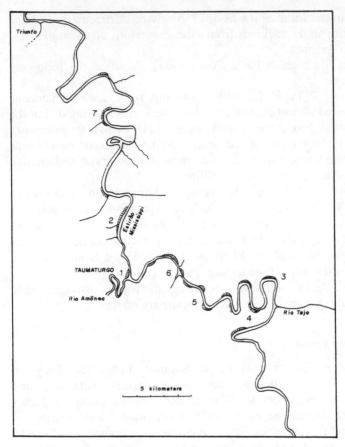

Figure 2. Occurrence of the principal Pleistocene outcrops described as Profiles on Figure 3. Dashed areas indicate the main sedimentary exposures. Coordinates are indicated in Table 1.

Table 1. Location of Profiles of Figure 3.

UNIT	PROFILES	Geographical position
Older Pleistocene Sediments	1	8°56.75 S; 72°47.05 W
Older Pleistocene Sediments	2	8°54.78 S; 72°46.71 W
Late Pleistocene Sediments	3	8°58.47 S; 72°43.24 W
Late Pleistocene Sediments	4	8°57.90 S; 72°44.07 W
Late Pleistocene Sediments	5	Locality 10 of Simpson
Late Pleistocene Sediments	6	8°57.43 S; 72°45.34 W
Holocene Sediments	7	8°50.87 S; 72°44.65 W
Holocene Sediments	8	8°25.12 S; 72°49.79 W
Holocene Sediments	9	8°24.18 S; 72°49.61 W
Holocene Sediments	10	8°04.35 S; 72°46.89 W
Holocene Sediments	11	7°51.38 S; 72°44.12 W

3.1 'Older' Pleistocene sediments

The Older Quaternary sediments, that we have ocassionally found, form a terrace level of nearly 15-20 m and were correlated with the Intermediate Terrace defined by Latrubesse (1992) and Latrubesse et al. (1997). The most clear exposure of this unit was recognized at the junction between the Juruá and Amonea rivers, near the town of Thaumaturgo. A detailed stratigraphical vertical section was studied at this site (Profile 1, Figs 2 and 3a). In other points this unit is not present in a very clear form. For example in 'Estirão de Mississippi', Simpson (in: Simpson & Paula Couto 1981) defined the exposures as 'Pseudo Puca type'. However in this area the stratigraphy is very complex. The Older Pleistocene sediments are lying in unconformity over the Mio-Pliocene Solimões Formation. On top of this outcrop the Late Pleistocene sediments occur. The Late Pleistocene sequence starts with a typical black conglomerate at the base (Profile 2, Figs 2 and 3a). The sediments outcropping at lat. 8°53.340 S and long. 72°47.228 W probably could belong to this older Quaternary sedimentary unit. The age of these deposits is unknown; radiocarbon dating indicates an infinite age (Table 2, Profile 1 of Fig. 3a).

Sedimentological and mineralogical analyses were performed in samples of sections 1 and 2 (Table 3, Fig. 4). The X-ray diffraction analyses indicate quartz is dominant, whereas feldspar is abundant and kaolinite, mica and calcite are scarce. Diffraction analysis of an organic bed of Profile 1 shows an existence of an abundant quantity of amorphous matter (organic matter), whereas quartz is dominant and feldspar is present.

3.2 Lower Terrace: Late Pleistocene and Holocene sediments

The Lower Terrace is formed by sediments of different cycles of erosion/deposition during the Late Pleistocene-Holocene. The sediments of this terrace include the PL Phase1 and PL Phase2 and Recent sediments, as described by Simpson (in: Simpson & Paula Couto 1981). The older sediments of this terrace are the Pl Phase 1 sediments. We do not include in this terrace the 'Pseudopuca type' facies as Simpson did, because this facies belong, in our opinion, to the Older Quaternary sediments and to the Intermediate Terrace.

3.2.1 The Late Pleistocene sediments
The sediment facies are typically a fining upward sequence with a basal conglomerate and related some sandy deposits to it. These sediments lie unconformably over the Tertiary sediments of the Solimões Formation.

The conglomerate facies outcrop mainly in the Juruá, upstream of the site of Triunfo, forming in some locations small waterfalls and rapids along the river. The bone-bearing conglomerate is vertically and laterally

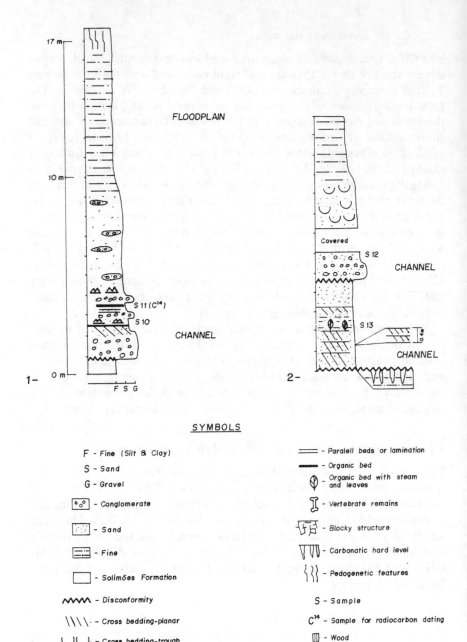

SYMBOLS

F - Fine (Silt & Clay)

S - Sand

G - Gravel

⬚ - Conglomerate

⬚ - Sand

⬚ - Fine

⬚ - Solimões Formation

〰️ - Disconformity

\\\\ - Cross bedding-planar

∪∪ - Cross bedding-trough

〜〜 - Cross lamination

═ - Paralell beds or lamination

▬ - Organic bed

Ⓥ - Organic bed with steam and leaves

I - Vertebrate remains

⌐F⌐ - Blocky structure

VVV - Carbonatic hard level

⌇⌇ - Pedogenetic features

S - Sample

C¹⁴ - Sample for radiocarbon dating

◫ - Wood

Figure 3. Dashed areas indicate the main sedimentary exposures. Coordinates are indicated in Table 1. a) Selected Profiles. Legend is the same for the next Figures 3b, c and d.

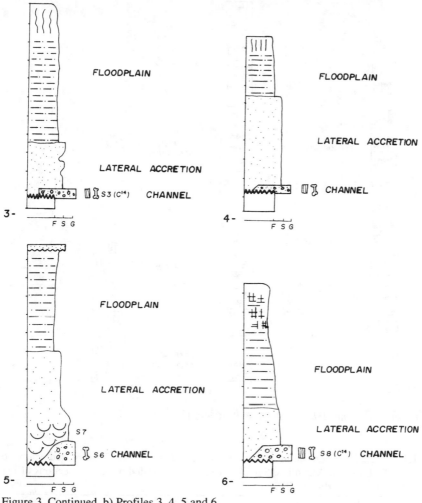

Figure 3. Continued. b) Profiles 3, 4, 5 and 6.

discontinuous. The prevailing colors range from black to red-brown, due to precipitation of iron oxides; most pebbles are composed of hard concretions and quartzites as well. The pebbles reach 10-15 cm in diameter and the matrix is sandy to clayey-sandy. The conglomerate is present as an indurated sediment, like a ferralithic crust. The X ray-diffraction analyses indicate that iron oxides, in the form of amorphous matter, are abundant. Quartz is abundant, goethite and hematite are largely present, but feldspar is scarce (Table 3).

In association with the conglomerate, sandy deposits can be found. The sands are coarse, red brown, and partly conglomeratic. Grain size analyses

35

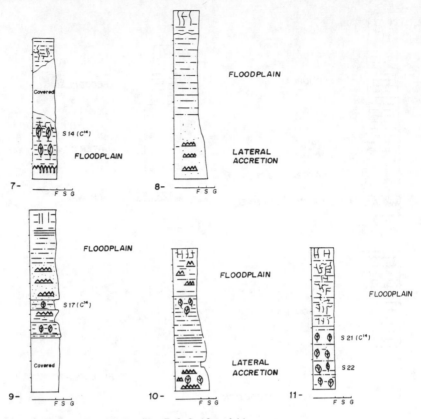

Figure 3. Continued. c) Profiles 7, 8, 9, 10 and 11.

Table 2. Radiocarbon dating performed in the Laboratorio de Tritio y Radiocarbono, LATYR, La Plata, Argentina. Sample locations is indicated in vertical sections of Fig. 3. Ages in radiocarbon yr BP.

Unit	Profile	Sample	Laboratory N[0]	Radiocarbon age
Older Pleistocene	1	S.9	LP-664	43900 ± 7000
Late Pleistocene	3	S.3	LP-634	29450 ± 1020
Late Pleistocene	6	S.8	LP-652	32300 ± 1600
Holocene sediments	7	S.14	LP-640	7960 ± 95
Holocene Sediments	9	S.17	LP-643	4510 ± 70
Holocene Sediments	11	S.21	LP-626	1990 ± 60

of selected samples are shown in Figure 4. The red colors are produced by iron oxidation. The sandy facies present trough cross bedding but planar cross-bedding and planar stratification may be found as well. Quartz is dominant and feldspar is scarce in the sand fraction (Table 3).

Table 3. Mineralogical composition determined by X-ray diffraction. Sample location is indicated in vertical sections of Figure 3. D = dominant; I = important; S = scarce; T = trace.

Unit	Sample	Q	F	Ill/Mi	K	Chl	P	H-G	S	C	Org. Matter	Iron Oxide Amorphous
OPS	S.11	D	I	S	S							
OPS	S.13	D	I	S	S	T	S	S			S	
OPS	S.9	D	I	S	S					I		
LPS	S.2	D	I	S	S	T						
LPS	S.6	I	S					I				D
LPS	S.4	I	S		T	T		S	S			D
HS	S.22	I	S		I					D		

References: Q. quartz, F. feldspar, Ill/Mi. illite/montmorillonite, K. kaolinite, Chl. Chlorite, P. pyrite, H-G. hematite-goethite, S. siderite, C. calcite.

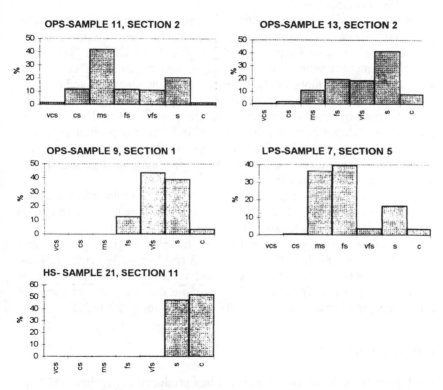

Figure 4. Grain size of sediments. Sample location is indicated in vertical sections of Figure 3. VCS. very coarse sand, CS. coarse sand, MS. medium sand, FS. fine sand, VFS. very fine sand, S. silt, C. clay, OQS. Old Quaternary Sediments, LPS. Late Pleistocene Sediments, HS. Holocene Sediments.

The conglomeratic and sandy deposits are typically channel deposits. We interpreted that the conglomerates indicate partly a channel lag facies and channel bars in other parts, as in Profile 5 (Fig. 3b). The sandy facies is interpreted as forming channel bars. Finer sediments such as silty sands are associated with conglomeratic and sandy deposits. They show a depositional dip indicating bed deposition by lateral accretion. This heterolithic stratification is covered by fine massive overbank deposits. The sediments of lateral accretion are reddish and mottled with limonithic staining and limonites plates along the contact of stratification beds. The more frequent colors are yellowish to reddish brown and, on the floodplain, finer grey to green-grey sediments can be found. In general, the sediments are unconsolidated or weakly consolidated. Selected vertical sections are shown in Fig. 3b (Profiles 3, 4, 5, 6). Two radiocarbon datings between 32 and 29 ka BP have been obtained in the Lower Terrace sediments (Table 2, Fig. 3b).

3.2.2 *Holocene Sediments*

We include in this category a set of sediments deposited during the Holocene. The younger deposits include part of the Recent sediments and Pleistocene Phase 2 of Simpson (in: Simpson & Paula Couto 1981).

The sediments are unconsolidated silty sands, clayey sands, clayey silts and silty clays. The more sandy deposits are yellowish brown and the finer are greenish-gray to gray. Black or dark organic beds up to one meter thick, formed by trunks and stems, are frequently found. The sandy deposits show a depositional dip of up to 15° indicating deposition by lateral accretion in point bars. The fine sediments are overbank deposits, mainly filling abandoned meanders. The sediments display a blocky structure and appear as reddish and mottled beds.

The facies of organic beds is very representative of this unit. Very good exposures of organic beds are found in the lower reaches of the Juruá River. Selected profiles are shown in Figure 3a, b, c and d.

Mineralogical composition and some examples of grain size distribution of the sediments are shown in Table 3 and Fig. 4. Mineralogical and grain size content of floodplain deposits, for example at Section 11, is dominated by a fine fraction, rich in organic matter. Three radiocarbon datings ranging between 8 and 2 ka BP were obtained (Table 2, Fig. 3c).

4 PALAEONTOLOGY

The Upper Juruá River is the area that has produced the earliest and largest quantity of Pleistocene mammals in Amazonia. Fossil mammals from the Juruá River are deposited at the Museu Paraense Emilio Goeldi (Belém), Museu Nacional (Rio de Janeiro), Departamento Nacional de Produção Mineral (Rio de Janeiro), American Museum of Natural History (New York) and Laboratório de Pesquisas Paleontológicas of the Universidade Federal do Acre (Rio Branco). These collections were all reviewed by Rancy (1991).

Until today, 24 genera have been recognized in Western Amazonia, of which 23 Pleistocene land mammal genera of Lujanan Land Mammal Age were recorded in the upper Juruá. The genus *Mylodon* has not been definitely recognized from the Juruá River. Among the 24 genera listed in Table 4, only five are extant in South America. An impressive feature of the

Table 4. List of Pleistocene land mammal genera from the Juruá River.

TAXA
EDENTATA/PILOSA
    Megatheriidae
        Ocnopus Reinhardt, 1875
        Eremotherium Spillmann, 1948
    Mylodontidae
        Glossotherium Owen, 1840
        Lestodon Gervais, 1855
        Scelidotherium Kraglievich, 1925
    Megalonychidae
        Megalonyx Harlan, 1823
EDENTATA/CINGULATA
    Pampatheriidae
        Pampatherium Ameghino, 1875
    Dasypodidae
        Euphractus Wagler, 1830
        Propraopus Ameghino, 1881
        Dasypus Linnaeus, 1758
    Glyptodontidae
        Hoplophorus Lund, 1839
        Neuryurus Ameghino, 1889
        Panochthus Burmeister, 1866
        Glyptodon Burmeister, 1898
PROBOSCIDEA
        Haplomastodon Hoffstetter, 1950
        Cuvieronius Osborn, 1923
ARTIODACTYLA
    Camelidae
        Lama Frisch, 1775
        Palaeolama Gervais 1867
    Tayassuidae
        Tayassu Fischer, 1814
NOTOUNGULATA
        Toxodon Owen, 1840
        Mixotoxodon van Frank, 1957
PERISSODACTYLA
        Tapirus Brisson, 1762
CARNIVORA
        Eira Hamilton Smith, 1842

Pleistocene fauna in western Amazonia is the diversity of mammalian species of very large size. There were two genera of each megatheriid ground sloths, toxodont notoungulates and gomphotheriid proboscidea. In addition, there was a variety of large edentates including large-shelled glyptodonts and pampatheres. Ground sloths include five genera that were of moderate size, compared to megatheriids. Particularly striking is the presence of two genera of Camelidae, *Palaeolama* and *Lama* (Rancy 1992).

Table 5 presents an inventory of the Pleistocene fossil mammals from the Upper Juruá River, which have been catalogued into the collection of the Laboratório de Pesquisas Paleontológicas of the Universidade Federal do Acre.

Table 5. Inventory of Pleistocene fossil mammals by family from the Juruá River, housed at the Laboratorio de Pesquisas Paleontologicas of the Universidade Federal do Acre, Rio Branco, Acre, Brazil.

EDENTATA/CINGULATA
Glyptodontidae (Osteoderms) – UFAC 022, UFAC 024, UFAC 025, UFAC 026, UFAC 1082, UFAC 1153, UFAC 1199, UFAC 1204.
Pampatheriidae (Osteoderms) – UFAC 023, UFAC 607, UFAC 861, UFAC 1131.

EDENTATA/PILOSA
Megatheriidae – UFAC 056 (left M^3), UFAC 065 (right M^4), UFAC 068 (right M^5), UFAC 096 (right metacarpus IV), UFAC 098 (left mandibular ramus), UFAC 099 (right tibia), UFAC 100 (left astragalus), UFAC 101 (distal portion of a left humerus), UFAC 108 (left M^5), UFAC 1118 (right radius).
Megalonychidae – UFAC 067 (molariform tooth), UFAC 074 (right M_2), UFAC 077 (molariform tooth).
Mylodontidae – UFAC 063 (left caniniform M^1), UFAC 082 (right M^5), UFAC 091 (left tibia), UFAC 102 (left humerus), UFAC 103 (left humerus).

NOTOUNGULATA
Toxodontidae – UFAC 142 (left I^2), UFAC 052 (right M_2), UFAC 085 (right I^1).

PROBOSCIDEA
Gomphotheriidae – UFAC 095 (right M_3), UFAC 104 (right M_2), UFAC 1212 (right M^2), UFAC 1213 (left mandibular ramus with M_3), UFAC 1214 (left M_2).

PERISSODACTYLA
Tapiridae – UFAC 034 (right mandibular ramus), UFAC 035 (right mandibular ramus-juvenile), UFAC 043 (right femur), UFAC 1150 (fragment of left mandibular ramus with P_3).

ARTIODACTYLA
Tayassuidae – UFAC 006 (right tibia), UFAC 008 (right radio-ulna), UFAC 011 (distal portion right femur), UFAC 016 (left tibia), UFAC 031 (left mandibular ramus), UFAC 032 (right mandibular ramus).
Camelidae – UFAC 061 (left P^4).

CARNIVORA
Mustelidae – UFAC 036 (right mandibular ramus)

40

5 FOSSILIFEROUS LEVELS

Two differents fossiliferous units can be differenciated in the Upper Juruá River. The older is the Solimões Formation, which has an Upper Miocene-Pliocene fauna (Huayquerian- Montehermosan SALMA). Unconformably over the Solimões Formation, the bone-bearing conglomerate of the Late Pleistocene sediments is found. The Pleistocene fossils were collected in the conglomerate facies of the Late Pleistocene sediments which was defined as the 'bone-bearing conglomerate' by Simpson & Paula Couto (1981). Some bone remains can be found as well in the sandy deposits related to the conglomerates.

The conglomerate has provided almost all of the Quaternary bone remains. Considering that the conglomerate is unconformably lying on the fossiliferous Solimões Formation, Miocene-Pliocene vertebrate remains can be found mixed with Quaternary vertebrates in the Late Pleistocene sediments. Trunks can be found in the bone-bearing conglomerate as well.

Older reworked trunks, probably coming from the Solimões Formation, are found mixed with the Pleistocene woods and trunks. The older trunks are totally silicified but the Pleistocene trunks contain iron oxide and some of them present a good state of preservation.

6 PALAEOENVIRONMENT RECONSTRUCTION DURING THE LATE PLEISTOCENE-HOLOCENE

The chronological and biostratigraphic record of the Upper Juruá River indicate a Late Pleistocene age for the fauna found in the bone-bearing conglomerate of the Late Pleistocene sediments.

The majority of the radiocarbon datings of the Late Pleistocene sediments indicate an infinite age or an age near the reliability limit of the method. However, a similar chronological record is found in other river basins of western Amazonia. In the Ucayali River, terrace sediments which are coarser than the present day sediments have been described. Trunks found in this terrace were deposited between 32 and more than 40 ^{14}C ka BP. In the Madre de Dios River, dating by Rasanen et al. (1990) in trunks indicates an age of 36-38 ^{14}C ka BP. In the Caquetá River, a lower terrace, composed of sand and coarse gravel of Cordilleran origin at its lower section, was dated between 55 ^{14}C ka BP and 30 ^{14}C ka BP (Van der Hammen et al. 1992a).

This sedimentation phase was correlated with the Middle Pleniglacial of the Last Glaciation by Van der Hammen et al. (1992a) in the Caquetá River and Dumont et al. (1992) in the Ucayali River. According to Van der Hammen & Absy (1994), Amazonia had a cooler or colder climate and relatively high values of rainfall during the Middle Pleniglacial changing to a colder climate, which was markedly drier, during the Late Pleniglacial (ca. 26 to 14-13 ^{14}C ka BP).

Considering the radiocarbon ages recorded in Ucayali, Madre de Dios and Caquetá rivers, we suggest that this phase that deposited the Late Pleistocene sediments in the Upper Juruá River can be tentatively correlated with the Middle Pleniglacial of the Last Glaciation.

However, our results indicate that the climatic deterioration in the lowlands of Southwestern Amazonia began previous to the Upper Pleniglacial of the Last Glaciation. At this time the dominant vegetation was of savanna type. The fauna recorded in the Juruá River indicates open country environments and the sedimentological record suggests dramatic changes in the hydrological behaviour of the lowland rivers.

Glyptodon, Eremotherium, Mylodon, Toxodon, Haplomastodon (Fig. 5) and *Cuvieronius* were well adapted groups for an open country, savanna or savanna-like habitat (Webb & Rancy 1996). The presence of *Lama* and *Palaeolama* (Fig. 6) in the Juruá River area are evidence of drier conditions in the region. *Tapirus* (Fig. 7) could indicate that the climate was tropical and that the aridity was not extreme. Redford & Eisenberg (1992; Map.9.1:227 and Map 10.4:235) suggest the geographic sympatry of *Tapirus* and *Lama* ('guanaco') on the border of Paraguay and Bolivia. During a field trip to the Paraguayan Chaco, in May 1997, we had the information about the sympatric occurrence of *Tapirus* and *Lama* on the same area. We infer that the same could have been possible in the region of the Juruá River at the border of Brazil and Peru, at least during some phases of the Pleistocene.

Considering the sedimentological record we can suggest that a hiatus exists between the Late Pleistocene sediments and the younger sediments

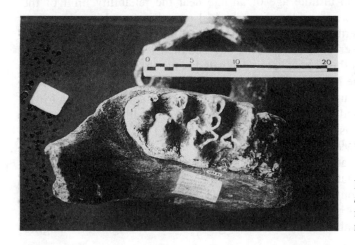

Figure 5. *Haplomastodon* – Left mandibular ramus with M$_3$.

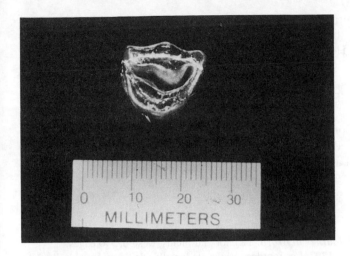

Figure 6. *Palaeo-lama* – Left tooth P[4], occlusal view.

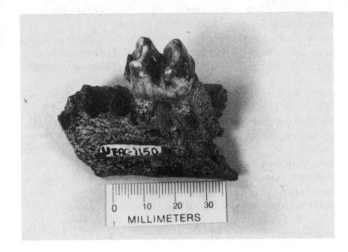

Figure 7. *Tapirus* – Left mandibular ramus with P_3.

of the Holocene. Until now, a sedimentological record does not appear in the fluvial belts of the Amazon rivers deposited during the Last Glacial Maximum times (ca. 24 ka to 14 ^{14}C ka BP). On the other hand, the sedimentological phase of deposition of the Latest Late Pleistocene (14-10 ka BP) that was recognized in the Ucayali, Caquetá and Negro rivers (Dumont et al. 1992, Van der Hammen et al. 1992b) and associated with the climatic changes produced during the deglaciation, was not identified in the Juruá River basin.

The sedimentological record of the Holocene is registered in the Juruá River since (at least) the last 8 ^{14}C ka BP, when the river was displaying a meandering pattern. At this time, the rain forest occuppied this area, which is confirmed by the large quantity of trunks and organic strata of leaves and stems in the Holocene sediments.

ACKNOWLEDGEMENTS

We thank Dr. Constance E. Campbell for English review of the manuscript and Dr. A. Rossi (University of Modena, Italy) by helping us in the sedimentological analyses.

REFERENCES

Agassiz, L. 1868. *A Journey in Brazil.* Ticknor and Fields, Boston, 540 pages.
Barbosa Rodrigues, J. 1892. Les reptiles fossiles de la vallée de l'Amazone. *Vellosia* 2: 41-56.
Chandless, W. 1866. Notes on the River Aquiry, the principal affluent of the River Purus. *Journal of the Geographical Society of London* 36: 119-128.
Dumont, J.F., F. García & M. Fournier 1992. Registros de cambios climáticos para depósitos y morfologías fluviales en la Amazonia Occidental. In: Ortlieb, L. & J. Macharé (eds), *Paleo ENSO records International Symposium,* Extended Abstracts, ORSTOM: 89-92.
Fawcett, P.H. 1909. Survey work on the frontier between Bolivia and Brazil. *Journal of the Geographical Society of London* 37: 181-185.
Frailey, C.D. 1986. Late Miocene and Holocene Mammals, exclusive of the Notoungulata of the Rio Acre Region, Western Amazon. *Contributions in Science,* Los Angeles County Museum 374: 1-46.
Frailey, C.D., E.L. Lavina, A. Rancy & J.P.P. de Souza Filho 1988. A proposal Pleistocene/Holocene Lake in the Amazon Basin and its significance to Amazonian Geology and Biogeography. *Acta Amazonica* 18: 119-143.
Gervais, P. 1876. Crocodile gigantesque fossile au Brèsil *(Dinosuchus terror). Journal de Zoologie* 3: 232-236.
Gervais, P.1877. Tortue gigantesque fossile au Brèsil. *Journal de Zoologie* 5-6: 283-285
Goeldi, E.P. 1906. Chelonios do Brasil. *Boletim do Museu Paraense* 4: 699-756. Brazil.
Gurich, G. 1912. *Gryposuchus jessei,* ein neues schmalschnauziges krokodil aus den jungeren ablagerungen de oberen Amazonas-Gebietes. *Jahrbuch der Hamburgischen Wissenschaftlichen Anstalten* 29: 59-71
Kraglievich, L. 1930. Reivindicación de *Carolibergia* de Mercerat, por Miranda Ribeiro. Crítica bibliográfica y descripción somera de un nuevo género. *Obras de Geología y Paleontología* 2: 473-474. La Plata. Argentina.
Kraglievich, L. 1931. Sobre *Trigonodops Lopesi* (Roxo) Kraglievich. *Obras de Geología y Paleontología* 2: 619-623. La Plata. Argentina.
Kronberg, D., R. Benchimol & M. Bird 1990. Evidence of aridity 50000 yr BP in western Amazonia. *Publicación Especial N^0 2. Proyecto IGCP 281,* Medellín, 7 pages. Colombia.

44

Latrubesse, E. 1992. El Cuaternario fluvial de la Cuenca del Purus en el Estado de Acre, Brasil. Ph.D. Dissertation, Universidad Nacional de San Luis, Argentina, 219 pages. Unpublished.

Latrubesse, E. & C. Ramonell 1993. El Neógeno de la Amazonia Sudoccidental brasilera. In: E. Franzinelli & E. Latrubesse (eds), *Resumos e Contribuições Científicas,* International Symposium on the Quaternary of Amazonia: 127-131. Universidade do Amazonas, Manaus. Brazil.

Latrubesse, E., J. Bocquentin, J. dos Santos & C. Ramonell 1997. Palaeoenvironmental Model for the Late Cenozoic of Southwestern Amazonia: Vertebrate Palaeontology and Geology. *Acta Amazonica,* 27(2): 103-118. Manaus, Brazil.

Marshall, L., A. Berta, R. Hoffstetter, R. Pascual., O. Reig, O. Bombin & A. Mones 1984. Mammals and stratigraphy: geochronology of the continental mammal bearing Quaternary of South America. *Paleovertebrata*: 1-76. Mémoire Extraordinaire. Montpellier.

Miranda Ribeiro, A. de 1938. *Plicodontinia mourai.* In: *Livro Jubilar do Professor Travassos:* 319-321. Instituto Oswaldo Cruz, Rio de Janeiro. Brazil.

Mook, C.C. 1921. *Brachygnathosuchus brasiliensis*, a new fossil crocodilian from Brazil. *Bulletin of American Museum of National History* 44: 43-49. New York.

Patterson,B. 1942. Two Tertiary mammals from northern South America. *American Museum Novitates* 1173: 1-7. USA. New York.

Paula Couto, C. de 1944. Noticia Preliminar sobre um novo Toxodonte do Cenozoico do Territorio do Acre. *Boletim do Museu Nacional,* Nova Série, Geologia 3: 1-4. Brazil.

Paula Couto, C. dc 1956. Mamíferos fósseis do Cenozoico da Amazônia. *Boletim do Conselho Nacional de Pesquisas* 3: 1-121. Brazil.

Paula Couto, C. de 1976. Fossil mammals from the Cenozoic of Acre, Brazil. 1. Astrapotheria. *Anais do XXVIII Congresso Brasileiro de Geologia* 1: 236-249. Brazil.

Paula Couto, C. de 1978. Fossil mammals from the Cenozoic of Acre, Brazil. 2. Rodentia, Caviomorpha Dinomydae. *Iheringia,* Série Geologia 5: 3-17.

Paula Couto, C. de 1981. Fossil mammals from the Cenozoic of Acre, Brazil. IV. Notoungulata, Notohippidae and Toxodontidae Nesodontinae. *Anais do II Congresso Latinoamericano de Paleontologia* 2: 461-477. Brazil.

Paula Couto, C. de 1982a. Fossil mammals from the Cenozoic of Acre, Brazil.V. Notoungulata, Nesodontinae (II) Haplodontheriinae and Litopterna, Pyrotheria and Astrapotheria (II). *Iheringia,* Série Geologia 7: 5-43.

Paula Couto, C. de 1982b. *Purperia,* a new name for *Megahippus* Paula Couto, 1981. *Iheringia,* Série Geologia 7: 69-70.

Paula Couto, C. de 1983a. Fossil mammals from the Cenozoic of Acre, Brazil. VI. Edentata Cingulata. *Iheringia,* Série Geologia 8: 101-120.

Paula Couto, C. de 1983b. Fossil mammals from the Cenozoic of Acre, Brazil. VII. Miscellanea. *Iheringia,* Série Geologia 8: 101-120

Price, L.I. 1957. Uma expedição geológica e zoológica ao alto Rio Juruá, Territorio Federal do Acre, realizada sob os auspicios do Instituto Nacional de Pesquisas da Amazônia. *Publicações Avulsas-INPA* 1: 1-5. Brazil.

Radambrasil, 1976. *Levantamento de Recursos Naturais, Folha SC 19 Rio Branco.* Volume 12. DNPM, Rio de Janeiro, 458 pages. Brazil.

Radambrasil, 1977. *Levantamento de Recursos Naturais, Folha SC 18 Javarí-Contamana.* Volume 13. DNPM, Rio de Janeiro, 413 pages. Brazil.

Rancy, A. 1981. Mamíferos fósseis do Cenozoico do Alto Juruá-Acre. Master Dissertation. Universidade Federal do Rio Grande do Sul, Porto Alegre, 122 pages. Brazil. Unpublished.

Rancy, A. 1991. Pleistocene Mammals and Paleoecology of the Western Amazon. Ph.D. Dissertation. University of Florida, Gainesville, 151 pages. Unpublished.

Rancy, A. 1992. The unexpected presence of Camelidae during the Pleistocene in the Western Amazon. *Journal of Vertebrate Paleontology* 12(3): 48-49.

Rasanen, M., J.S. Salo, H. Jungnert & L. Romero Pitman 1990. Evolution of western Amazon lowland Relief: impact of Andean foreland dynamics. *Terra Nova* 2: 320-332.

Redford, K.H. & J.F. Eisenberg 1992. *Mammals of the* Neotropics – *The Southern Cone*. University of Chicago Press, 430 pages.

Roxo, M.G. de O. 1921. *Note on a new species of Toxodon Owen, T. lopesi Roxo*. Rio de Janeiro: Empresa Brasil Editora, 12 pages.

Roxo, M.G. de O. 1937. Fósseis pliocenos do Rio Juruá, Estado do Amazonas. Departamento de Geologia e Mineralogia. *Notas Preliminares e Estudos* 9: 4-10. Brazil.

Simpson, G.G. & C. de Paula Couto 1981. Fossil mammals from the Cenozoic of Acre Brazil. III. Pleistocene Edentata Pilosa, Proboscidea, Sirenia, Perissodactyla and Artiodactyla. *Iheringia*, Série Geologia 6: 11-73.

Spillmann, F. 1949. Contribución a la paleontología del Perú. Una mamifauna fósil de la región del Río Ucayali. *Publicaciones de Museo de Historia Natural 'Javier Prado'*, Universidad Mayor de San Marcos, Serie Geología y Paleontología 1: 1- 40.

Van der Hammen, T. & M.L. Absy 1994. Amazonia during the Last Glacial. *Palaeogeography, Palaeoclimatology, Palaeoecology* 109: 247-261.

Van der Hammen, T., J.F. Duivenvoorden, J.M. Lips, L.E. Urrego & N. Espejo 1992a. The Late Quaternary of the Middle Caquetá River area (Colombian Amazonia). *Journal of Quaternary Sciences* 7(1): 45-55.

Van der Hammen, T., L.E. Urrego, N. Espejo, J.F. Duivenvoorden & J.M. Lips 1992b. Late-glacial and Holocene sedimentation and fluctuations of river water level in the Caquetá area (Colombian Amazonia). *Journal of Quaternary Sciences* 7(2): 57-67.

Webb, S.D. & A. Rancy 1996. Late Cenozoic Evolution of the Neotropical Mammal Fauna. In Jackson, J.B.C., A.B.Budd & A.G.Coates (eds), *Evolution and Environment in Tropical America*: 335-358. The University of Chicago Press, Chicago.

Determination of water palaeotemperature in the Beagle Channel (Argentina) during the last 6000 yr through stable isotope composition of *Mytilus edulis* shells

4

BOGOMIL OBELIC
Rudjer Boskovic Institute, Zagreb, Croatia

AURELI ÁLVAREZ & JUDIT ARGULLÓS
Universitat Autònoma de Barcelona, Departamento de Geología, Bellaterra, Spain

ERNESTO LUIS PIANA
Centro Austral de Investigaciones Científicas (CADIC), CONICET, Ushuaia, Tierra del Fuego, Argentina

ABSTRACT: This paper presents surface water palaeotemperature data of the Beagle Channel in the last 6000 yr, using the isotopic composition ($\delta^{18}O$) of *Mytilus edulis* shells from archaeological shell middens. The accumulation of these 30-40 mm shells is the result of the exploitation of marine resources by aborigine people in the Fuegian archipelago. The absolute age of shells was determined by radiocarbon measurement of associated charcoal. The isotopic composition of the present shells collected in the period 1988-1993 from the marine waters near shell middens is used in palaeotemperature calculation. The results obtained do not indicate any significant changes of temperature during the studied period. A slightly lower temperature (0.8°C) of superficial sea water compared with the present temperatures was observed during the studied period.

RESUMEN: Este artículo presenta las paleotemperaturas del agua superficial del canal de Beagle en los últimos 6000 años a través de la composición isotópica ($\delta^{18}O$) obtenida de las valvas de *Mytilus edulis* recogidas de yacimientos arqueológicos (concheros). La acumulación de estas valvas, de 30-40 mm de longitud, es el resultado de la explotación de recursos marinos por parte de los antiguos aborígenes adaptados a la vida litoral en los canales magallánico-fueguinos. Su edad se determinó mediante la datación por ^{14}C en carbón vegetal asociado perfectamente a las valvas. También se han obtenido valores isotópicos de valvas actuales (1988-1993) extraídas del agua cercana a los sitios arqueológicos que usamos para calcular la paleotemperatura. Una temperatura más baja del agua superficial (0.8°C), en comparación con la temperatura actual en el agua del canal, ha sido observada para el período estudiado.

1 INTRODUCTION

The Beagle Channel (lat. 54°53'S, long. 67°00"–68°30'W) is about 180 km long and communicates the Atlantic and Pacific oceans, separating Isla Grande de Tierra del Fuego from the southern islands of the Fuegian archipelago (Fig. 1). This uttermost part of the world is the nearest to Antarctica, and – according to the present knowledge – both water temperature and natural radioactive (^{14}C Reservoir Effect) isotopes depend on Antarctic ice melting. Within the global warming framework it was of primary importance to determine whether climatic fluctuations or stable conditions are reflected in isotopic relations for the last 8000 yr, i.e. after the opening of the Beagle Channel (Rabassa et al. 1986), as a new insight to the main question: is this a process of fluctuation or a change? The

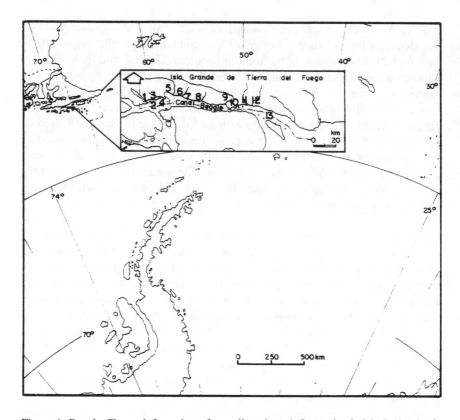

Figure 1. Beagle Channel. Location of sampling sites: 1. Lapataia; 2. Isla Redonda; 3. Ensenada; 4. Punta Occidental; Bahía Golondrina; 5. Lomada Alta Olivia; 6. Túnel, Lancha Packewaia, Encajonado; 7. Remolino or Shamakush; 8. Almanza, Bahía Brown; 9. Isla Gable; 10. Harberton, Bahía Relegada; 11. Bahía Varela, Rancho Tambo; 12. Cambaceres; 13. Islas Becasses.

question is whether the present isotopic relationships in the Beagle Channel are within the fluctuations range of those in the past. In this regard, the Beagle Channel nearness to the Antarctic Peninsula and its hyperoceanic climate (Tukhanen 1992) must be born in mind.

For the calculation of palaeotemperature in the area of interest it is necessary to quantify stable isotope values. The most suitable carbonate-secreting organism for this purpose is the mollusk *Mytilus edulis* because it has its ubiquity in the area and because it was used from early ages as a food resource by indigenous people. These people, whose descendants were ethnographically known as *Yámanas* or *Yaghans*, developed a nomadic canoer tradition based on successful exploitation of marine resources since 6500 yr ago up to the beginning of this century when their cultural behavior was finally disrupted and they faced almost total extinction (Orquera & Piana 1987). Their numerous campsites can be still seen along the beaches as rounded mounds consisting mainly of shells, thrown away after consumption.

The archaeological samples were obtained from about 20 archaeological sites in the region along the northern coast of the Beagle Channel, between Lapataia and Moat bays (Fig. 1). The samples were classified and dated by the ^{14}C method by using wood charcoal remains, always found in shell middens, and perfectly associated with the shell samples.

The present-day shells have been collected along the same littoral zone (from Lapataia, Ensenada, Túnel, Bahía Golondrina and Punta Occidental) and similar sea depths to the ones where the archaeological samples were originally gathered. To follow the same conditions in which archaeological samples were treated, the mollusks were opened on the same way as the *Yámanas* did when cooking them: Placing the mollusk directly on the hot coal and picking them up immediately after opening.

2 THEORETICAL BACKGROUND

The deposition of shell carbonates in water is accompanied by the isotope exchange, i.e. the exchange of atoms between chemical components or phases resulting in an equilibrium state between different isotopic composition involved. The reason for this exchange between atoms of an element is the isotopic fractionation caused by their mass differences. The isotopic composition of deposited carbonates depends on two main factors: (1) The isotopic content of the water and (2) The temperature at which the deposition takes place, the latest being a consequence of the temperature dependence on the fractionation between water and calcite (Mook 1971). This dependency was very well established for the ratio of oxygen isotopes ^{18}O and ^{16}O in the system calcite-H_2O, which describes the precipitation of shells and skeletons in surrounding sea water (Epstein & Mayeda

1953; Craig 1965). On the other hand, $^{13}C/^{12}C$ ratio is controlled not only by one parameter varying over narrow limits (temperature), but is the function of several additional factors, such as the origin of bicarbonates (dissolved inorganic carbon) and food and therefore cannot be used for palaeotemperature calculation.

To avoid the accurate determination of the absolute amount of isotopes in every compound through routine analyses, an expression was introduced for geochemical purposes. The relative differences in the isotopic composition of a sample with respect to a standard value is given by the δ-value, defined as:

$$\delta \equiv \frac{R_{sample} - R_{standard}}{R_{standard}} \cdot 1000 \, (\%o) \tag{1}$$

(Epstein & Lowenstam 1953), where R represents the ratio between the heavier (less abundant) isotope, in our case ^{18}O, and the lighter (more abundant) isotope, in our case ^{16}O.

Since shells were precipitated in chemical equilibrium with the surrounding sea water, the following palaeotemperature equation was introduced by Epstein & Mayeda (1953) and improved by Craig (1965):

$$t \, (^{\circ}C) = 16.9 - 4.2 \cdot (\delta_c - \delta_w) + 0.13 \cdot (\delta_c - \delta_w)^2 \tag{2}$$

where both δ_c and δ_w are expressed as $^{18}O/^{16}O$ isotope ratios (Eq. 1). δ_c is defined as CO_2 gas obtained from shell $CaCO_3$ by the same technique as the PDB standard (defined as $\delta^{18}O$ value of CO_2 obtained from rostrum of a Cretaceous belemnite – *Belemnitella americana* from the Peedee Formation in South Carolina – by reaction with 100% H_3PO_4 at 25°C) and is related to SMOW (Standard Mean Ocean Water) as:

$$\delta_c \equiv (\delta^{18}O_c)_{PDB} = 0.97006 \cdot (\delta^{18}O_c)_{SMOW} - 29.94 \tag{3}$$

δ_w is expressed as CO_2 equilibrated with Standard Mean Ocean Water (SMOW) at 25°C, and is represented on PDB scale as (Craig 1961):

$$\delta_w \equiv (\delta^{18}O_w)_{PDB} = 0.9978 \cdot (\delta^{18}O_w)_{SMOW} - 0.22 \tag{4}$$

Temperatures derived from Equation 2 depend critically on the isotopic composition of ocean water (δ_w.). This composition varied in the past in response to the amount and the isotopic composition of excess continental ice which is largely depleted in heavy isotopes compared to SMOW. The melting of Antarctic ice caps with an average oxygen isotope content of − 33‰ could cause the decrease in oxygen content of sea water, which consequently causes a decrease in the isotopic composition of the shells for the given temperature. The estimated decrease in $\delta^{18}O$ in water between

50

the glacials and interglacials was 1.23‰ (Dansgaard & Tauber 1969), but such a large decrease should not be expected within the last 6000 yr.

The other problem in determining palaeotemperatures is that the regions of the sea where mollusk shells are normally found, especially coastal waters and estuaries, are often influenced by fresh water (brackish water) which is generally depleted in ^{18}O (relative to ocean water) because they are influenced by precipitation which is generally depleted in heavy isotopes. Long-term measurements of oxygen isotopes in precipitation at Ushuaia, as a part of a IAEA-WMO program, gave $\delta^{18}O = -10.53$‰ for the period 1981-1985 (Rozanski et al. 1993). Depending on the local degree of freshwater admixture, the ^{18}O content of coastal water is consequently lower, causing correspondingly lower $\delta^{18}O$ of the shell carbonate. If this factor were ignored, it would result in erroneous, i.e. too high temperatures.

Biological factors also play an important role in the recording of temperature in calcareous shells. Marine animals may deposit shell over only a portion of the total temperature range, rather than over the entire ambient range (Epstein & Lowenstam 1953). Therefore, the analyses of such shell material will yield different temperatures for different species. If an entire shell were to be prepared for isotope measurements, the 'average' temperature obtained would be heavily biased toward maximum growth temperature, usually higher.

However, many organisms or parts of organisms in contact with water produce $CaCO_3$ which is not in equilibrium with the surrounding water. This so called 'vital effect' (Urey et al. 1951) is species-specific and time invariant. Various mechanisms are responsible for this effect, such as physiological control of isotopic composition or kinetic isotope effects. The degree of disequilibrium is generally smaller for oxygen than for carbon, because there exists a larger pool of oxygen (i.e. water) and a smaller one of carbon (mostly dissolved bicarbonates) which interacts with the various metabolic products associated with shell formation (Wefer & Berger 1991). Any deviation from equilibrium would change the isotopic composition of shells and this would lead to wrong temperatures obtained by Equation 2. In fact, such vital effects have been observed, but not yet properly understood.

3 MATERIALS AND METHODS

Archaeological shells are always associated with organic material, mostly charcoal remains. These 'fossil' valves have been collected from shell middens that were never affected by the variations in the sea level. Samples from layers that ever suffered any special temperature change (such as hearths or wood fire on them) have been discarded (Orquera & Piana

1992). The mean size of the majority of shells is relatively small, however the smallest of them indicate the growing interval of at least two yr (Silva 1996). Medium size of archaeological shells is from 30 mm to 40 mm, and dimension of present-day shells is 57.4±7.2 mm (Orquera & Piana 1994).

The ages of fossil shells which we used were obtained by measuring the ^{14}C content in the associated charcoal remains, mostly of *Nothofagus* (the southern beech, the dominant genus in the Fuegian forests) origin. Although the Reservoir Effect of the zone is known (Albero et al. 1986), charcoal was preferred than shells themselves to avoid modifications due to freshwater income in specific points (Albero et al. 1987). Radiocarbon activities were corrected to δ^{13}C value of –25‰.

In order to eliminate as much as possible the distorting influence of excessive high temperature, the present-day shells were thermally tested. Changes in color and other alterations helped us to eliminate those shells that underwent excessive heat when cooked. Any sign of thermal alteration – even doubtful ones – in a shell was enough to discard it.

3.1 *Mineralogical analyses*

The mollusks synthesize simultaneosly two polymorphs of $CaCO_3$, i.e. calcite (rhomboedric system) and aragonite (orthorhombic system). Since isotopic enrichment factors for aragonite follow different rules than those for calcite, causing the fractionation of calcite in relation to aragonite to be 0.6‰ at 25°C (Tarutani et al. 1969) and about 1.2‰ at 7°C (Sommer & Rye 1978), it was necessary to know the exact proportion of these minerals in the shells. In the valves of the *Mytilus* shells the outer layer consists of calcite and the inner layer of aragonite, and their ratio varies with temperature (Eisma et al. 1976). The percentage of calcite is inversely proportional to the mean temperature of water, i.e. higher temperature favors the formation of aragonite (Lowenstam 1954, Dodd 1963, 1964). This inverse relation, according to the same authors, also applies to salinity.

In order to know how aragonite and calcite are distributed within the shells we used the Feigel solution (Hutchinson 1974). The stained thin section of the *Mytilus* shells was observed with a petrographic microscope: aragonite was basically located on the mother-of-pearl part of the shell, though it could also be found on the hinge. However, on the external part of the individual shells mainly calcite was found. According to these results we took the samples from the external part of the shells and then ground them in an agate mortar.

Nonetheless, to maintain control over the following measurements, all samples were characterized by X-ray diffraction. Some samples have a small percentage of aragonite present in some samples. To calculate this percentage we prepared some comparative standards made by known ratios of pure calcite and pure aragonite and assessed them with X-ray analysis. The diffraction counts corresponding to known aragonite-calcite ratios enabled us to establish a regression curve which was used afterwards for samples containing the unknown mineralogic ratio. This is shown in Table 1 for modern samples and in Table 2 for archaeological samples.

Table 1. The measured and corrected $\delta^{18}O$ values of modern shells due to the presence of aragonite. The percentage of calcite in samples (Column 5) has been calculated on the basis of the relation $C=A+B*R$, where R is the count ratio (Column 4), and coefficients are $A=88.75$ and $B=0.06$. In the last two columns, the ^{18}O contents in pure calcite are calculated according to Equation 5, taking into account the isotopic fractionation between calcite and aragonite ε as 0.6‰ (at 25°C) and 1.2‰ (at 7°C), respectively.

Sample code	Calcite (counts)	Aragonite (counts)	Counts ratio	Calcu-lated (%)	$\delta^{18}O_{(PDB)}$ (mixture)	$\delta^{18}O_{(PDB)}$ (calcite, ε=0.6)	$\delta^{18}O_{(PDB)}$ (calcite, ε=1.2)
1	2145	44	488	91.8	−0.04	−0.09	−0.14
2	2030	0	*	100.0	−0.81	−0.81	−0.81
3	1651	41	40.3	91.3	0.09	0.04	−0.02
4	2079	62	33.5	90.9	0.51	0.46	0.40
5	2060	57	36.1	910	0.39	0.34	0.28
6	4511	103	43.8	91.5	−0.36	−0.41	−0.46
7	4523	93	48.6	91.8	−0.26	−0.31	−0.36
8	4236	159	26.6	90.4	0.15	0.09	0.04
9	3577	129	27.7	905	−0.14	−0.20	−0.25
10	3808	47	81.0	93.8	0.27	0.23	0.20
11	4167	103	40.5	913	0.60	0.55	0.50
12	3968	157	25.3	90.3	−0.06	−0.12	−0.18
13	4028	128	31.5	90.7	0.42	0.36	0.31
14	4252	126	33.8	90.9	−0.04	−0.10	−0.15
15	3842	122	31.5	90.7	0.73	0.67	0.62
16	1485	90	16.5	89.8	0.22	0.16	0.10
17	1875	80	23.4	90.2	0.32	0.26	0.20
18	1907	83	23.0	90.2	0.38	0.32	0.26
19	1815	113	16.1	89.8	−0.09	−0.15	−0.21
20	1268	63	20.1	90.0	0.33	0.27	0.21
21	1256	85	14.8	89.7	−0.05	−0.11	−0.17
22	1999	0	*	100.0	0.67	0.67	0.67
23	1959	100	19.6	90.0	−0.74	−0.80	−0.86
24	1933	107	18.1	89.9	−0.85	−0.91	−0.97
25	1888	0	*	100.0	0.52	0.52	0.52
26	1713	93	18.4	89.9	−0.74	−0.80	−0.86
27	2067	68	30.4	90.6	0.27	0.21	0.16
28	1863	90	20.7	90.1	0.07	0.01	−0.05
29	2308	28	82.4	93.9	0.43	0.39	0.36
30	2079	117	17.8	89.9	0.81	0.75	0.69
31	1554	0	*	100.0	−0.13	−0.13	−0.13
32	1874	75	25.0	90.3	0.45	0.39	0.33
33	3703	180	20.6	90.0	0.05	−0.01	−0.07

Table 1. Continued.

Sample code	Calcite (counts)	Aragonite (counts)	Counts ratio	Calcu-lated (%)	$\delta^{18}O_{(PDB)}$ (mixture)	$\delta^{18}O_{(PDB)}$ (calcite, ε=0.6)	$\delta^{18}O_{(PDB)}$ (calcite, ε=1.2)
34	3477	193	18.0	89.9	−0.23	−0.29	−0.35
35	3603	220	16.4	89.8	−0.16	−0.22	−0.28
36	3842	120	32.0	90.8	−0.20	−0.26	−0.31
37	4311	150	28.7	90.6	−0.06	−0.12	−0.17
38	4075	188	21.7	90.1	0.10	0.04	−0.02
39	3934	158	24.9	90.3	0.47	0.41	0.35
40	3826	249	15.4	89.7	−0.10	−0.16	−0.22
41	3703	213	17.4	89.8	−0.68	−0.74	−0.80
42	4005	178	22.5	90.2	−0.17	−0.23	−0.29
43	3739	246	15.2	89.7	−0.46	−0.52	−0.58
44	3923	195	20.1	90.0	−0.23	−0.29	−0.35
45	4008	167	24.0	90.3	0.23	0.17	0.11
46	3802	175	21.7	90.1	−0.42	−0.48	−0.54
47	4123	192	21.5	90.1	−0.43	−0.49	−0.55
48	4047	108	37.5	91.1	0.86	0.81	0.75
49	4012	136	29.5	90.6	0.32	0.26	0.21
50	4242	146	29.1	90.6	−0.56	−0.62	−0.67
51	4104	121	33.9	90.9	0.38	0.33	0.27
52	4123	101	40.8	91.3	0.62	0.57	0.52
53	4375	95	46.1	91.6	0.40	0.38	0.33
54	3845	162	23.7	90.2	0.10	0.04	−0.02
55	4455	57	78.2	93.7	0.22	0.18	0.14
56	4301	109	39.5	91.2	−0.39	−0.44	−0.50
57	4410	86	51.3	92.0	0.04	−0.01	−0.06
58	4079	0	*	100.0	−0.08	−0.08	−0.08
59	4474	70	63.9	92.8	0.23	0.19	0.14

Table 2. The measured and corrected $\delta^{18}O$ values of archaeological shells due to the presence of aragonite. The percentage of calcite in samples (Column 5) has been calculated on the basis of the relation $C=A+B*R$, where R is the count ratio (Column 4), and coefficients are A=88.75 and B=0.06. In the last two columns, ^{18}O contents in pure calcite are calculated according to Equation 5, taking into account isotopic fractionation between calcite and aragonite ε as 0.6‰ (at 25°C) and 1.2‰ (at 7°C), respectively.

Sample code	Calcite (counts)	Aragonite (counts)	Counts ratio	Calcu-lated (%)	$\delta^{18}O_{(PDB)}$ (mixture)	$\delta^{18}O_{(PDB)}$ (calcite, ε=0.6)	$\delta^{18}O_{(PDB)}$ (calcite, ε=1.2)
60	1870	63	29.7	90.6	1.06	100	0.95
61	1790	44	40.7	91.3	0.25	0.20	0.15
62	2140	18	118.9	96.2	0.52	0.50	0.48
63	1600	25	64.0	92.8	0.93	0.89	0.84
64	2270	28	81.1	93.8	0.32	0.28	0.25
75	1840	18	102.2	95.2	0.82	0.79	0.76
66	4168	64	65.2	92.8	0.66	0.62	0.57
67	4259	47	90.6	94.4	0.83	0.80	0.76
68	3924	0	*	100.0	1.04	1.04	1.04

Table 2. Continued.

Sample code	Calcite (counts)	Aragonite (counts)	Counts ratio	Calcu-lated (%)	$\delta^{18}O_{(PDB)}$ (mixture)	$\delta^{18}O_{(PDB)}$ (calcite, ε=0.6)	$\delta^{18}O_{(PDB)}$ (calcite, ε =1.2)
69	3786	63	60.1	92.5	0.39	0.35	0.30
70	3960	35	113.1	95.9	1.31	1.29	1.26
71	4390	0	*	100.0	0.98	0.98	0.98
72	3910	0	*	100.0	1.00	1.000	1.00
73	–	–	*	100.0	0.98	0.98	0.98
74	3920	0	*	100.0	0.99	0.9	0.99
75	4680	0	*	100.0	0.68	0.68	0.68
76	3930	0	*	100.0	0.69	0.69	0.69
77	3560	0	*	100.0	0.99	0.99	0.99
78	4300	40	107.5	95.5	0.81	0.78	0.76
79	3600	0	*	100.0	1.21	1.21	1.21
80	3650	0	*	100.0	0.77	0.77	0.77
81	4120	42	98.1	94.9	0.49	0.46	0.43
82	3650	0	*	100.0	0.72	0.72	0.72
83	4070	0	*	100.0	0.76	0.76	0.76
84	4300	0	*	100.0	0.88	0.88	0.88
85	–	–	–	–	–	–	–
86	4180	0		100.0	0.43	0.43	0.43
87	3570	34	105.0	95.3	0.83	0.80	0.77
88	3590	0	*	100.0	1.10	1.10	1.10
89	4550	0	*	100.0	0.83	0.83	0.83
90	3780	0	*	100.0	0.80	0.80	0.80
91	4250	0	*	100.0	0.57	0.57	0.57
92	3680	0	*	100.0	0.59	0.59	0.59
93	4330	0	*	100.0	0.68	0.68	0.68
94	3410	0	*	100.0	0.58	0.58	0.58
95	4280	0	*	100.0	0.39	0.39	0.39
96	4180	0	*	100.0	0.84	0.84	0.84
97	3970	0	*	100.0	1.02	1.02	1.02
98	4540	0	*	100.0	0.83	0.83	0.83
99	4730	0	*	100.0	0.73	0.73	0.73
100	4300	30	143.3	97.7	0.75	0.74	0.72
101	4040	0	*	100.0	0.40	0.40	0.40
102	4350	0	*	100.0	–0.14	–0.14	–0.14
103	4200	34	123.5	96.5	0.75	0.73	0.71
104	3920	0	*	100.0	0.64	0.64	0.64
105	4210	0	*	100.0	0.28	0.28	0.28
106	4150	0	*	100.0	0.57	0.57	0.57
107	4530	0	*	100.0	0.18	0.18	0.18
108	4180	0	*	100.0	–0.96	–0.96	–0.96
109	3300	0	*	100.0	0.61	0.61	0.61
110	4040	0	*	100.0	0.58	0.58	0.58
111	3600	0	*	100.0	0.88	0.88	0.88
112	3960	0	*	100.0	0.70	0.70	0.70
113	3930	31	126.8	96.7	0.78	0.76	0.74
114	4170	0	*	100.0	0.66	0.66	0.66

Table 2. Continued.

Sample code	Calcite (counts)	Aragonite (counts)	Counts ratio	Calculated (%)	$\delta^{18}O_{(PDB)}$ (mixture)	$\delta^{18}O_{(PDB)}$ (calcite, ε=0.6)	$\delta^{18}O_{(PDB)}$ (calcite, ε =1.2)
115	4690	0	*	100.0	0.61	0.61	0.61
116	3573	55	65.0	92.8	0.57	0.53	0.49
117	3986	53	75.2	93.5	0.26	0.22	0.18
118	3593	101	35.6	91.0	–0.15	–0.21	–0.26
119	4217	0	*	100.0	0.00	–0.00	–0.00
120	3591	103	34.9	90.9	0.07	0.02	–0.04
121	–	–	–	–	–	–	–
122	3626	0	*	100.0	0.35	0.35	0.35
123	4280	0	*	100.0	0.55	0.55	0.55
124	4337	0	*	100.0	0.45	0.45	0.45
125	3616	0	*	100.0	0.45	0.45	0.45
126	3520	76	46.3	91.7	–0.11	–0.16	–0.21
127	3891	0	*	100.0	–0.06	–0.06	–0.06
128	4008	0	*	100.0	–0.41	–0.41	–0.41
129	4052	0	*	100.0	–0.10	–0.10	–0.10
130	3811	93	41.0	91.3	0.29	0.24	0.19
131	3532	139	25.4	90.4	–0.43	–0.49	–0.55
132	3824	76	50.3	91.9	0.41	0.37	0.32
133	3840	0	*	100.0	–0.07	–0.07	–0.07
134	3548	121	29.3	90.6	0.40	0.34	0.28
135	3929	0	*	100.0	–0.28	–0.28	–0.28
136	4249	44	96.6	94.8	0.33	0.30	0.27
137	4143	0	*	100.0	0.85	0.85	0.85
138	3948	54	73.1	93.3	0.26	0.22	0.18
139	4234	0	*	100.0	0.22	0.22	0.22
140	3558	0	*	100.0	0.11	0.11	0.11
141	4779	0	*	100.0	0.42	0.42	0.42
142	4980	49	101.6	95.1	0.33	0.30	0.27
143	5204	126	41.3	91.3	0.38	0.32	0.27
144	5302	0	*	100.0	0.68	0.60	0.60
145	4838	0	*	100.0	0.55	0.55	0.55
146	4840	0	*	100.0	0.51	0.51	0.51
147	5021	0	*	100.0	0.47	0.47	0.47
148	4252	31	137.2	97.4	0.41	0.39	0.38
149	4843	34	142.4	97.7	–0.01	–0.02	–0.04
150	4813	55	87.5	94.3	0.48	0.45	0.41

3.2 Present-day temperature and salinity

The present-day temperature data have been collected in the past few yr by the CADIC staff. Sea surface temperature in the Beagle Channel has been measured since 1985 by CADIC with a circular belt thermograph installed in 1985 at the site called 'Muelle de Combustibles', located at the piers of the port of Ushuaia. Data are weekly quality controlled with an immersion thermometer. For the present-day period (1989-1993) the temperature data from the Ushuaia bay are the only available. Data are comparable to the average temperature of sea water in Beagle Channel (Iturraspe et al. 1989) at 2 m depth which is 6.5°C, with a maximum in February being 10°C and a minimum in August being 3°C.

The sea water salinity was also controlled at the sites where present-day mollusk were sampled. The salinity data helped determining locations with possible influence of fresh water. A contribution of fresh water could change the isotopic content because of the influence of low ^{18}O water and the organic material from Isla Grande of Tierra del Fuego (Eisma et al. 1976). Therefore, both, archaeological and present-day shells were collected by dividing them into two groups: Those with the freshwater influence and those without it.

The salinity in the bay of Ushuaia reaches its highest values in July (31 to 35 g/L) and diminishes in spring and summer (Fig. 2) due to the influence of the Río Olivia – one of the most important sources in the area – whose drainage depends on ice melting. The freshwater contribution into the bay of Ushuaia is most changeable in November and December, reaching minimal values of 23 g/L.

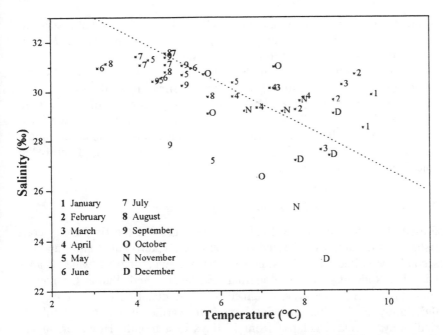

Figure 2. The correlation of salinity with temperature in the bay of Ushuaia measured within the period 1971-1987. Coefficients of linear fitting to the curve *Salin.*= *A+B*temp.* are *A*= 32.63 ± 0.47 and *B*= -0.40 ± 0.07, respectively, with *R*= -0.67. For calculation only samples denoted by * were taken into account.

3.3 ^{18}O and ^{13}C isotopic analyses

In addition to calcium carbonate the shells contain a small amount of organic material which is important for the growth of calcite crystals (Donner & Nord 1986). Therefore, powdered samples were pretreated in a plasma reactor, model (Fisons) PT 7150 (RF Plasma Bareel Etcher), which requires 10 mg as the maximal quantity of the sample was used. The samples have been afterwards measured in the mass spectrometer in order to obtain the ratio of isotopes $^{18}O/^{16}O$ and $^{13}C/^{12}C$ (the analyses were made by SIRA 10 and Delta S spectrometers).

The internal precision of the laboratory is:

$$\Delta\delta^{13}C = \pm0.02$$

$$\Delta\delta^{18}O = \pm0.06$$

The samples show a minor percentage of aragonite due to the difficulty to separate it from the calcite (Table 1 and Table 2). The correction due to the influence of the aragonite to the results has been made by using the following formula:

$$\delta^{18}O_c = \delta^{18}O_m + \varepsilon \cdot (p-1) \tag{5}$$

where $\delta^{18}O_c$ = $\delta^{18}O$ value in calcite, $\delta^{18}O_m$ = measured $\delta^{18}O$ value of sample (mixture of calcite and aragonite), p = percentage of calcite, ε = enrichment factor between aragonite and calcite.

This correction was made under the assumption that at about 7°C the aragonite is 1.2‰ richer in ^{18}O than calcite formed under the same conditions (Sommer & Rye 1978). Due to the minor percentage of aragonite in most samples (maximally 10%) the difference between the measured and calculated $\delta^{18}O$ values (related to pure calcite) is relatively small causing a decrease of no more than 0.1‰, which does not influence significantly the calculated values of temperature.

3.4 Present-day isotopic composition of sea water

Present-day mollusks were divided into two groups, the first group consisting of the samples collected from pure sea water (Túnel site), and the second one collected at the sites with the influence of fresh water (Lapataia, Isla Redonda-Bahía Golondrina, Ensenada, and Punta Occidental sites; Fig. 1). Samples were obtained from the distal portion of the shell, thus implying the final growing period. In warmer water, this 'final period' may represent a lapse from 15 days to a month. However, in the Beagle Channel the growth pattern is uneven, depending on coastal micromorphological features and the growing rhythm is much slower (Silva 1996). So, to be cautious, the 'final growing' period included in the sampled portion was regarded as a three months lapse. This lapse was used as temperature values (fourth column of Tables 3 and 4).

Present-day temperature data, along with the mollusk stable isotope data ($\delta^{18}O_c$) were used in Equation 2 in order to obtain the isotopic content of sea water ($\delta^{18}O_w$), which is necessary as the standard for the calculation of palaeotemperatures obtained from archaeological samples. It was not possible to determine exactly the last growing period for each sample and thus take into consideration the respective water temperatures for calculation of stable isotope content of sea water ($\delta^{18}O_w$). Therefore we considered two extreme cases: the first one uses for calculation temperatures of the water corresponding to the last month before the sample collection, and the second one the mean temperature in the last year before the sample was collected.

Results of calculation are presented in Tables 3 and 4. The resulting isotopic composition of water is presented relative to PDB and relative to SMOW, according to Equation 4 (Rye & Sommer 1980).

Table 3. Calculation of $\delta^{18}O_w$ from modern shell samples from pure sea water by using Equation 2. Monthly temperatures refer to the last month before sample collection; yearly temperatures refer to mean values of one year before the sample collection

Sample code	$\delta^{13}C$ (‰ PDB)	$\delta^{18}O_{calcite}$ (‰ PDB)	t (°C) monthly	t (°C) annual	$\delta_c–\delta_w$ ‰ (monthly)	δ_w	$\delta_c–\delta_w$ ‰ (annual)	δ_w
11	−0.38	0.50	8.5	6.5	1.99	–	2.53	−2.04
12	0.11	−0.18	8.6	6.4	1.96	–	2.56	−2.74
13	−0.42	0.31	8.6	6.5	1.96	–	2.53	−2.23
14	−0.34	−0.15	5.5	6.5	2.82	–	2.53	−2.68
15	−0.11	0.62	8.3	5.9	2.04	–	2.70	−2.08
16	0.61	0.10	–	–	–	–	–	–
17	−0.45	0.20	8.1	6.5	2.10	–	2.53	−2.33
18	−0.13	0.26	5.0	6.1	2.96	–	2.65	−2.38
19	0.21	−0.21	5.7	5.9	2.76	–	2.70	−2.92
20	−0.18	0.21	5.0	6.1	2.96	–	2.65	−2.44
21	−0.37	−0.05	8.6	6.5	1.96	–	2.53	−2.58
27	0.43	0.16	8.5	6.5	1.99	–	2.53	−2.38
28	0.27	−0.05	8.5	6.5	1.99	–	2.53	−2.58
29	−0.16	0.36	5.5	6.5	2.82	–	2.53	−2.18
30	−0.44	0.69	4.5	6.5	3.10	–	2.53	−1.85
31	−0.04	−0.13	8.6	6.5	1.96	–	2.53	−2.66
32	0.02	0.33	8.6	6.5	1.96	–	2.53	−2.20
38	0.09	−0.02	5.7	5.9	2.76	–	2.70	−2.72
39	−0.29	0.35	8.6	6.1	1.96	–	2.65	−2.29
‰ PDB	−0.08	0.17	7.2	6.3	2.34	–	2.58	−2.41
Errors (1σ)	0.31	0.27	1.7	0.3	0.46	0.52	0.07	0.29
‰ SMOW						–		−2.19

Table 4. Calculation of $\delta^{18}O_w$ from modern shell samples from the sea influenced by freshwater by using Equation 2. Monthly temperatures refer to the last month before sample collection; yearly temperatures refer to mean values of one year before the sample collection.

Sample code	$\delta^{13}C$ (‰ PDB)	$\delta^{18}O_{calcite}$ (‰ PDB)	t (°C) monthly	t (°C) annual	$\delta_c-\delta_w$ ‰ (monthly)	δ_w	$\delta_c-\delta_w$ ‰ (annual)	δ_w
1	–0.38	–0.09	8.4	6.5	2.02	–	2.53	–2.62
2	–1.49	–0.81	8.4	6.5	2.02	–	2.53	–3.34
3	0.22	–0.02	5.2	6.5	2.90	–.	2.53	–2.55
4	0.34	0.40	4.5	6.5	3.10	–	2.53	–2.13
5	–0.42	0.28	4.5	6.5	3.10	–	2.53	–2.25
6	–1.03	–0.46	–	6.5	–	–	2.53	–3.00
7	–1.68	–0.36	–	6.5	–	–	2.53	–2.89
8	–0.59	0.04	8.4	6.5	2.02	–	2.53	–2.50
9	–0.07	–0.25	5.0	6.1	2.96	–	2.65	–2.90
10	–0.24	0.20	6.5	6.1	2.53	–	2.65	–2.45
22	–0.73	0.67	8.4	6.5	2.02	–	2.53	–1.86
23	–0.40	–0.86	8.4	6.5	2.02	–	2.53	–3.39
24	–0.91	–0.97	5.2	6.5	2.90	–	2.53	–3.51
25	–0.33	0.52	4.5	6.5	3.10	–	2.53	–2.01
26	–0.76	–0.86	–	6.5	–	–	2.53	–3.40
33	–0.53	–0.01	8.5	6.1	1.99	–	2.65	–2.66
34	–1.07	–0.29	9.0	6.1	1.86	–	2.65	–2.94
35	–0.62	–0.22	4.0	6.1	3.25	–	2.65	–2.87
36	–0.32	–0.26	5.7	5.9	2.76	–	2.70	–2.96
37	–0.52	–0.17	4.1	5.9	3.22	–	2.70	–2.88
40	–0.50	–0.22	5.2	6.1	2.90	–	2.65	–2.87
41	0.12	–0.80	8.6	6.4	1.96	–	2.56	–3.36
42	0.05	–0.29	8.0	6.4	2.12	–	2.56	–2.85
43	–0.21	–0.58	6.7	6.4	2.48	–	2.56	–3.15
44	–0.18	–0.35	4.5	6.4	3.10	–	2.56	–2.91
45	–0.31	0.11	4.6	6.4	3.08	–	2.56	–2.45
46	–0.37	–0.12	4.7	6.4	3.05	–	2.56	–2.68
47	–0.36	–0.12	5.6	6.4	2.79	–	2.56	–2.68
48	–0.35	–0.11	6.7	6.4	2.48	–	2.56	–2.67
49	–0.32	0.21	7.9	6.4	2.15	–	2.56	–2.36
50	0.10	–0.67	8.9	6.4	1.88	–	2.56	–3.24
51	0.28	–0.11	8.6	6.4	1.96	–	2.56	–2.67
52	0.29	0.52	8.0	6.4	2.12	–	2.56	–2.05
53	0.25	–0.10	6.7	6.4	2.48	–	2.56	–2.66
54	0.37	–0.02	5.8	* 6.4	2.73	–	2.56	–2.58
55	0.03	0.18	4.5	6.4	3.10	–	2.56	–2.38
56	0.11	–0.50	4.4	6.4	3.13	–	2.56	–3.06
57	0.04	–0.10	4.6	6.4	3.08	–	2.56	–2.66
58	–1.00	–0.08	4.7	6.4	3.05	–	2.56	–2.64
59	–0.64	0.14	7.9	6.4	2.15	–	2.56	–2.42
‰ PDB:	–0.35	–0.16	6.4	6.4	2.58	–	2.57	–2.74
Errors (1σ):	0.45	0.38	1.7	0.2	0.56	0.64	0.31	0.48
‰ SMOW:						–		–2.52

60

In both cases $\delta^{18}O_w$ values are lower than those reported for high latitude sea water. According to Epstein & Mayeda (1953) and Roether (1970) this value should be -0.85 ± 0.05 ‰, but introducing this value for the water isotopic composition into the Equation 2 and using isotope data obtained from *Mytilus edulis* shells would result in the extremely high temperature $12.27 \pm 1.07°C$. This is non real and is much above the summer temperature values in the Beagle Channel which do not exceed $10°C$.

As we had no other isotopic data of the sea from the Magellan–Fuegian area at our disposal, the results of calculated $\delta^{18}O_w$ were used as the parameter in Equation 2 in order to calculate the sea temperature in the past by using the measured isotope content in archaeological shells of *Mytilus edulis*.

4 RESULTS OF PALAEOTEMPERATURE CALCULATION

Knowing the stable isotope results of the archaeological shells and the oxygen isotope content of the sea water ($\delta^{18}O_w$), it is possible to calculate the temperature of the mollusks formation in the past. The mollusk samples were again divided into two groups. The first one corresponding to those collected from the pure sea water and the second one collected from the sites with the influence of fresh water. The sites are also presented on Figure 1. Using of monthly temperature data for the past while ^{14}C dates errors cover many yr would be meaningless. Therefore annual, or more than annual, temperature had to be looked for. For this reason, in the case of archaeological shells larger portions were sampled. Therefore the samples prepared for stable isotope analyses of archaeological shells were taken from all parts of the shells, which correspond almost to the total life of the bivalve. Therefore we supposed that obtained stable isotope results reflect at least one year in the life of each shell, and the derived temperature should be the mean annual temperature of sea water.

Stable isotopes content of both, modern and archaeological samples is presented in Figure 3. Scales for oxygen ($\delta^{18}O$) and carbon ($\delta^{13}C$) records are conventionally reversed (Williams et al. 1988), because – to a first approximation – carbonate formed in warm water has a more negative isotopic value than carbonate from colder water (change of -0.22‰ per $1°C$ in $\delta^{18}O$). Distribution of ^{18}O and ^{13}C results in archaeological shells on the time scale is presented in Figure 4.

Lower isotopic composition of samples from sites with freshwater influence (brackish waters) comparing to those originating from pure sea water is the result of the influx of river water which has lower concentration of the heavier oxygen and carbon isotopes (Keith et al. 1964). This is valid for both, archaeological and modern samples. However, modern samples from pure sea water are offset by $\delta^{13}C = -0.90$‰ and $\delta^{18}O =$

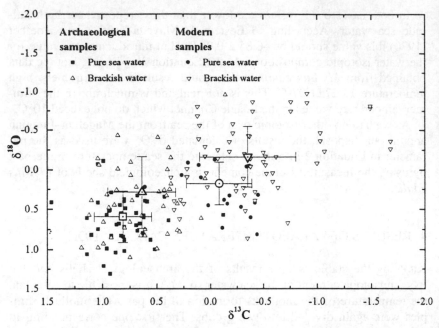

Figure 3. Stable isotope content in modern and fossil *Mytilus edulis* shells. Mean values of each group, denoted by error bars are: (1) *solid square*: archaeological samples from pure sea water ($\delta^{13}C$= +0.81 ± 0.26‰, $\delta^{18}O$= +0.59 ± 0.32‰); (2) *solid circle*: present-day samples from pure sea water ($\delta^{13}C$=–0.08 ± 0.31‰, $\delta^{18}O$= +0.17 ± 0.27‰); (3) *open up triangle*: archaeological samples from brackish water ($\delta^{13}C$= +0.64 ± 0.31‰, $\delta^{18}O$= +0.29 ± 0.45‰); (4) *open down triangle*: present-day samples from brackish water ($\delta^{13}C$= –0.35 ± 0.45‰, $\delta^{18}O$= –0.16 ± 0.38‰).

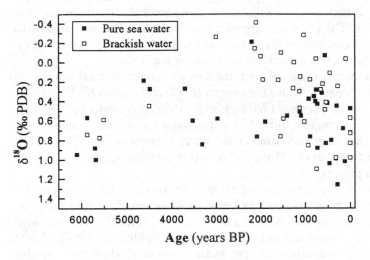

Figure 4. Distribution of ^{18}O isotopes in archaeological *Mytilus* shells.

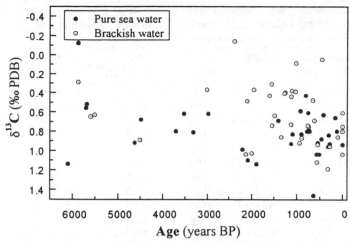

Figure 4. Continued. Distribution of ^{13}C isotopes in archaeological Mytilus shells.

−0.42‰ compared to the archaeological ones, while those from brackish water are offset by δ^{13}C = −0.98‰ and δ^{18}O= −0.28‰.

δ^{18}O$_w$ values calculated from oxygen isotopic content in modern shells according to Equation 2, together with δ^{18}O$_c$ values measured from archaeological samples, were introduced again into Equation 2 in order to calculate the surface sea temperature corresponding to the age of *Mytilus edulis* shells. Values for oxygen composition of sea water (δ^{18}O$_w$) are based on monthly temperatures (Tables 3 and 4). Results of calculation are listed in Tables 5 and 6, and presented on Figure 5. The ages of particular samples were obtained by ^{14}C dating of associated organic material and span the period from about 6000 yr BP up to now. Samples which are modern according to radiocarbon dating obviously denote those stemming from the end of the last century, which corresponds to the end of the canoer tradition due to the extinction of the *Yámana* people. In order to avoid the excessive influence of some values which exceed the average isotopic course and which cannot be considered as representative, we have introduced in Figure 5 the fast Fourier transform smoothing.

The mean temperature obtained for the period of last 6000 yr, based upon archaeological samples of *Mytilus edulis* originating from pure sea water is 5.74 ± 1.13°C, while the mean temperature based on samples originating from brackish water is 4.80 ± 1.74°C. The division on samples into the groups with and without freshwater influence was only estimated, because the influence of freshwater to the sites where shells were growing before they were collected by the indigenous people often changed due to the retreat of the sea during the last 6000 yr. The larger standard deviations for temperatures obtained by *Mytilus edulis* samples from sites influ-

Table 5. Results for archaeological samples from pure sea water. $\delta^{18}O$ isotope values are corrected according to Equation 5. Temperature is calculated according to Equation 2 with $\delta^{18}O_w = -2.16\ \%_0$ (PDB), as shown in Table 3.

Sample Code	^{14}C laboratory sample code	Age (BP)	$\delta^{13}C$ $\%_0$ (PDB)	$\delta^{18}O_{calcite}$ $\%_0$ (PDB)	t (°C)
62	AC–1282	0	0.94	0.48	6.2
100	AC–1286	0	1.04	0.72	5.3
97	AC–1296	124 ± 109	0.80	1.02	4.3
93	AC–1306	161 ± 81	0.66	0.68	5.4
70	AC–1307	284 ± 112	0.93	1.26	3.4
74	AC–1301	317 ± 128	0.96	0.99	6.2
124	AC-1355	305 ± 144	0.84	0.45	4.4
90	AC–1308	422 ± 77	0.63	0.80	5.0
68	AC–1304	464 ± 127	0.88	1.04	4.2
141	AC–1374	511 ± 113	1.04	0.42	6.3
133	AC–1375	565 ± 62	1.04	–0.07	4.9
89	AC–1284	545 ± 123	0.92	0.83	8.1
132	AC–1363	633 ± 108	1.46	0.32	6.7
81	AC–1309	731 ± 79	0.79	0.43	4.7
111	AC–1373	727 ± 127	0.80	0.88	6.3
134	AC–1366	754 ± 104	0.61	0.28	6.8
101	AC–1275	775 ± 116	0.80	0.40	6.4
122	AC–1365	806 ± 94	0.43	0.35	6.6
67	AC–1281	895 ± 138	0.83	0.76	5.1
148	AC–1302	920 ± 134	0.59	0.38	6.5
123	AC-1353	1093 ± 107	0.83	0.55	5.9
146	AC–1290	1123 ± 101	0.93	0.51	6.0
145	AC–1289	1403 ± 111	0.69	0.55	5.9
109	AC–1372	1897 ± 115	1.14	0.61	5.7
65	AC–1283	2082 ± 126	1.10	0.76	5.1
126	AC-1352	2199 ± 150	0.99	–0.21	8.6
94	AC–1273	2974 ± 130	0.62	0.58	5.8
96	AC–1274	3317 ± 120	0.81	0.84	4.9
144	AC–1288	3527 ± 163	0.62	0.60	5.7
143	AC–1285	3703 ± 125	0.80	0.27	6.9
130	AC–1369	4627 ± 125	0.92	0.19	7.2
142	AC-1376	4485 ± 414	0.68	0.27	6.9
84	AC–236	5700 ± 100	0.56	0.88	4.3
72	AC–1272	5684 ± 196	0.52	1.00	4.7
91	AC–1397	5872 ± 147	–0.12	0.57	5.8
60	BETA-2819	6140 ± 130	1.14	0.95	4.5
		Mean values:	0.81	0.59	5.7
		Errors (1σ):	0.26	0.32	1.1

Table 6. Results for archaeological samples from brackish water. $\delta^{18}O$ isotope values are corrected according to Equation 5. Temperature is calculated according to Equation 2 with $\delta^{18}O_w=-2.71‰$ (PDB), as shown in Table 4.

Sample Code	^{14}C laboratory sample code	Age (BP)	$\delta^{13}C$ ‰ (PDB)	$\delta^{18}O_{calcite}$ ‰ (PDB)	t (°C)
66	AC–1271	0	0.75	0.57	3.9
73	AC–1277	0	1.04	0.98	2.5
98	AC–1280	0	0.61	0.83	3.0
99	AC–1300	0	0.80	0.73	3.4
149	Archaeological	100	0.86	–0.04	6.0
64	MC–1062	280 ± 85	0.96	0.25	5.0
71	AC–1276	321 ± 120	1.19	0.98	2.5
140	AC–1368	457 ± 102	0.05	0.11	5.5
150	AC–1295	541 ± 100	0.94	0.41	4.4
116	AC–1349	561 ± 165	1.12	0.49	4.2
105	AC–1370	654 ± 103	0.48	0.28	5.1
139	AC–1367	615 ± 191	0.69	0.22	4.9
120	AC–1360	759 ± 103	0.74	–0.04	2.1
88	AC–1278	745 ± 160	0.75	1.10	6.0
138	AC–1357	892 ± 96	0.87	0.18	5.3
137	AC–1356	933 ± 101	0.92	0.85	3.0
115	AC–1343	1018 ± 105	0.09	0.61	3.8
135	AC–1358	1029 ± 94	0.39	–0.28	6.9
147	CSIC–311	1120 ± 50	0.44	0.47	5.4
61	AC–1291	1106 ± 109	0.38	0.15	4.2
69	AC–1305	1126 ± 111	0.73	0.30	4.8
119	AC–1347	1251 ± 139	0.40	0.00	5.9
131	AC–1354	1340 ± 111	0.86	–0.55	4.9
136	AC–1345	1268 ± 250	0.41	0.27	7.9
110	AC–1377	1487 ± 130	0.64	0.58	3.9
86	AC–1267	1552 ± 70	0.31	0.43	6.2
129	AC–1371	1541 ± 109	0.74	–0.10	4.4
117	AC–1350	1602 ± 141	0.43	0.18	5.2
107	AC–1341	1950 ± 123	0.37	0.18	5.2
127	AC–1362	1996 ± 167	1.03	–0.06	6.1
102	AC–1346	2129 ± 103	1.04	–0.14	7.4
128	AC–1364	2101 ± 176	0.49	–0.41	6.4
118	AC–1342	3904 ± 408	0.37	–0.26	6.8
125	AC–1361	4505 ± 602	0.89	0.45	4.3
87	AC–1164	5600 ± 125	0.65	0.77	3.8
92	AC–1299	5510 ± 219	0.63	0.59	3.2
113	AC–1397	5872 ± 147	0.29	0.74	3.3
		Mean values:	0.66	0.22	4.7
		Errors (1s):	0.29	0.41	1.6

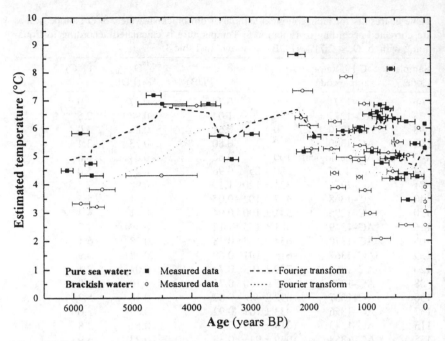

Figure 5. Surface sea water temperature of the Beagle Channel calculated from oxygen isotopic composition in archaeological *Mytilus edulis* shells according to Craig (1965) by using the sea water isotopic content $\delta^{18}O_w = -2.16\%_o$ PDB for samples from pure sea water and $\delta^{18}O_w = -2.71\%_o$ PDB for samples from brackish waters. The curve corresponding to the samples proceeding from sites without brackish water influence (pure sea water) should be considered as valid.

enced by brackish waters are understandable, because the exact amount of freshwater influence to those sites, especially in the past, is not known. This could significantly change the isotopic content of the water, influencing thus the isotopic content in calcite during the formation of mollusks. Therefore only the values corresponding to pure sea water should be considered as valid.

5 DISCUSSION

Calculation of palaeotemperatures in a calcite-water system, according to Equation 2, requires the knowledge of the oxygen isotope content of the water in which the shells were formed. This value is unknown and at present it can not be estimated exactly for the past. There are two possibilities to obtain a figure of the oxygen isotope content of water: (a) to calculate it from modern shells, by using the actual temperature where the mollusks

66

have grown or (b) to measure the present-day oxygen content of pure sea water and to extrapolate this value to the past.

The first method was already done in this research project, resulting in the absolute palaeotemperature values shown in Figure 5. For the second method, a first measurement of a sample of Beagle Channel sea water with no freshwater influence, taken in February 1997, was recently done. This isotopic analysis gave the value $-1.6‰$ vs. SMOW (salinity being some 30 ‰), which is not far from the value we used in calculations ($\delta^{18}O_w= -2.16‰$, or $-1.92‰$ vs. SMOW) based on present-day shells. This method can not be applied for samples growing in brackish waters because they are not homogeneous, so a mean value cannot be estimated. Mean temperature for the period studied, based on the first method gave $5.74 \pm 1.13°C$, whereas when the second method was used, it yielded 6.97 ± 1.16 °C. Since the first method was based on actual (measured) sea surface temperature, it is not possible to compare results of both methods with present-day one, because it would be tautological.

Quantitative processing of our measurements shows that the mean temperature, based on the samples from pure sea water, is $5.74°C$ for the entire 6000 year time period. This value is lower than today's mean surface temperature in the Beagle Channel measured at 2 m depth ($6.5°C$), but within its standard deviation ($\pm 1.13°C$). Estimated temperature variation in all this period is lower than other data available from the literature. According to Heusser & Streeter (1980), standard deviation of surface temperature in southern Patagonia and Tierra del Fuego in the Holocene is ca. $1.9°C$.

Calculated temperature obtained from the archaeological samples dated at the end of the 19th century is $5.7°C$, and this is lower by $0.8°C$ than the yearly mean temperature value in Beagle Channel as obtained by instrumental measurements in the last decades. Although higher than expected, this difference is in agreement with the global (combined land and ocean) warming by $0.45 \pm 0.15°C$ in the Southern Hemisphere, recorded in the period from 1861 to 1989 (Folland et al. 1991). The so-called 'Little Ice Age' period, from the 15th to 19th centuries, is also identifiable in Figure 5, as well as the warming period of the Middle Ages. The maximum at 2200 yr BP cannot be confirmed, due to the only one offset datum. Moreover, this sample was collected at the Moat Channel near the eastern mouth of the Beagle Channel, 150 km east of Ushuaia. Therefore, it may not reflect the same sea water composition. The obtained temperatures show a depletion at about 3000-3500 yr BP, followed by an increase at 4000-4500 yr BP. Lower temperature values for the period 5000-6000 yr BP correspond to those measured by Panarello (1987).

Measurements of palaeotemperature in the Beagle Channel by Panarello (1987) were based on several shell species from two shell middens dated to 3030 ± 100 BP and 5630 ± 120 BP, respectively, as well as from modern shells. He used the isotopic value for sea water $\delta^{18}O_w= -0.85‰$, as suggested by Epstein & Mayeda (1953) and Roether (1970). According to his results, the greatest deviations from expected temperatures were obtained just from *Mytilus edulis* shells, giving 11.6°C for present-day sea water surface temperature. This is much higher than today's average temperature, and even higher than the summer maximum (10°C). It is interesting that about 4°C higher temperatures than expected were also obtained by Donner & Nord (1986) on the basis of *Mytilus edulis* samples, and this was attributed to the mixing with fresh water.

However, *Mytilus* shells analyzed by Panarello (1987) were composed mostly of aragonite. It is not known if the values presented in his paper were already corrected or not for aragonite–calcite (Cornu et al. 1993), because the relation (2) is valid only for pure calcite samples. This could additionally increase the error in the obtained temperature. On the other hand, temperatures which he obtained from *Balanus* sp. and *Trophon* sp. (mostly calcite) are realistic and within the expectations. Generally lower temperatures for the period of ca. 5630 BP, which he obtained from all species and which coincide with our results, were explained by this author as due to local processes, such as the approaching of a cold marine current to the Channel, and not to global climatic trends.

Recent isotopic measurements made by Gordillo (1996) on *Hiatella solida* (Sowerby) shells indicate a temperature maximum at 5900 yr BP (3°C higher than present) and a cooler period at 4400 yr BP (1°C lower than present). This last assessment does not agree with our data.

6 RESEARCH IN PROGRESS

The main problem which we faced during the calculation of palaeotemperatures was the lack of information on the salinity and isotopic composition in the Beagle Channel surface sea water (particularly the value $\delta^{18}O_w$). During the planning of the project it was supposed that isotopic results from modern shells would give enough information, enabling the calculation of parameters necessary for palaeotemperature calculation. This investigation is still in progress. At the moment we have obtained only one mean value of $\delta^{18}O$ of the Beagle Channel sea water. In the future samples of water will be collected at the same time and from the same place as the mollusk samples.

However, possible changes in oxygen isotopic composition in sea water due to the melting of the mountain glaciers in the Darwin Cordillera (western Tierra del Fuego) and even Antarctic ice during the last 6000 yr cannot be excluded. In order to assess $\delta^{18}O_w$ values in the past some indirect methods could improve our knowledge. Collecting of modern shell samples should be accompanied always with salinity and temperature measurements *in situ* (the data which we had at our disposal were obtained from only one site in Ushuaia). This could establish a correlation between the isotopic content in water and salinity, similar to those made for warmer oceans (Fairbanks et al. 1992, in: Cornu et al. 1993).

In order to compare results obtained by palaeotemperature calculation by using the calcite–water system (Equation 2) with other methods we are trying to use the method based on the calcite-aragonite system. For this reason, aragonite was extracted from inner parts of nine *Mytilus edulis* shells dated from 317 ± 128 yr BP to 5510 ± 219 yr BP in 500-year sequence and were subjected to isotopic analyses. The measurements are in progress. Although this method is more sensitive than the water–calcite method, and therefore sometimes more inaccurate, it is independent of the isotopic content of sea water which is still unknown for the past. If reasonable results are obtained, more aragonite samples will be prepared.

ACKNOWLEDGMENTS

This study was made within the frame of the project 'Beagle Channel Marine Resources Prior to the Industrial Exploitation', funded by the European Community (CI1*-CT93-0015). Radiocarbon analyses were made at the Instituto de Geocronología y Geología Isotópica (INGEIS), CONICET, Buenos Aires, and the stable isotope analyses at the Unity for Geochemical Analyses of the Servicio Científico Técnico de la Universitat de Barcelona. The authors are much indebted to Josep Elvira, Felicià Plana and Ignasi Queralt, Instituto de Ciencias de la Tierra 'Jaume Almera' (CSIC) for X-rays diffractometric analyses and to Pilar Teixidor and Eva Aracil for stable isotopes analyses.

REFERENCES

Albero, M.C., F.E. Angiolini & E.L. Piana 1986. Discordant ages related to reservoir effect of associated archaeological remains from Túnel site (Beagle Channel, Argentine Republic). *Radiocarbon* 28, No. 2, 748-753.
Albero, M.C., F.E. Angiolini & E.L. Piana 1987. Holocene ^{14}C reservoir effect at Beagle Channel (Tierra del Fuego, Argentine Republic). *Quaternary of South America and Antarctic Peninsula* 4: 59-72. Rotterdam: Balkema.
Cornu, S., J. Pätzold, E. Bard, J. Meco & J. Cuerda-Barcelo 1993. Palaeotemperature of the last interglacial period based on $\delta^{18}O$ of *Strombus bubonius* from western Mediterranean sea. *Palaeogeography, Palaeoclimatology, Palaeoecology* 103: 1-20.
Craig, H. 1961. Standard for reporting concentrations of deuterium and oxygen-18 in natural waters. *Science* 133: 1833-1834.
Craig, H. 1965. The Measurement of oxygen isotope palaeotemperatures. In Tongiorgi, E. (ed.), *Stable Isotopes in Oceanographic Studies and Palaeotemperatures* 3: 1-24,

Spoleto, July 26-30, 1965. Consiglio Nationale delle Richerche, Laboratorio di Geologica Nucleare, Pisa.

Dansgaard, W. & H. Tauber 1969. Glacier oxygen-18 content and Pleistocene ocean temperatures. *Science* 166: 500-502.

Dodd, J.R. 1963. Paleoecological implication of shell mineralogy in two pelecypod species. *Journal of Geology* 71: 1-11.

Dodd, J.R. 1964. Environmentally controlled variation in the shell structure of a pelecypod species. *Journal of Paleontology* 71: 1-11.

Donner, J. & A.G. Nord 1986. Carbon and oxygen stable isotope values in shells of *Mytilus edulis* and *Modiolus modiolus* from Holocene raised beaches at the outer coast of the Varanger peninsula, north Norway. *Palaeogeography, Palaeoclimatology, Palaeoecology* 56: 35-50.

Eisma, D., W.G. Mook & H.A. Das 1976. Shell characteristics, isotopic composition and trace-element contents of some euryhaline mollusks as indicators of salinity. *Palaeogeography, Palaeoclimatology, Palaeoecology* 19: 39-62.

Epstein, S. & H.A. Lowenstam 1953. Temperature-shell growth relations of Recent and interglacial Pleistocene shoal-water biota from Bermuda. *Journal of Geology* 61: 424-438.

Epstein, S. & T. Mayeda 1953. Variations in [18]O content of eaters from natural sources. *Geochimica et Cosmochimica Acta* 27(4): 213-224.

Folland, C.K., T.R. Karl, K.Y.A. Vinnikov 1991. Observed climate variations and change. In *Climate Change – The IPCC scientific assessment*. WMP-UNEP Intergovernmental panel on climate change, Cambridge Univ. Press, 2nd Ed.

Gordillo, S. 1996. Subfossil and living *Hiatella solida* (SOWERBY) from the Beagle Channel, South America. *Quaternary of South America and Antarctic Peninsula* 9: 183-204. Rotterdam: Balkema.

Heusser, C.J. & S.S. Streeter 1980. A temperature and precipitation record of the past 16,000 yr in southern Chile. *Science* 210: 1345-1347.

Hutchinson, Ch.S. 1974. *Laboratory handbook of petrographic techniques*. New York: John Wiley & Sons.

Iturraspe, R. R. Sottini, C. Schroeder & J. Escobar 1989. Hidrología y variables climáticas del Territorio de Tierra del Fuego. *Contribución Científica del Centro Austral de Investigaciones Científicas* 7: 1-196. Ushuaia.

Keith M.L., G.M. Anderson & R. Eichler 1964. Carbon and oxygen isotopic composition of mollusk shells from marine and fresh-water environments. *Geochim. Cosmochim. Acta* 28: 1878-1816.

Lowenstam, H.A. 1954. Factors affecting the aragonite-calcite ratios in carbonate secreting marine organisms. *Journal of Geology*. 62: 284-322.

Mook, W.G. 1971. Palaeotemperatures and chlorinities from stable carbon and oxygen isotopes in shell carbonate. *Palaeogeography, Palaeoclimatology, Palaeoecology* 9: 245-263.

Orquera, L.A. & E.L. Piana 1987. Human littoral adaptation in the Beagle Channel region: maximum possible age. *Quaternary of South America and Antarctic Peninsula* 4: 133-162. Rotterdam: Balkema.

Orquera, L.A. & E.L. Piana 1992. Un paso hacia la resolución del palimpsesto. In: Borrero L.A. & J.L. Lanata (eds), *Análisis espacial en la arqueología patagónica*: 21-52. Edit. Ayllu, Buenos Aires.

Orquera, L.A. & E.L. Piana 1994. Análisis de conchales de la costa del Canal Beagle. In *Actas y Memorias del XI Congreso Nacional de Arqueología Argentina*. Abstract, *Revista del Museo de Historia Natural de San Rafael* XIV (1/4): 308-310. Mendoza.

Panarello, H.O. 1987. Oxygen-18 temperatures on present and fossil invertebrated shells from Túnel Site, Beagle Channel, Argentina. *Quaternary of South America & Antarctic Peninsula* 5: 83-91. Rotterdam: Balkema.

Rabassa, J., C.J. Heusser & R. Stuckenrath 1986. New data on Holocene sea transgression in the Beagle Channel, Tierra del Fuego, Argentina. *Quaternary of South America and Antarctic Peninsula* 4: 291-309. Rotterdam: Balkema.

Roether, W. 1970. Water CO_2 exchange set up for the routine oxygen-18 assay of natural waters. *International Journal of Applied Radioactivity and Isotopes* 21: 379-387.

Rozanski, K., L. Araguás-Araguás & R. Gonfiantini 1993. Isotopic patterns in global precipitation. In P.K. Swart, K.C. Lohmann, J. McKenzie & S. Savin (eds), *Climate Change in Continental Isotopic Records, Geophysical Monograph* 78. American Geophysical Union.

Rye, D.M. & M.A. Sommer II 1980. Reconstructing palaeotemperature and palaeosalinity regimes with oxygen isotopes. In D.C. Rhoads & R.A. Lutz (eds), *Skeletal growth of aquatic organism:* 169-202. Plenum Publishing Co.

Silva, M.R. 1996. Patrón de asentamiento larval y crecimiento de *Mytilus edulis chilensis* en el canal de Beagle. Dirección de Ciencia y Tecnología de la provincia de Tierra del Fuego. Junio. Unpublished report, Ushuaia, June 1996.

Sommer II, M.A. & D.M.Rye 1978. Oxygen and carbon isotope internal thermometry using benthic calcite and aragonite foraminifera pairs. *Short papers of the Fourth International Conference 'Geochronology, Cosmochronology, Isotope Geology'*, U.S.G.S. Open File Rep. 78-701: 408-410.

Tarutani, T., R.N. Clayton & T.K. Mayeda 1969. The effect of polymorphism and magnesium substitution on oxygen isotope fractionation between calcium carbonate and water. *Geochimica et Cosmochimica Acta* 33: 987-996.

Tukhanen, S. 1992. The Climate of Tierra del Fuego from a vegetation geographical point of view and its ecoclimatic counterparts elsewhere. *Acta Botanica Fennica* 145: 1-64.

Urey, H.C., H.A. Lowenstam, S. Epstein & C.R. McKinney 1951. Measurements of palaeotemperatures and temperatures of the Upper Cretaceous of England, Denmark and the Southeastern United States. *Bulletin of the Geological Society of America* 62: 399-416.

Wefer, G. & W.M. Berger 1991. Isotope paleontology: growth and composition of extant calcareous species; *Marine Geology* 100/101: 207-248.

Williams, D.F., I. Lerche & W.E. Full 1988. *Isotope Chronostratigraphy: Theory and Methods.* San Diego, CA: Academic Press, Inc.

Holocene palaeoenvironments of the northern coastal plain of Rio Grande do Sul, Brazil, reconstructed from palynology of Tramandaí lagoon sediments

5

MARIA LUISA LORSCHEITTER
Departamento de Botânica, Instituto de Biociências, Universidade Federal do Rio Grande do Sul (UFRGS), Brazil – CNPq. researcher.
SÉRGIO REBELLO DILLENBURG
Centro de Estudos de Geologia Costeira e Oceânica (CECO), Universidade Federal do Rio Grande do Sul (UFRGS), Brazil

ABSTRACT: The palynological analysis of sediments of Tramandaí lagoon (long. 50°07'16"-50°11'08"W and lat. 29°56'46"-30°00'00"S) has as its objective the palaeoenvironmental reconstruction of the last millennia of the Rio Grande do Sul northeastern coast, in relation to the evolving geological pattern of Tramandaí lagoon described by Dillenburg (1994).

Nineteen samples of a lagoon core (T-03) were analysed. The results showed a maximum relative sea level transgression at about 5000 yr BP and another less intense transgression at about 1800 yr BP.

The palynological analysis also showed highly unstable environments during the eustatic oscillations, significantly changing the patterns of the coast vegetation. High temperatures and humidity around 5000 yr BP seem to have contributed greatly to the vegetation expansion.

RESUMEN: A análise palinológica de sedimentos da laguna de Tramandaí (long. 50°07'16"-50°11'08"W e lat. 29°56'46"-30°00'00"S) tem com objetivo a reconstituição paleoambiental dos últimos milênios do litoral nordeste do Rio Grande do Sul, comparando com o modelo geológico evolutivo da laguna de Tramandaí, elaborado por Dillenburg (1994).

Foram analisadas 19 amostras de um testemunho da laguna (T-03). Os resultados evidenciaram um máximo transgressivo do nível relativo do mar há cerca de 5000 anos AP e uma outra transgressão, menos intensa, há cerca de 1800 anos AP.

A análise palinológica mostrou também ambientes altamente instáveis durante as oscilações eustáticas, alterando significativamente os padrões da vegetação litorânea. Temperatura e umidade elevadas, em torno de 5000 anos AP, parecem ter contribuído em muito para a expansão vegetal.

1 INTRODUCTION

Tramandaí lagoon (long. 50°07'16"-50°11'08"W and lat. 20°56'46"-30°00'00"S) in the northern coastal plain of Rio Grande do Sul, the southernmost state of Brazil (Fig. 1), consists of two bodies of water – Lagoa de Tramandaí and Lagoa do Armazém – respectively situated north and south of a sand spit stretching northwest-southeast. The total area of the lagoon, including the drainage channel, measures 18.8 km^2. Its average depth is 1 m in the lagoon body (maximum depth 1.4 m) and, in the channel, 4 m (maximum 6 m). The formation of this lagoon is linked to the evolution stages of the depositional system lagoon/barrier IV, defined by Villwock (1984). According to Dillenburg (1994), the development of system IV in the Tramandaí lagoon region began at about 7000 yr BP.

Tramandaí lagoon is the only natural outflow to the Atlantic Ocean of a drainage system, including a set of northern and southern lakes of the coastal plain, inter-connected by natural channels. This lagoonal system of 1800 km^2, which lies parallel to the coastline, draws a large amount of river drainage, at the north of the lagoon area.

The ample rainfall in this region causes a large and continuous flow of continental waters into the lagoon, chiefly through the Tramandaí River. On the other hand, eventual drought periods may permit a larger amount of marine water to enter through the channel, with a disastrous effect to the lagoon margin vegetation as well as at the borders of the neighbouring inter-connected lakes, both at the north and south (B.E. Irgang, pers. comm.).

The present depositional system of Tramandaí lagoon follows the classical pattern of sediment distribution: sandy at the margin, due to waves caused by local winds, gradually becoming thinner (silt and clay) towards the greater depths, where there is a lower energy level. The intra-lagoon delta of the Tramandaí River carries a rather large amount of clay into the Lagoa de Tramandaí. Lagoa do Armazém is dominated by essentially sand sediments.

Tomazelli (1990) and Dillenburg (1994) indicate that the present system of lakes inter-connected by channels was a rather recent occurrence, resulting from a lowering of the sea level which happened after a maximum transgressive of 5100 ^{14}C yr BP.

The objective of this paper is to compare palynological data with the evolving geological pattern of Tramandaí lagoon described by Dillenburg (1994) and to study the history of the vegetation and climate of the latter millennia of this microregion (lagoon and adjacent areas). Thus, a core taken from the sand spit of the lagoon margin was analysed in detail.

74

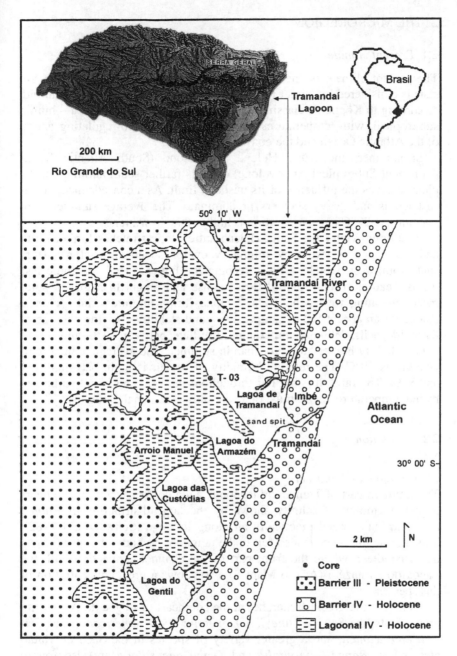

Figure 1. Tramandaí lagoon localization and chosen core (T-03).

2 THE MICROREGION

2.1 *Present climate*

The present climate of the northern coastal plain of Rio Grande do Sul state is considered by Nimer (1979) as mild mesothermic and very humid. According to Köppen's classification, the regional climate is CFa – humid sub-tropical, with temperatures softened by the thermo-regulating action of the Atlantic Ocean and the coastal lake bodies.

In summer, the Santa Helena anticyclone (South Atlantic Semi-permanent Subtropical Anticyclone) extends farther south and the coastal plain receives the influence of its unstable limit. As a consequence, north-east winds and heavy showers predominate. The average summer temperature is >22°C and evaporation is higher than in winter.

In autumn-winter the Santa Helena anticyclone stretches farther north and the region is swept by migratory cyclones and anticyclones from south-southwest, which, while sweeping the hotter surfaces become less intense, causes frontal rainfall (less intense and longer than that in summer) generating west winds. The cyclones bring migratory anticyclones, that cause an abrupt fall of temperatures and stability after the passage of the cold spells. Winter temperatures oscillate between -3°C and +8°C, with a slightly higher humidity than in summer. The average annual temperature is 20°C. The relative humidity of the air is high, between 76% and 86%. The rainfall is regular throughout the year and is not surpassed by the temperature curve, thus, there are only occasional droughts.

2.2 *Vegetation*

2.2.1 *Lagoa do Armazém*

The southern part of Tramandaí lagoon (Lagoa do Armazém, Fig. 1) bears a greater amount of salinity shown by the large distribution of *Ruppia maritima*, submerged practically all along its edge (Schwarzbold 1982). This marine influence is also felt in the neighboring body of water (Lagoa das Custódias), where the front part of the channel connection with the Lagoa do Armazém (Arroio Manuel) also shows a great concentration of this species.

Linked to the brackish marshes at the southeast of Lagoa do Armazém, grow plants like *Scirpus Olneyi*, *S. maritimus* and *S. americanus* and *Crinum americanum*, and a greater variety of species in the Saco do Ratão area, where *Rapania parvifolia* and *Typha angustifolia* are also found, which resist the saline stress (Würdig 1984). *R. parvifolia* may also be found to a great extent along the margin of Lagoa do Armazém.

An extense creeping vegetation of *Salicornia gaudichaudiana* indicates brackish and salt soils at the margin nearest to the sea link channel, at the southeast part of the Lagoa de Tramandaí. Resistance to brackish soils is a characteristic of *Paspalum vaginatum* Sw., which grows close to the Lagoa do Armazém, near the Arroio Manuel (Schwarzbold 1982).

The greater influence of fresh water marshes in Lagoa do Armazém is felt on the sand margin of the spit which separates it from Lagoa de Tramandaí, with an extense creeping field of *Scirpus californicus. Nitella* sp. grows abundantly submerged in the flooded and brackish areas of the remaining spit.

2.2.2 *Lagoa de Tramandaí*

Lagoa de Tramandaí is very much influenced by the fluvial outflow of the Tramandaí River, which is constant due to the high regional rainfall (Fig. 1). Yet there is always a certain regional marine influence close to the intra-lagoon delta, where varied brackish soils cause a diversified vegetation subject to edaphic instability in transitional soil zones between continental and oceanic waters.

Close to the delta, the higher and dry sandy areas, with greater deposits of fluvial sediments are characterised by a significant presence of *Paspalum vaginatum*, a pioneer plant in monospecific stretches (Ramos 1977). In the more humid depressions, mainly during rainy seasons, species like *Ludwigia elegans*, *Eleocharis flavescens* and *Bacopa monnieri* appear. The latter resists emersions and immersions for several months, even in brackish waters (Cordazzo & Seeliger 1988).

The greatest parts of the lower dry fields of the delta receive a constant inflow of sand material transported by the river with a permanent fluctuation of the water table, giving rise to *Gley* type soils (Ramos 1977). A mosaic of increasing humidity may be found towards the interior parts of the delta. Thus, the drier and sandier soils are characterised by *Ischaemum urvilleanum* and *Eleocharis minima*. Corresponding to a gradual increase of humidity in the soils *Axonopus affinis*, *Paspalidium paludivagum* and *Panicum gouini*, *Scirpus olneyi* and *Cyperus polytachyus*, *Juncus dichotomus* and *Eleocharis flavescens* appear as principal species here.

However, in the flooded depressions of the interior of the delta, the peat-soil presents *Bacopa monnieri*, *Eleocharis flavescens*, *Pontederia lanceolata* f. *lanceolata*, *Paspalidium paludivagum*, *Scirpus olneyi* and *Panicum gouini* as characteristic species. On the flooded margins of the delta *Scirpus californicus*, here a pioneer plant, appears in monospecific patches.

A monospecific carpet of *Scirpus giganteus* from 1.5 to 2 m in height covers almost all the western part of the delta, also peat-soil, with vast low flooded soils. In this area, among the patches of *S. giganteus*, other species like *Achrostichum danaeifolium*, *Crinum americanum*, *Nymphoides*

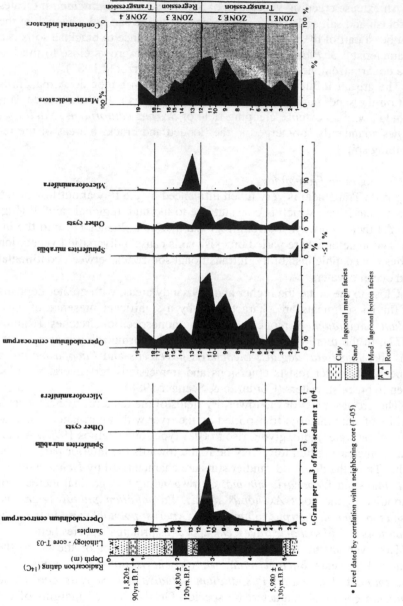

Figure 2. Palynological concentration and percentage diagrams of marine indicators.

humboldtianum, *Polygonum punctatum*, *Eryngium pandanifolium*, *Hydrocotyle bonariensis* and *Panicum helobium* may be found in abundance.

At the mouth of the Tramandaí River and southeast of the delta, where the greatest fluviodeltaic deposits occur, grow monospecific groups of *Typha angustifolia* among fields of *Scirpus californicus* and *S. giganteus* (Ramos 1977).

Close to the Tramandaí river channel there is an ample growth of fresh water species, like *Eichornia crassipes*, *Pontederia lanceolata*, *Ludwigia longifolia*, *Hydrocotyle ranunculoides*, *Salvinia auriculata*, *Leersia hexandra*, besides *Chara sp. Scirpus* species predominate on the south, southwest and west margins of the Lagoa de Tramandaí.

Coastal forests consisting of a large number of tropical species, mainly of Myrtaceae, grow in the most sheltered areas.

3 MATERIAL AND METHODS

The core extraction (T-03 core, Fig. 2) was made in the lagoon margin over to the sand spit (Fig. 1). Percussion was used, with penetration of a 6 meter-long × 7.5 cm wide PVC tube.

The sediment core brought out was 5.40 m long. The sediment analyses were made by laboratories of CECO (Centro de Estudos de Geologia Costeira e Oceânica) and the palynological analyses by the Quaternary Palynology Laboratory, both at Universidade Federal do Rio Grande do Sul.

In the core, the two samples of mud facies from the lagoon bottom – 4.75 m deep (base) and 2.19 m deep (top) – were dated by ^{14}C at the Beta Analytic Inc. lab, Miami, Florida (USA). By stratigraphic correlation with a core situated at a distance of 2 km (T-05, Dillenburg 1994), the sand-mud interval of the core, at a depth of 0.70 m, was estimated at 1820 ± 90 ^{14}C yr BP (*Beta* 59296).

The palynological analysis reached a total of 19 samples of T-03 core, ranging from the bottom to the top. The chemical processing was according to Faegri & Iversen (1989), using hydrofluoric and hydrochloric acids, boiling in KOH at 10%, acetolysis and filtering in a 250 µm net. The slides were prepared with glycerol-jelly (Salgado-Labouriau 1973).

The qualitative analysis consisted of the botanical determination of the palynomorphs. When this was impossible, the material was determined only by morphology (ex: granulate trilete, other triletes, other tricolporates). A number after the taxon name separates distinct groups within itself (ex: *Pediastrum* 1, *Pediastrum* 2). The word 'type' before the name was used when the botanical determination could not be fixed which could refer to one genus or to another one related to it, within the same family.

The countings were based on a minimum number of 500 grains per sample, corresponding to angiosperm and gymnosperm pollen + pteridophyte and bryophyte spores.

The other palynomorphs were counted separately, while trying to reach the minimum number of 500 grains. The quantitative analysis took into consideration the relative frequency (%) and absolute frequency (grain concentration per cm^3 of fresh sediments), through the previous introduction of *Lycopodium clavatum* tablets into the samples (Stockmarr 1971).

The percentage diagrams were drawn up based on the following calculations:

1. Diagrams of angiosperms, gymnosperms, pteridophytes and bryophytes: Percentages calculated over the total of the minimum number counted.

2. Diagrams of fresh water algae: percentages calculated over the total of the minimum number counted + total of fresh water algae.

3. Diagrams of marine indicators: percentages calculated over the total of the minimum number counted + total of marine indicators.

4. Grouping diagrams: Σ of the percentages of dry field indicators, forest indicators and pteridophytes, calculated over the total of the minimum number counted; Σ of the percentages of fresh water indicators, calculated over the minimum number counted + total of fresh water algae (not including *Botryococcus* and *Pediastrum,* due to excessive representation).

5. Composite diagrams: dry field × forest, calculated over the total indicators of these two groups; marine × continental, calculated over the total of the minimum number counted + total of marine indicators.

The pollen + spore total concentration was calculated over the total of the minimum number counted.

4 RESULTS AND DISCUSSION

4.1 *Marine indicators*

Dinoflagellates *Operculodinium centrocarpum, Spiniferites mirabilis* and other cysts); microforaminifera.

4.2 *Continental indicators*

Dry fields: *Alternanthera, Baccharis* type, *Amaranthus*-Chenopodiaceae type, *Plantago, Gomphrena, Eryngium, Gnaphalium* type, *Ephedra, Vernonia* type, Caryophyllaceae and Poaceae (part). All these pollen grains were included in the of the percentages of dry field indicators, except Poaceae, which were analysed separatedly, due to their high representativity and because marshy elements may also be included among them.

Forests: *Alchornea triplinervia, Trema, Celtis, Urticales, Anacardiaceae, Myrtaceae, Ilex, Acacia* type, *Podocarpus, Araucaria*, Sapindaceae, Meliaceae, *Roupala, Mimosa, Mimosa lepidotae* series, *Drimys and Rapanea*. All these pollen grains were included in the Σ of the forest indicator percentages.

Pteridophytes: *Lycopodiella alopecuroides, Huperzia, Osmunda, Azolla filiculoides, Isoetes, Blechnum* type, *Microgramma*, Cyatheaceae, *Dicksonia* type, *Dryopteris* type, *Anogramma* type, *Ceratopteris* type, *Marattia* type, granulate monolete, echinate trilete, indeterminate triletes, indeterminate monoletes and other pteridophytes. All these spores were included in the Σ of pteridophyte percentages.

Fresh water: *Azolla filiculoides, Isoetes, Myriophyllum, Eichhornia*, Alismataceae and algae (*Botryococcus, Pediastrum* 1, *Pediastrum* 2, *Spirogyra, Zygnema, Mougeotia, Debarya, Pseudoschizaea rubina*). All these palynomorphs were included in the Σ of aquatic fresh water environment indicators, excepting *Botryococcus, Pediastrum* 1 and *Pediastrum* 2, which were analysed separately due their high representativity.

Fresh water marshes: Poaceae (parte), Cyperaceae, *Typha, Lycopodiella alopecuroides, Huperzia, Osmunda, Phaeoceros, Anthoceros, Sphagnum, Ludwigia*, Eriocaulaceae and *Gunnera*.

Varied environments: Liliaceae, Melastomataceae, *Relbunium, Polygonum*, Rosaceae, *Cuphea*, Scrophulariaceae, *Polytrichum, Mutisia* type, *Pamphalea* type, Acanthaceae, *Chrysophyllum* type, Polygalaceae, Fabaceae, *Valeriana*, Apiaceae, Malvaceae, Rhamnaceae, tricolpates, stephanocolporates, pantoporates, tricolporates and indeterminates.

Exotic pollen: *Nothofagus* and *Alnus* (not considered in the palaeoenvironmental analyses).

Fungi: *Tetraploa, Rhizofagus, Athelia*, other spores and hyphae. Although fungi were taken into consideration in the parallel countings, they did not show significant data and thus were not considered in the palaeoenvironmental analyses.

Other palynomorphs: Arthropod jaws, other arthropod debris and platyhelminthes eggs. Like the fungi, the above were not considered in the palaeoenvironmental analysis.

81

4.3 Pollen assemblage zones

4.3.1 Zone 1 (samples 1 to 5)

The very fine sand (with rare shell debris) identified at the base of this zone (Fig. 2), corresponds to one facies of the lagoon margin environment, produced during the initial stages of the Tramandaí lagoon formation. In its upper portion, this facies seems to have been partially reworked by a high energy event which is registered by fine sand covered by shell debris. The facies succession identified in this zone culminates with the beginning of a lagoon bottom mud facies deposition, with a high concentration of organic matter whose base was dated at 5980 ± 130 ^{14}C yr BP (*Beta* 56511).

Samples 1 and 2, collected from the sandy facies of the lagoon margin (lower portion of this zone), showed a very low pollen concentration, compared with the samples of the muddy facies of lagoon bottom (upper portion of the zone, samples 3 and 5). Possibly this is due not to climatic influences, but to the kind of sediment (better conservation of palynomorphs in mud sediments). Because of this, the pollen analysis corresponding to the sandy intervals of zone 1, and its posterior comparison with the other facies of the core was based on the percentage diagrams, leaving out pollen concentration diagrams.

The presence of marine elements in the base facies of the lagoon margin shows a phase of marine transgression already reaching this region at about 6000 yr BP (Fig. 2).

According to the development pattern of this region, presented by Dillenburg (1994), this transgression phase was controlled by the rise of relative sea level (RSL).

With the continuation of RSL elevation, the volume of marine water increases in the lagoon and, while spreading over the land, it covers the higher marginal portions, beginning the mud-facies deposition of lagoon bottom at the core site. This initial mud facies deposition (upper portion of zone 1) is also significantly characterised by marine palynomorphs, thus strenghtening the hypothesis of a transgressive sea, controlled by RSL elevation.

In the upper portion of the zone there is an increase in concentration and percentage of dry field indicators (Figs 3, 4 and 5). Some of these components may represent halophytes, like *Salicornia gaudichaudiana*, *Rapanea parvifolia* and *Paspalum vaginatum*, plants which at present grow in brackish soils close to the lagoon. Others may be related to psammophytes which beginning at this phase increase in representativity, probably due to the dry sand soil expansion of the adjacent lagoon areas. *Ephedra,* although at low frequency, is best represented in zone 1 (Fig. 5).

Progressive brackish soils also seem to become evident in the analysis of fresh water marsh and swamp indicators (including the Σ of freshwater percentage elements). Some of these, formerly well represented, now

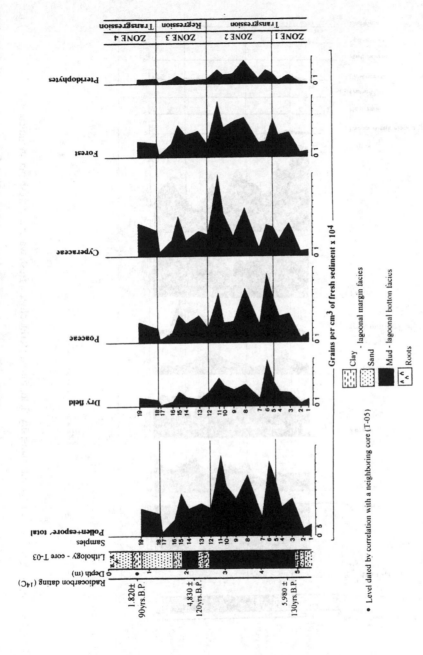

Figure 3. Palynological concentration diagrams of main indicator groups.

83

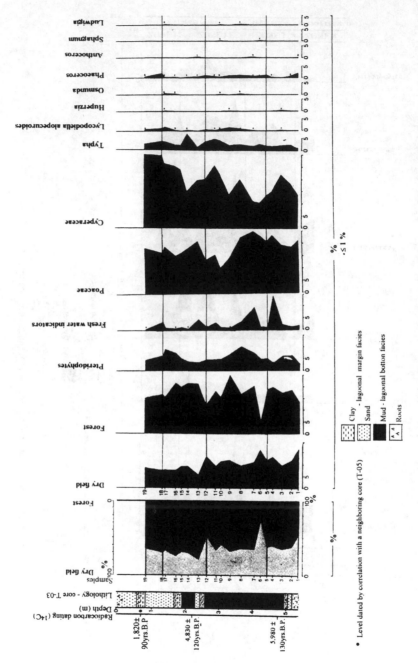

Figure 4. Palynological percentage diagrams: dry field, forest, pteridophytes, fresh water and fresh marsh indicators.

84

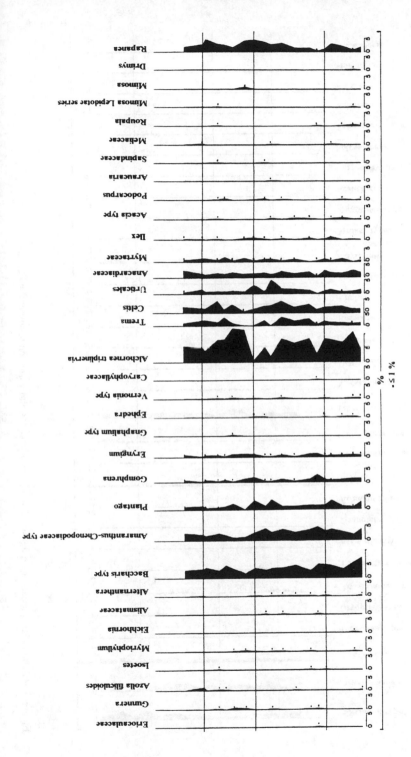

Figure 5. Palynological percentage diagrams (cont.): fresh marsh and fresh water indicators, dry field and forest components.

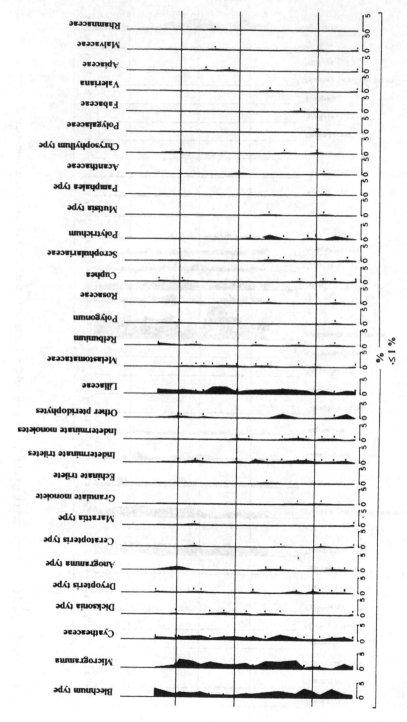

Figure 6. Palynological percentage diagrams (cont.): pteridophytes and varied environment indicators.

86

show significantly lower percentages in the upper portion and/or at the top of the zone (Figs 3, 4, 5 and 7). With the exception of sample 1, where *Typha* appears in a lower frequency, it continues in similar percentages in all of the zone (Fig. 4). This may indicate the presence of *Typha domingensis*, which is more adapted to brackish waters, and at present grows in the marshy coast system (Cordazzo & Seeliger 1988).

The tendency of an increase of halophytes and psammophytes and the decrease of fresh water indicators coincide with a significant representativity of marine origin elements (Fig. 2), which show the growing marine transgression over the region which must have enlarged the sandy lands and elevated the soil level salinity.

With the exception of sample 1, forest elements are better represented than those of dry fields in the whole of zone 1, both in concentration and in percentage, a tendency which is accentuated in the upper and top portions of this sequence (Figs 3 and 4). Part of *Rapanea* may represent arboreal species in non-brackish soils (Fig. 5).

This higher representativity of forest elements compared with dry field elements seems to indicate forests located farther inward, on lands still protected from transgressive marine influence. With the progressive temperature and humidity elevation at about 6000 yr BP (Cordeiro & Lorscheitter 1994; Neves & Lorscheitter 1995), the forest vegetation could develop as a whole on these more sheltered sites, increasing the number of its indicators in the upper portion of zone 1, where pioneer plants like *Alchornea triplinervia*, *Trema* and *Celtis* (Fig. 5) are well represented.

Pteridophytes are, generally, well represented in the zone (Σ of frequencies in the concentration and percentage diagrams, Figs 3, 4 and 6), decreasing significantly in the upper portion (contrary to the forest elements), already showing their sensitivity to environment alterations caused by the gradual elevation of RSL (Fig. 6).

4.3.2 *Zone 2 (samples 5 to 12)*

Contrary to the former zone, the whole depositional interval of zone 2 is represented by organic mud sediments. In the lower portions, the mud consists of silt and clay, rare layers and lenses of sand and scarce shell debris. The sand influence is a little higher in the uppermost portions with sand, silt and clay sediments, intercalated sand layers and lenses, and very few shell debris (Dillenburg 1994) (see Fig. 2).

Samples taken only in the mud portions made the comparison of distinct pollen concentration possible, contrary to zone 1, which presents more heterogeneous sediments.

The RSL elevation, already found in zone 1, increases constantly at about 5700-5500 ^{14}C yr BP (samples 6 to 8). In these samples the Σ of marine palynomorphs increases in percentage and concentration diagrams. However, the concentration diagrams show the gradual marine elevation

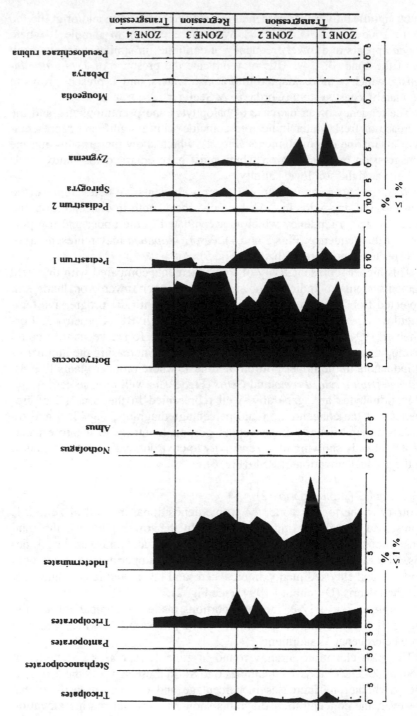

Figure 7. Palynological percentage diagrams (cont.): varied environment indicators, exotic pollen and algae.

88

better. *Operculodinium centrocarpum* and other cysts are the most sensitive indicators of RSL elevation in the interval between samples 6 and 8 (Fig. 2).

At about 5700 ^{14}C yr BP (sample 6) the transgression must have expanded its effects yet farther into the previously still protected areas, destroying the coppices and replacing the arboreal vegetation by species from more brackish and drier sand soil environments.

This environment alteration is shown in the quantitative pollen spectrum analysis, with significant changes on the frequency diagrams. Thus, sample 6 presents one of the highest total pollen + spore concentrations of the core, coinciding with the highest concentration and percentage of dry and/or brackish field elements of the whole profile (Figs 3 and 4). There is in this sample an increase in the frequencies of *Baccharis* type, *Amaranthus*-Chenopodiaceae type, *Gomphrena*, Caryophyllaceae and *Eryngium*, besides Poaceae, which also show their highest concentration in the core and a rather large percentage (Figs 3, 4 and 5). On the other hand, the arboreal pollen frequency is much lower in sample 6, to such an extent that this sample is the only core level where forest indicators are lower than those of dry fields (Σ of concentration and percentage diagrams – Figs 3, 4 and 5).

Pteridophytes show a low percentage in sample 6, following the former zone tendency (Figs 4 and 6).

Also following this tendency, the fresh water marsh elements are found in reduced numbers in sample 6 (Figs 4 and 7).

Everything indicates that this higher representation of halophytes and psammophytes, coinciding with the lessening of other elements, reflects alterations caused by a transgressive sea over the coastal plain (increase of marine indicators) not being the result of a dry phase, reducing the river inflow. This opinion agrees with other papers which mention a hot and humid climate on the Rio Grande do Sul coast approaching 5000 yr BP (Cordeiro & Lorscheitter 1994, Neves & Lorscheitter 1995).

The rapid and accentuated development of dry field plants must have required a temporary stabilization of RSL at about 5700 ^{14}C yr BP, favoring the vegetation expansion over the sandy soils.

At about 5500 ^{14}C yr BP (sample 7) a still greater transgression must have covered the dry sandlands, destroying a large area of herb vegetation while, in the more distant areas, arboreal species were trying to get back to their former stage of development. The lowest pollen + spore concentration of the whole mud interval is found in this sample (Fig. 3). The total of the percentage of dry field elements also decreases greatly, as well as the percentage of their most typical components (Figs 4 and 5).

Though at a lower concentration, the percentage of forest elements begins to grow during this phase, surpassing, as a whole, the dry field percentage elements and showing with the passing of time, a tentative adap-

89

tation to new sheltered sites (Fig. 4). Thus, pioneer plants like *Alchornea triplinervia* and *Trema,* as well as *Celtis,* Urticales, Anacardiaceae, *Ilex, Acacia* type and *Rapanea* show again increased frequencies (Fig. 5).

It is clear that the beginning of the forest development was made easier by the climate amelioration, with temperature and humidity increase. This too seems evident in the percentage increase of pteridophytes spores in sample 7, though still in a low concentration (Figs 4 and 6). The frequency elevation of fresh water indicators, both in the Σ of percentages and in individualized percentages in samples 7 and/or 8 also indicate higher river inflow into the region, likewise a probable result of a higher rainfall (Figs 4, 5 and 7). In sample 8, in spite of the fact that evidence continues to indicate the gradual marine transgression the indication of climate amelioration seems clear in the increase of pollen concentration, especially of arboreal components (Fig. 3).

The field formation begins to be restored in the sandy and dry areas and those of the forests follow their own development (sample 8). Pteridophytes here present their highest indices of the core in percentage and concentration, showing a quick response to the temperature and humidity increase (Figs 3, 4 and 6). It is possible that the percentage decrease of Poaceae, besides percentage of dry field and forest elements decrease in sample 8 (contrary to that indicated in the concentration diagrams) is due to this higher representativity of both pteridophytes and Cyperaceae, the latter showing the fresh water and/or brackish marshes expansion (Fig. 4).

Among the algae, only the increase of *Botryococcus* and *Spirogyra* show continental water influence in sample 8, indicating a greater brackish marsh expression (Fig. 7).

After this there is a rapid elevation of RSL in samples 9, 10 and 11, culminating in a maximum transgression at about 5000 yr BP (sample 11, Fig. 2).

At this phase there was a single and large lagoon which at present appears split into a big series of smaller interconnected lakes (Tomazelli, 1990, Dillenburg, 1994). These results agree with those already obtained for the coast of Rio Grande do Sul (Villwock et al. 1986, Villwock & Tomazelli 1989, Cordeiro & Lorscheitter 1994) and for other places of the Brazilian coast (Suguio et al. 1985), which seem to confirm an event of great extension.

In this phase, still under RSL stable conditions, sediment lagoonal processes, controlled by waves and currents, began the formation of the sand spit which at present partially separates the Tramandaí lagoon (Dillenburg 1994) (see Fig. 1).

The climate amelioration with a still greater elevation of temperature and humidity, allowed the pollen concentrations in the uppermost levels of the zone considerably (sample 9 to 11). The maximum concentration of continental origin elements in the whole mud interval can be seen in sample 11, coinciding with maximum transgression. Here Cyperaceae and forest elements reach the highest concentration in the whole core also with a significant increase in dry field element concentration (Fig. 3). The high frequencies of Cyperaceae reduce the Σ of the percentage of forest, dry field and pteridophyte elements in spite of their high concentrations (Fig. 4).

With a higher rainfall, the drainage system had to pour a greater volume of continental waters into the region, contributing to the expansion and diversification of marshes and swamps with varying grades of salinity (Figs 4, 5 and 7). A greater expansion of marshes is also indicated by the percentage decrease of *Botryococcus* and percentage increase of *Pediastrum*, since the latter lives preferentially in waters with more vegetation (Fig. 7).

Coppices and meadows, conditioned by distinct edaphic formations originated by the degree of salinity and kind of sediment, must have spread (samples 9 and 11). Thus, in sample 9, forest elements are more evident, showing at this level the highest Σ of percentage in the core (Figs 4 and 5).

Among psammophytes and halophytes (these possibly now less expressive due to high fluvial inflow) are also significant in samples 9 to 11, indicators like *Baccharis* type, *Amaranthus*-Chenopodiaceae type, *Plantago*, *Gomphrena*, *Eryngium*, followed by *Alternanthera*, *Vernonia* type and *Ephedra* (Fig. 5). In this way the development of the herbaceous coast vegetation becomes evident. The pteridophytes concentrations and percentages, though lessened by environment alterations, are still significant (Figs 3, 4 and 6).

Therefore, an accentuated vegetal diversification took place at about 5000 ^{14}C yr BP with temperature and humidity elevation, compensating the effects of the highest Holocene marine transgression over the Rio Grande do Sul coastal plain.

Likewise at about 5000 yr ^{14}C BP (sample 12) the regression began. This is shown by a significant reduction of concentration and percentage of marine origin elements (Fig. 2).

Coinciding with the marine regression, the pollen concentration as a whole decreases considerably in sample 12, as well as the concentration and percentage of Cyperaceae, Poaceae likewise the total of arboreal pollen, while pteridophytes continue scarcely represented in very low percentages and concentrations (Figs 3 and 4). On the contrary, the total percentage of dry field elements increases considerably. These data, as a whole, seem to show important environment changes caused especially by new edaphic conditions (degree of salinity, level of water table, etc.). Perhaps there was also a decrease of fluvial inflow in this phase as a result of the lower local rainfall (noticeable decrease of concentration and percentage of Cyperaceae and in the percentage of *Typha*, *Lycopodiella alopecu-*

roides, Botryococcus, Pediastrum, Zygnema and *Debarya*; Figs 3, 4 and 7).

4.3.3 *Zone 3 (samples 12 to 18)*

The base sediments of zone 3 consist of two narrow sand belts of moderate selection, with clay, showing a level of shell debris, which seem to indicate a short alteration phase in the sediment sequence. Then there comes a new mud interval, dated at 4830±120 ^{14}C yr BP (*Beta* 56512), with sand-silt-clay and intercalated sand lenses, representing a probable continuation of the deposition interval of lagoonal bottom of zone 2. A short sand-clay interval follows and another longer interval with fine selection sand and rare mud blades and lenses (Fig. 2), interpreted by Dillenburg (1994) as a lagoonal margin facies.

In the whole of zone 3 the indicators continue showing the marine regression process, probably controlled by a lowering of RSL, with a gradual coast emergence (great decrease in the concentration and percentage of marine origin elements). Only microforaminifera show a significant frequency in sample 13 (Fig. 2).

At about 4800 ^{14}C yr BP (sample 14), the depth of the lagoon water level probably had already been lowered, due to the gradual decrease of RSL. Thus the mud interval of lagoonal bottom deposition came to an end, giving rise to a sand deposition of lagoonal margin.

Samples 13 and 14 show a transition phase as related to zone 2 (sample 12). The corresponding pollen diagrams seem to indicate that the vegetation tried to adapt to the altered new environments but do not show the principal tendencies very clearly. However, forest indicators show again much higher frequencies as compared to dry field elements, pointing to a new forest development. *Celtis* and *Alchornea triplinervia*, pioneer tree plants, are more evident, the latter with the highest percentages of the whole core in these two samples (Fig. 5). The pteridophytes, on the contrary, still keep their reduced frequencies (Fig. 4).

Marsh expansion, now with less salinity is indicated by a significant representativity of Poaceae, Cyperaceae (especially in sample 13) and *Typha* (with the highest core percentages in sample 14; Fig. 4). Fresh water algae, such as *Botryococcus, Spirogyra* and *Zignema,* are well represented in sample 13 (Fig. 7). In this sample the reverse relation between *Pediastrum* and *Botryococcus* (better represented than *Pediastrum*) seems now to indicate a lesser influence of aquatic macrophytes (only *Gunnera, Myriophyllum, Phaeoceros* and *Anthoceros* are slighty more evident among marsh and swamp elements; Figs 4 and 5).

The interval between samples 14 and 17 shows signs of a still greater marine regression, and in sample 17 the maximum regression is registered (Fig. 2). This maximum regression is estimated by Dillenburg (1994) at 3500 ^{14}C yr BP.

The tentative to conquer new emergent and less saline environments seems to have been successful, showing in sample 15 a more diversified vegetation with a balanced development of fields and forests, the latter being well represented (Figs 3 and 4).

Pteridophytes, which since the former interval kept their low frequencies, increase their percentages (samples 15 to 17), which shows that this group is more sensitive to environment alterations, developing only after the establishment of more stable environment conditions (Fig. 4). Cyperaceae have very high percentages in this interval showing, together with Poaceae (part), the now less saline marsh expansion (Fig. 4). The greater influence of the fluvial network in the region, with marine regression, is also evident (in samples 15 to 17) by the presence of marsh and fresh water components; Figs 4, 5, and 7). *Botryococcus* have very high frequencies since sample 13. Even so, its inverse relation with *Pediastrum* (now in a greater number) suggests an aquatic vegetation development.

With the gradual lowering of RSL and progressive less saline soils, ever more favorable conditions appeared for the settlement of new continental lands, which is chiefly indicated by the increase of the pteridophytes (samples 15 to 17; Fig. 4). This increase reduces somewhat the percentages of dry field and forest elements, though signs still indicate the significant presence of these two vegetal formations in the region: forests in the more withdrawn and sheltered places and fields in the sandy dry portions and/or still under some saline influence.

At the top of zone 3 (sample 18) the first signs of a second RSL ascension appear, shown by a slight increase of the Σ of marine indicators (Fig. 2). As a whole, there are no significant vegetation changes, only an accentuated increase in pollen concentration compared with sample 17, which shows the continuity of vegetation development still caused by the expansion of emergent areas due to previous marine regression (Fig. 3).

The forest continues to be better represented than the field in this phase, the latter practically unchanged (Fig. 4). Extense marshes and swamps continue being indicated by Poaceae (part) and chiefly by Cyperaceae which here have the highest core percentages, contributing to the reduction of the relative frequency of pteridophytes and arboreal pollen (Fig. 4), though both have increased concentrations in relation to the previous sample (other marsh elements more evident in this sample are *Typha, Lycopodiella alopecuroides* and *Phaeoceros*; Fig. 4).

Continental waters continue being indicated especially by the high frequency of *Botryococcus* and increased percentages of *Pediastrum, Spirogyra, Zygnema, Debarya* and *Azolla filiculoides* (Figs 5 and 7).

4.3.4 *Zone 4 (samples 18 to 19)*

The lowest portion of zone 4 consists of fine selected sand with scarce lenses and blades of mud, corresponding to the lagoon margin facies.

There follows a different interval, also with fine sand, yet with clear inter-calated mud blades dated, by correlation with a neighboring core (T-05), at 1820 ± 90 ^{14}C yr BP. Above the sequence becomes again clearly sandy with well selected fine sand and a few roots at the top, corresponding to the end of the sand spit formation (Dillenburg 1994).

The pollen sequence finishes in sample 19, collected in the sand inter-val with intercalated mud blades. It corresponds to the new transgression over the Rio Grande do Sul coastal plain, already shown in sample 18, which took place at about 1800 yr BP, when the sea may have risen up to 1.5 m above the present level (Dillenburg 1994). This second transgres-sion was mentioned for the first time for Rio Grande do Sul, and because it is based on only one dating, its chronology is not yet completely sure. Yet the transgression is proved by the increase of all of the marine indi-cators (Fig. 2), not reflecting an occasional event.

The results about sea level oscillations after 5000 yr BP are not yet very clear as to the Brazilian coast, although researches are already underway (Suguio et al. 1985, Villwock & Tomazelli 1989, Angulo & Suguio 1992).

Suguio et al. (1985) mention two transgressions for some parts of the Brazilian coast after the maximum transgression of 5100 yr BP: for Salva-dor (Bahia State), Paranaguá (Paraná State) and Itajaí-Laguna (Santa Catarina State). For Salvador, where detailed studies were made, a trans-gression at about 3600 yr BP and another one at about 2500 yr BP are well defined. It is possible that the Tramandaí transgression at about 1800 yr BP is related to the one of 2500 yr BP. Only a greater number of datings can clarify this problem. Not depending on chronological precision, only one transgression was detected for Rio Grande do Sul after the maximum transgression of 5100 yr BP, diverging from the results which indicate three Holocene transgressions over the Brazilian coast.

Following the tendency of the previous sample (18) pteridophytes once more show their sensitivity to environment alterations in this last sample (19), with reduced frequencies even in mud sediments (Figs 3 and 4). The changes seem to affect, as a whole, also forest and fresh water vegetation, decreasing their pollen component percentages in the sediments (Fig. 4). On the contrary dry field and/or salt soil vegetation increases, showing a tendency of similar behavior of the vegetation to the one at the beginning of the previous transgression (Fig. 4).

5 CONCLUSIONS

The evidences of only two marine transgressions based on the pollen analysis of Tramandaí lagoon sediments, at about 5000 ^{14}C yr BP and 1800 ^{14}C yr BP, corroborate the results obtained by Dillenburg (1994), related to the development of the lagoon/barrier IV at the northern region of the coastal plain of Rio Grande do Sul. No signs were found of a third Holocene transgression, detected in other parts of the Brazilian coast (Suguio et al. 1985), which might be the object of future investigations and debates.

The maximum transgression at about 5000 ^{14}C yr BP has already been mentioned for the coastal plain of Rio Grande do Sul and for other parts of the Brazilian coast, thus the present paper reinforce these results.

The second transgression (1800 ^{14}C yr BP), because detected for the first time and proceeding from an only sample (Dillenburg 1994), deserves more geochronologic research, though the sediment and pollen data agree as to the RSL elevation during this phase. Therefore it does not seem to be a simple occasional event, and may be related to the transgression of 2500 ^{14}C yr BP, detected at other parts of the Brazilian coast.

The palaeoenvironment reconstruction of the last millennia in the Tramandaí lagoon microregion revealed highly unstable coast environments, originated by the sea level oscillations. Although the vegetation distribution patterns in the coastal plain were similar to the present, the greater marine influence, especially at about 5000 ^{14}C yr BP, removed the distinct vegetation belts towards the interior of the continent, partially due to land submersion and partially due to edaphic alterations, resulting from sea proximity.

The halophytic and psammophytic belt was enlarged during this phase. Yet the hot and humid climate favored the vegetation diversification and the full coast forest development over the most protected places of marine influence. The following RSL regression and the progressive lower salinity in the soils caused new plant settlements over the previously conquered areas, affecting the vegetation stretches once more. New transgression evidences, controlled by RSL elevation, at about 1800 ^{14}C yr BP are linked to a vegetation behavior similar to the one of the previous transgression.

The Tramandaí lagoon therefore proved to be an excellent site for palaeoenvironment analysis, and may be indicated for further research to clarify questions which have not yet been well understood, especially as to the number and chronology of the Holocene oscillations of RSL in southern Brazil.

ACKNOWLEDGEMENTS

The authors wish to express their deep recognition to Stella Ursula Tesche, for her help in the final version of this paper. This research was supported by grants from Conselho Nacional de Desenvolvimento Científico e Tecnológico (CNPq. – Brazil).

REFERENCES

Angulo, R.J. & K. Suguio 1992. Reavaliação dos máximos da curva de variação do nível do mar durante o Holoceno no Estado do Paraná. *Anais 36 Congresso Brasileiro de Geologia,* SBG 1: 82-83. São Paulo.

Cordeiro, S.H. & M.L. Lorscheitter 1994. Palynology of Lagoa dos Patos sediments, Rio Grande do Sul, Brazil. *Journal of Paleolimnology* 10: 35-42.

Cordazzo, C.V. & V. Seeliger 1988. *Guia ilustrado da vegetação costeira no extremo sul do Brasil.* Editora da FURG, Rio Grande. 275 pages.

Dillenburg, S.R. 1994. A Laguna de Tramandaí: Evolução geológica e aplicação do Método geocronológico da termoluminescência na datação de depósitos sedimentares lagunares. Tese de Doutorado, Curso de Pós-graduação em Geociências, Universidade Federal de Rio Grande do Sul. 113 pages. Brazil.

Faegri, K. & J. Iversen 1989. *Textbook of pollen analysis.* New York: John Wiley & Sons. 328 pages.

Hasenack, H. & L.W. Ferraro 1989. Considerações sobre o clima da região de Tramandaí, RS. *Pesquisas* 22: 53-70.

Neves, P.C.P. & M.L. Lorscheitter 1995. Upper Quaternary palaeoenvironments in the Northern Coastal Plain of Rio Grande do Sul, Brazil. *Quaternary of South America and Antarctic Peninsula* 9: 39-67. Rotterdam: Balkema Publishers.

Nimer, E. 1979. *Climatologia do Brasil.* Superintendência de Recursos Naturais e Meio Ambiente (SUPREN), IBGE, Rio de Janeiro, 421 pages.

Ramos, R.F. 1977. Composição florística e ecologia do delta do Rio Tramandaí - RS. Dissertação de Mestrado, Curso de Pós-graduação em Botânica, Universidade Federal de Rio Grande do Sul. 131 pages. Brazil.

Salgado-Labouriau, M.L. 1973. *Contribuição à palinologia dos Cerrados.* Academia Brasileira de Ciências. 291 pages.

Schwarzbold, A. 1982. Influência da morfologia no balanço de substâncias e na distribuição de macrófitos aquáticos nas lagoas costeiras do Rio Grande do Sul. Dissertação de Mestrado, Curso de Pós-graduação em Ecologia, Universidade Federal de Rio Grande do Sul. 91 pages. Brazil.

Stockmarr, J. 1971. Tablets with spores used in absolute pollen analysis. *Pollen et Spores* XIII(4): 615-621.

Suguio, K., L. Martin, A.C.S.P. Bittencourt, J.M.L. Dominguez, J.M. Flexor & A.E.G. Azevedo 1985. Flutuações do nível do mar durante o Quaternário Superior ao longo do Litoral Brasileiro e suas implicações na sedimentação costeira. *Revista Brasileira de Geociências,* 15: 273-286.

Tomazelli, L.J. 1990. Contribuição ao Estudo dos Sistemas Deposicionais Holocênicos do Nordeste da Província Costeira do Rio Grande do Sul – Com Ênfase no Sistema Eólico. Tese de Doutorado. Curso de Pós-graduação em Geociências, Universidade Federal do Rio Grande do Sul. 270 pages. Brazil.

Villwock, J.A. 1984. Geology of the Coastal Province of Rio Grande do Sul, Southern Brazil. A Synthesis. *Pesquisas,* 16: 5-49.

Villwock, J.A., L.J. Tomazelli, E.L. Loss, E.A. Dehnhardt, N.O. Hörn F°, F.A. Bachi & B.A. Dehnhardt 1986. Geology of the Rio Grande do Sul Coastal Province. *Quaternary of South America and Antarctic Peninsula* 4: 79-97. Rotterdam: Balkema Publishers.

Villwock, J.A. & L.J. Tomazelli 1989. Sea-level changes and Holocene evolution in the Rio Grande do Sul Coastal Plain, Brazil. In: *International Symposium on Global*

Changes in South America during the Quaternary. Past-Present-Future: 192-196 (Special Publication). São Paulo.

Würdig, N.L. 1984. Ostracodes do sistema lagunar de Tramandaí, RS, Brazil – sistemática, ecologia e subsídios à paleoecologia. Tese de Doutorado. Curso de Pós-graduação em Geociências, Universidade Federal de Rio Grande do Sul, 300 pages. Brazil.

Evidence of a forest free landscape under dry and cold climatic conditions during the last glacial maximum in the Botucatú region (São Paulo State), Southeastern Brazil

6

HERMANN BEHLING
University of Amsterdam, Hugo de Vries-Laboratory, Department of Palynology and Paleo/Actuo-ecology, The Netherlands Centre for Geo-ecological Research, Amsterdam

MARTIN LICHTE
Nordenham, Germany

ATTILA WOLINSK MIKLOS
Departamento de Ciencia do Solo, Escola Superior de Agricultura 'Luiz de Queiroz', Universidade de São Paulo, Piracicaba, Brazil

ABSTRACT: Late Quaternary vegetational and climatic environments have been interpreted from ca. 30,000 ^{14}C yr BP organic rich headwater area deposits from the Botucatú region in São Paulo State (Southeastern Brazil). The pollen and charcoal records indicate mainly grassland (campos) vegetation cover with little stands of subtropical forest, reflecting cold and dry climatic conditions during the recorded glacial period from about 30,000 to ca. 18,000 ^{14}C yr BP. Fires occurred in the campos vegetation during glacial times. Probably climatic conditions during the early Holocene may have caused decomposition of the deposits of glacial age, and a sedimentation gap from ca. 18,000 to 6000 ^{14}C yr BP. Pollen assemblages from peaty swamp deposits indicate aboriginal land use by deforestation and plantation of *Zea mays* and *Manihot* since at least 2900 ^{14}C yr BP.

RESUMEN: Ambientes vegetacionales y climáticos del Cuaternario tardío desde ca. 30,000 años ^{14}C AP han sido interpretados a partir de depósitos de cabeceras de agua ricos en materia orgánica de la región de Botucatú, en el Estado de São Paulo (Sudeste de Brasil). Los registros de polen y carbón indican principalmente una cubierta vegetacional de 'campos' con escasos espacios de bosque subtropical, reflejando condiciones climáticas frías y secas durante el período glacial, registrado desde ca. 30.000 a 18.000 años ^{14}C AP. Ocurrieron incendios en los 'campos' durante épocas glaciales. Probablemente, las condiciones climáticas durante el Holoceno temprano podrían haber causado la descomposición de los depósitos de edad glacial y una interrupción en la sedimentación desde ca. 18.000 a 6000 años ^{14}C AP. Conjuntos polínicos de depósitos turbosos indican el uso de la tierra por los aborígenes, por medio de desforestación y siembra de *Zea mays* y *Manihot* desde por lo menos 2900 años ^{14}C AP.

1 INTRODUCTION

There have been long and controversial discussions about the Late Quaternary vegetational and climatic history of Southeastern Brazil. Recently, a few publications on palynological surveys improved our understanding of the palaeoenvironmental history of this area. In a report on a palynological study (Behling & Lichte 1997) we were able to define a clear vegetational and climatic sequence for the site Catas Altas in Minas Gerais, Brazil (lat. 20°05' S, long. 43°22' W; see Fig. 1): the last glacial landscape (recorded period: ca. 18,000 to >48,000 yr BP) was covered by extensive areas of subtropical grasslands and small areas of gallery forests along the rivers. The data indicate that in the lower highland region of Minas Gerais a marked drier and colder climate with temperatures of 5-7 °C below the present value predominated, with strong frosts during the winter months. A similar climate at Morro de Itapeva, in the Serra da Mantiqueira mountains in Southeastern Brazil, has been reconstructed (Behling 1997a). In contrast to the Catas Altas and Morro de Itapeva region, records from Salitre (Ledru 1993, Ledru et al. 1994; Ledru et al. 1996) and Serra Negra and Lagoa dos Olhos (De Oliveira 1992) were interpreted as cold and wet climatic conditions for glacial times.

Interesting is the difference in reconstructed palaeoclimatic conditions using the pollen records of both mentioned areas. We continued our search more to the Southwest from the sites mentioned before into the São Paulo State. The point in question is which type of palaeoclimate can be found in the newly studied site near the town of Botucatú. The situation in Botucatú resembles the one found in Catas Altas: in both locations dark organic rich headwater deposits occur, which are in general overlain by colluvial accumulations.

2 STUDY AREA

The studied site (lat. 22°48' S, long. 48°23' W) is located close to the town of Botucatú, about 230 km west of the city of São Paulo in São Paulo State, Southeastern Brazil (Fig. 1), at 770 m a.s.l. The study region of Botucatú is characterized by a soft rolling landscape at elevations of about 500 to 800 m a.s.l. Small patches of black organic rich deposits can be found in most of the headwater areas in that region. The headwater area of Corrego Santana is the one studied in this paper.

The major vegetation forms in Southeastern Brazil are the Atlantic rain forests along the coast, the semideciduous forests in the hinterland and different 'cerrado' forms in the northwestern regions (IBGE 1993). *Araucaria* forest is distributed on the Southeastern Brazilian highlands only in form of small isolated areas between lat. 18° and 24° S at elevations mainly

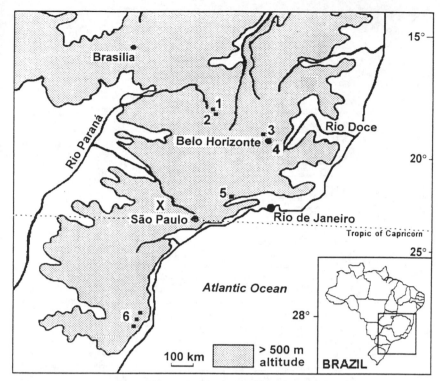

Figure 1. Sketch map showing location of the study site (X) Botucatú and other mentioned sites Serra Negra (1), Salitre (2), Lagoa dos Olhos (3), Catas Altas (4), Morro de Itapeva (5) in SE Brazil and Serra da Boa Vista, Morro da Igreja and Serra do Rio Rasto in Southern Brazil (6).

between 1400 and 1800 m a.s.l. (Hueck 1953). The primary vegetation around Botucatú can be characterized as a mosaic of ca. 20 m tall and dense semideciduous forest and cerrado vegetation, mainly cerradão (10 to 15 m tall dry forest with a close canopy) (IBGE 1993, Hueck 1966). In the Botucatú region, cerrado vegetation occurs in its southernmost extension (Silberbauer-Gottsberger et al. 1977, Hueck 1956). The primary vegetation has almost been destroyed by formation of pastures. Only small patches of secondary forests remain in the study area (Miklos 1992).

Tropical semideciduous forests in Southeastern Brazil are found in regions with annual precipitation between 1000 and 1500 mm and a dry season of between 3 and 5 months. The average annual precipitation in the main part of the cerrado region is between 1000 and 1750 mm, the average annual temperature is 20-26°C, and the length of the dry season is 5-6 months (Nimer 1989).

The closest meteorological station to the area of study is Piracicaba,

101

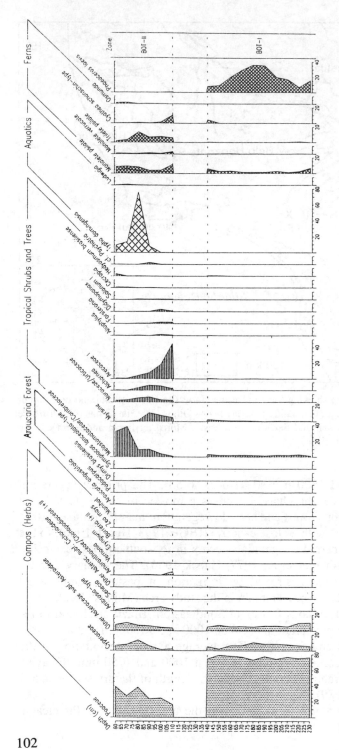

Fig. 2a. Percentage pollen diagram of the most frequent pollen and spore taxa from Botucatú.

102

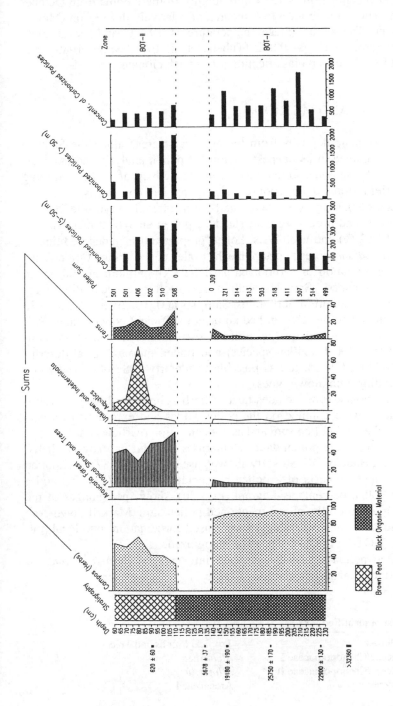

Fig.2b. Summary pollen diagram showing groups of pollen taxa representing different vegetation types. In addition, radiocarbon dates, stratigraphical characteristics, pollen concentration, percentages and concentration of carbonized particles, and pollen zones from the core Botucatú are shown.

103

about 80 km east of Botucatú and at 556 m elevation. The recorded average annual precipitation is 1314 mm, mainly in the months from October to March. There are about 3 to 5 months of relatively dry season (May to September). The annual mean temperature is 19.4°C and the absolute minimum temperature is -0.8°C (Tubelis et al. 1971, Nimer 1989). According to the Koeppen classification it is a Cwb climate.

3 MATERIAL AND METHODS

Excavated organic deposits from the headwater areas along the Santana river were sampled in its deepest section for pollen analysis and radiocarbon dating. All samples were collected in plastic bags and, after returning from the field, stored under cool (ca. +8°C) and dark conditions.

Samples of 0.5 cm^3 were taken at 10 cm intervals along the 260 cm thick deposits and processed with standard pollen analytical methods, using HF (47-52 %) and acetolysis. Pollen preparation included addition of exotic *Lycopodium* spores to determine the pollen concentration (grains/cm^3) and the concentration of carbonized particles. A minimum of 500 pollen grains was counted. Samples with a low pollen concentration were counted to a minimum of 300 grains. Samples at 0, 10, 20, 30, 40, 50, 120 and 130 cm and below 235 cm had no or not sufficient pollen grains. Pollen identification relied on the author's own reference collection (containing about 1400 Brazilian species) and pollen morphological descriptions in Behling (1993). It was possible to identify 108 pollen and spore types including 5 unknown types.

The total pollen sum includes taxa of herbs, shrubs and trees. Carbonized particles were counted on the pollen slides and expressed as a percentage of the total pollen sum and as concentration (particles/cm^3).

For plotting of the pollen data, calculations and cluster analysis TILIA, TILIAGRAPH and CONISS software was used (Grimm 1987). The pollen diagram, which consists of a pollen percentage diagram (Fig. 2a) and a summary pollen percentage diagram (Fig. 2b), shows percentages of the major taxa based on the above mentioned pollen sum. Marked down-core changes in the pollen assemblages were used for zonation: two local pollen zones (BOT-I and BOT-II) can be recognized.

Identified pollen taxa were grouped into campos (herbs), *Araucaria* forest, and tropical shrubs and trees (Table 1).

Table 1. List of identified taxa.

Campos (Herbs)	Tropical Shrubs and Trees
Amaranthaceae/Chenopodiaceae I	*Alchornea*
Amaranthaceae/Chenopodiaceae II	*Allophylus*
Ambrosia-type	Arecaceae I

Table 1. Continued.

Apium-type	*Aspidosperma*
Baccharis-type	*Astrocaryum aculeatissimum*-type
Borreria I	*Cardiospermum*
Borreria II	*Cecropia*
Convolvulaceae	*Celtis*
Cuphea	cf. *Psychotria*
Cyperaceae	*Clethra*-type
Ericaceae	*Croton*
Eryngium	*Didymopanax*
Gomphrena/Pfaffia	*Dodonaea*
Hyptis-type	*Ephedra tweediana*-type
Iridaceae I	*Eucalyptus*-type
Iridaceae II	*Euphorbia*
Jungia/Holocheilus-type	*Eorsteronia*
Lamiaceae	*Gallesia*
Malvaceae	*Hedyosmum brasiliense*
Manihot	*Hedyosmum brasiliense*
Other Asteraceae subf. Asteroideae	*Hyeronima*
Other Asteraceae subf. Cichorioideae	*Ilex*
Poaceae	*Jacobina/Justicia*-type
Senecio	*Luehea*
Valeriana	Malpighiaceae
Vernonia-type	*Mandevilla*-type
Vicea/Lathyrus	*Matayba*
Xyris	Melastomataceae/Combretaceae
Zea mays	*Meliaceae*
	Meliosma
Araucaria Forest	*Mimosa*
Araucaria angustifolia	Moraceae/Urticaceae
Podocarpus	*Myrsine*
Myrtaceae	*Drimys brasiliensis*
Symplocos lanceolata-type	Papilionaceae I
Symplocus tenuifolia-type	Papilionaceae III
	Pera
Aquatics	*Piper*
Echinodorus	*Prockiacrucis*-type
Hydrocotyle	*Rhipsalis*
Ludwigia	Sapindaceae
Typha domingensis	Sapotaceae
Utricularia	*Schinus/Lithraea*-type
	Sebastiania brasiliensis
Unknown	*Sloanea*
Type 1 - 5	*Solanum* I
	Solanum II
	Styrax
Ferns	*Trema*-type
Cyathea schanschin-type	Weinmannia
Cyathea-type	*Zanthoxylum*
Lycopodium alopecuroides-type	

Table 1. Continued.

Lycopodium clavatum-type
Lycopodium sp.
Monolete psilate < 50μm
Monolete psilate > 50μm
Monolete verrucate < 50μm
Monolete verrucate > 50μm
Osmunda
Pteris-type
Trilete psilate
Trilete verrucate indet.
Moss
Phaeoceros laevis
Fungal Remains
Gelasinospora (Sordariaceae)
Algae
Debarya (Zygnemataceae, Chlorophyceae)

4 RESULTS

4.1 *Stratigraphy*

The studied outcrop starts with a thin grass layer at the surface. Organic deposits are in the thickest section 200 cm thick and located below 60 cm thick yellow to reddish erosional slope and soil accumulations. The sediments were described in the field as follows:

0-60 cm	Yellow – reddish erosional slope and soil accumulations
60-110 cm	brown peat
110-260 cm	dark gray – black compact completely decomposed organic material, fine sandy
>260 cm	sandy subsurface

4.2 *Radiocarbon dates*

The 6 radiocarbon dates (Table 2), including 3 AMS dates (at 133, 186 and 226 cm depth) indicate that the lower part of the organic-rich deposits between ca. 140 and 230 cm represents the interval of about ca. 30,000 to ca. 18,000 ^{14}C yr BP. The radiocarbon dates with >32,360 ^{14}C yr BP (246-250 cm depth) and 25,750 ^{14}C yr BP (186 cm depth) may suggest that the date of 22,900 ^{14}C yr BP at 226 cm depth is not correct.

Accumulations between 140 and 110 cm depth, which do not con-

106

Table 2. Radiocarbon dates of sediment samples of the Botucatú core.

Lab. Number	Depth (cm)	^{14}C yr BP	^{13}C/^{12}C
Hv-20822	90-92	620 ± 60	−22.3
UtC-5544	133	5678 ± 37	−20.8
Hv-20823	150-152	19180 ± 190	−14.9
UtC-5545	186	25750 ± 170	−14.9
UtC-5546	226	22900 ± 130	−16.2
Hv-20824	246-250	>32360	−16.9

tain sufficient pollen grains, are apparently of Holocene age (one radio-carbon date with 5678 ^{14}C yr BP at 133 cm depth). There is a gap in the pollen recovery, due to corrosion, that represents the period of ca. 18,000 ^{14}C yr BP to 6000 ^{14}C yr BP. Sediments of the uppermost section (0-110 cm depth) are of late Holocene age and represents the period from ca. 2900 ^{14}C yr BP (interpolated age) to the present.

4.3 Description of the Botucatú pollen diagram (Figs. 2a and 2b)

4.3.1 Zone BOT-I (230 - 140 cm, ca. 30,000 - ca. 18,000 ^{14}C yr BP, 10 samples)

This zone is characterized by abundant pollen grains of campos vegetation (90-95%), primarily Poaceae, followed by Cyperaceae, Asteraceae and taxa such as *Eryngium, Borreria,* which occur in low percentages. Pollen percentages of taxa of *Araucaria* forest are very low (<0.5 %): only some single grains of *Araucaria, Podocarpus, Drymis brasiliensis* and *Symplocos* have been found. Pollen grains of taxa representing tropical shrubs and trees (3-6%), such as Melastomataceae/Combretaceae, *Myrsine,* Moraceae/Urticaceae and Arecaceae (palms) are found only in low percentages. Representation of spores from ferns is low: between 3 and 11%. Moss spores of *Phaeoceros laevis* are frequent. The percentages and concentrations of carbonized particles are high in this zone.

4.3.2 Zone BOT-II (110-60 cm, 2910 ^{14}C yr BP to Recent, 6 samples)

Abundant herb pollen (35-60%), dominated by Poaceae, Cyperaceae Asteraceae and Amaranthaceae/Chenopodiaceae characterize this zone. Pollen grains of *Zea mays* and *Manihot* appear for the first time and are only present in the lower part of this zone. Taxa of *Araucaria* forest are also rare in this zone (> 1%). Representation of tropical shrubs and trees is high (40-65%), primarily *Myrsine,* Arecaceae in the lower part and Melastomataceae/Combretaceae in the upper part of the zone. Moraceae/Urticaceae are more frequent than in zone BOT-I. *Alchornea, Allophylus, Forsteronia, Didymopanax* and *Solanum* appear for the first time and they are more frequent in the lower part of the zone than in the

107

upper part. Aquatics are mainly represented by *Typha domingensis* (0-75%) in the middle part of the zone. Spores of ferns (13-35%) show markedly higher percentages than in zone BOT-I. Carbonized particles are frequent in this zone.

5 INTERPRETATION OF THE RESULTS

The pollen assemblages indicate an almost treeless landscape of campos vegetation in which fires occurred frequently during the glacial period between ca. 30,000 and ca. 18,000 yr BP. Only very few stands of subtropical *Araucaria* forest may have occurred along water currents with sufficient moisture. Arboreal species of the family Melastomataceae and *Myrsine* might have been during that time part of subtropical gallery forests. *Araucaria angustifolia* and *Podocarpus* itself were very rare.

The largely treeless landscape reflects cold and dry climatic conditions in the Botucatú region at that time, whereas today semideciduous forest and tropical seasonal climate exists.

The very rare presence of *Araucaria angustifolia* in the Botucatú pollen record indicates during the glacial time a temperature which is similar to Catas Altas (Behling & Lichte 1997) of at least 5-7°C lower than today. The modern minimum temperature for this region is –0.8°C (meteorological station of Piracicaba). *Araucaria* can grow in areas of an absolute minimum temperature of –10 °C (Golte 1978), and stronger frosts may have affected the stands of *Araucaria* in the Botucatú region.

The gap in the pollen recovery between zone BOT-I and BOT-II cannot be definitely explained, but may have been caused by dry climatic conditions, probably during the early Holocene. Dry climates have been observed in different sites in Southern Brazil during early and mid Holocene periods (Behling 1995, 1997b).

The following Late Holocene period, from 2910 yr BP (interpolated age) to probably modern age, is represented in the core by a 50 cm thick peaty deposits (pollen zone BOT-II). High percentages of *Typha* are indicative of swamp conditions, possibly by the damming of the river by human activity.

High percentages of Poaceae pollen and carbonized particles in the zone BOT-II is indicative of forest disturbance and deforestation, related to human presence in the area of Botucatú, at least since 2900 yr BP. Pollen grains of *Zea mays* and *Manihot,* primarily at the beginning of the zone, are indicative of corn and *Manihot* plantations, suggesting agricultural activities by aboriginals in the study area. The decrease of several forest taxa, such as palms, *Myrsine*, *Alchornea* and others indicate a change of the main vegetation form of secondary forests and agriculture to pasture land with Melastomataceae shrubs and less intensive agriculture. The transformation into pasture land of the surroundings of the study site may have caused soil erosion leading to accumulation of this erosion product over the peaty deposits.

6 DISCUSSION AND CONCLUSION

The pollen and charcoal record of Botucatú shows what the main vegetation was in the last glacial landscape with large areas of grassland and small stands of forests on the Southeastern Brazilian highland at lower elevations, indicating cold and dry climatic conditions. Palaeofires were frequent during glacial times. Similar results have been obtained from the record of Catas Altas (Behling & Lichte 1997) about 650 km northeast of Botucatú.

This study is another proof of grassland expansion from Southern to Southeastern Brazil and evidence of drier and colder climatic conditions in Southeastern Brazil during glacial times. The Botucatú pollen record supports the idea that the results from Catas Altas (Behling & Lichte 1997) and Morro de Itapeva (Behling 1997a) cannot be seen only as local environmental conditions. The results have to be interpreted as a general regional feature for the Southeastern Brazilian highlands during the glacial times, as far as indicated in the pollen records.

These results may suggest that the pollen records from the mountain region of Serra Negra at 1170 m elevation (De Oliveira 1992) and Salitre (Ledru et al. 1996) reflect a locally different climate (wetter between 40,000 to 27,000 yr BP) than in the lower highland regions. The sedimentation gap in the Salitre record and perhaps also in the Serra Negra record (P.E. De Olivcira, pers. comm.) during the last glacial maximum may be related to dry climatic conditions.

ACKNOWLEDGMENTS

We thank Soil Science Department of the ESALQ in Piracicaba, Universidade Estadual de São Paulo (USP) for infrastructure assistance during the field work. Our thanks are to Dr. Henry Hooghiemstra for providing a constructive review. We thank the Deutsche Forschungsgemeinschaft (DFG) for support of travel expenses and radiocarbon dating, the Institute of Palynology and Quaternary Sciences (University of Göttingen) for the kind permission to process the pollen samples in their laboratory. We thank the Niedersächsisches Landesamt für Bodenforschung (Hannover) and the University of Utrecht (Klaas Van der Borg) for radiocarbon dating.

REFERENCES

Behling, H. 1993. Untersuchungen zur spätpleistozänen und holozänen vegetations- und Klimageschichte der tropischen Küstenwälder und der Araukarienwälder in Santa Catarina (Südbrasilien). *Dissertationes Botanicae* 206, J. Cramer, Berlin Stuttgart, 149 pp.

Behling, H. 1995. Investigations into the Late Pleistocene and Holocene history of vegetation and climate in Santa Catarina (Southern Brazil). *Vegetation History and Archaeobotany* 4: 127-152.

Behling, H. 1997a. Late Quaternary vegetation, climate and fire history from the tropical mountain region of Morro de Itapeva, SE Brazil. *Palaeogeography, Palaeoclimatology, Palaeoecology* 129: 407-422.

Behling, H. 1997b. Late Quaternary vegetation, climate and fire history in the *Araucaria* forest and campos region from Serra Campos Gerais (Paraná), South Brazil. *Review of Palaeobotany and Palynology* (in press).

Behling, H. & M. Lichte 1997. Evidence of dry and cold climatic conditions at glacial times in tropical Southeast Brazil. *Quaternary Research* (in press).

De Oliveira, P.E. 1992. A palynological record of late Quaternary vegetational and climatic change in southeastern Brazil. Ph.D. Dissertation, The Ohio State University, Columbus, 238pp. Unpublished.

Golte, W. 1978. Die südandine und die südbrasilianische Araukarie. *Ein ökologischer Vergleich, Erdkunde*, Band 32: 279-296.

Grimm, E.C. 1987. CONISS: A Fortran 77 program for stratigraphically constrained cluster analysis by the method of the incremental sum of squares. *Pergamon Journals* 13: 13-35.

Hueck, K. 1953. Distribuição e habitat natural do Pinheiro do Paraná (*Araucaria angustifolia*). *Boletim Facultade de Filosofia e Ciências, Botanica* 10:1-24. Univ. São Paulo. Brazil.

Hueck, K. 1956. Die Ursprünglichkeit der brasilianischen 'Campos cerrados' und neue Beobachtungen an ihrer Südgrenze. *Erdkunde* 11: 193-203.

Hueck, K. 1966. *Die Wälder Südamerikas*. Stuttgart: Fischer. 422 pp.

IBGE 1993. *Mapa de Vegetação do Brasil*. Rio de Janeiro.

Ledru, M.P. 1993. Late Quaternary environmental and climatic changes in central Brazil. *Quaternary Research* 39: 90-98.

Ledru, M.P., H. Behling, M. Fournier, L. Martin & M. Servant 1994. Localisation de la forêt d'*Araucaria* du Brésil au cours de l'Holocène. Implications paléoclimatiques. *C. R. Acad. Sci. Paris*, 317: 517-521.

Ledru, M.P., P.I. Soares Braga, F. Soubiés, M. Fournier, L. Martin, K. Suguio & B. Turcq 1996. The last 50,000 years in the Neotropics (Southern Brazil): evolution of vegetation and climate. *Palaeogeography, Palaeoclimatology, Palaeoecology* 123: 239-257.

Miklos, A.A.W. 1992. Biodynamique d'une couverture pedologique dans la region de Botucatú – SP, Brésil. Ph.D. Dissertation. Université de Paris VI, Vol. I, 247 p. Unpublished.

Nimer, E. 1989. *Climatologia do Brasil*. Rio de Janeiro: IBGE. 421pp.

Silberbauer-Gottsberger, I., W. Morawetz & G. Gottsberger 1977. Frost damage of cerrado plants in Botucatu. *Biotropica* 253-261. Brazil.

Tubelis, A., F.J.L. Nasciment & L.L. Foloni 1971. *Parámetros climáticos de Botucatú*. Universidade de Piracicaba. Botucatú, 25 p.

Chronostratigraphic and Palynozone chronosequences charts of Napostá Grande Creek, Southwestern Buenos Aires Province, Argentina.

7

MIRTA E. QUATTROCCHIO, SILVIA C. GRILL &
CARLOS A. ZAVALA
CONICET & Departamento de Geología, Universidad Nacional del Sur. Bahía Blanca, Argentina

ABSTRACT: Changes in Late Pleistocene and Holocene vegetation, including sea level fluctuations in the Middle Holocene, were detected by means of pollen analysis and marine palaeomicroplankton of three sediment profiles from the Napostá Grande creek and one profile from layer with the 'fossil footprints' at Monte Hermoso I, southwestern Buenos Aires Province. Changes in sedimentary facies associated with changes in pollen assemblages allowed the inference changes in depositional environments and climate. Fifteen pollen zones and four subzones (one interval zone) were determined. Arid to semiarid climatic conditions were inferred for the Late Pleistocene (Profile 1) represented from the abundance of Chenopodiaceae-Amaranthaceae, and Asteraceae (Compositae) or Cruciferae + Chenopodiaceae-Amaranthaceae in Profile 2. Both zones are overlain by truncated palaeosoils followed by erosion. The early Holocene interval (Profile 4: Monte Hermoso I), dominated by Chenopodiaceae-Amaranthaceae, and Gramineae is dated at 7125 ± 75 and 7030 ± 100 ^{14}C yr BP. This vegetation community is characteristic of coastal dunes associated with lacustrine conditions, with slight marine influence suggests the existence of local humid conditions. The Middle Holocene is palynologically barren in Profile 1. While the Pollen Zone Chenopodiaceae-Amaranthaceae, Gramineae (presence), of Profile 3, dated at $6000/5580 \pm 100$ ^{14}C yr BP, reflects the destruction of the littoral environment by the rise of the sea level as evidenced by appearance of dinocysts and acritarchs. Between 3560 ± 100 yr BP and 3000 yr BP, grass steppe replaced the littoral environment. The uppermost levels dated at 2610 ± 60 ^{14}C yr BP are dominated by Chenopodiaceae-Amaranthaceae, Cruciferae, Compositae and Gramineae, reflecting semiarid conditions comparable to the present ones.

RESUMEN: Se han detectado cambios en la vegetación del Pleistoceno tardío y Holoceno, incluidas las fluctuaciones del nivel del mar en el

Holoceno medio, a través del análisis de polen y el microplancton marino de tres perfiles sedimentarios y un perfil del nivel con pisadas fósiles en Monte Hermoso I, en el Sudoeste de la provincia de Buenos Aires. Cambios en las facies sedimentarias asociados a cambios en conjuntos polínicos permitieron inferir cambios en ambientes depositacionales y clima. Se determinaron quince zonas polínicas y cuatro subzonas (una zona de intervalo). Se infirieron condiciones climáticas áridas a semiáridas para el Pleistoceno tardío (Perfil 1) representadas por la abundancia de Chenopodiaceae-Amaranthaceae y Asteraceae (Compositae) o Cruciferae + Chenopodiaceae-Amaranthaceae en el Perfil 2. Paleosuelos truncados seguidos de erosión sobreyacen ambas zonas. El intervalo Holoceno temprano (Perfil 4: Monte Hermoso I), dominado por Chenopodiaceae-Amaranthaceae y Gramineae está datado en 7125 ± 75 y 7030 ± 100 ^{14}C años AP. Esta comunidad vegetal es característica de las dunas costeras asociadas con condiciones lacustres con escasa influencia marina, que sugiere la existencia de condiciones locales húmedas. El Holoceno medio es palinológicamente estéril en el Perfil 1. Mientras la Zona polínica Chenopodiaceae-Amaranthaceae, Gramineae (presencia) del Perfil 3, datada en $6000/5580 \pm 100$ ^{14}C años AP, refleja la destrucción del ambiente litoral por el ascenso del nivel del mar, tal como lo evidencia la aparición de dinocistos y acritarcos. Entre 3560 ± 100 años AP y 3000 años AP, la estepa de gramíneas reemplazó el ambiente litoral. Los niveles superiores datados en 2610 ± 60 ^{14}C años AP están dominados por Chenopodiaceae-Amaranthaceae, Cruciferae, Compositae y Gramineae, lo cual refleja condiciones semiáridas comparables a las actuales.

1 INTRODUCTION

The present paper attempts to improve the knowledge of the stratigraphy and palaeoenvironments of the Late Pleistocene and Holocene of southern Buenos Aires province, Argentina, based on the analysis of pollen grains, spores and marine microplankton.

The aim of this work is the recognition of plant communities, their fluctuations through time and the specification of pollen assemblage zones. This paper also purports to determine the space-time correlation of the recorded events (chronosequences chart) and how changes of the vegetation interact with palaeoclimatic fluctuations.

The studied area (Fig. 1) comprises the Naposta Grande creek basin (Profiles 1, 2 and 3) and the layer with 'fossil footprints' at Monte Hermoso I (Profile 4). Deposits are distinctly continental (fluvial, lacustrine and aeolian) or transitional and marine.

Previous palynological research in southern Buenos Aires province are those of Guerstein & Quattrocchio (1984), Quattrocchio et al. (1988,

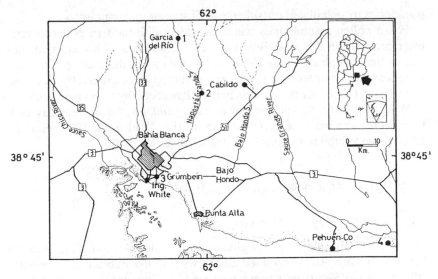

Figure 1. Location map. The four profile studied are represented with numbers 1, 2, 3, 4.

1993, 1995), Prieto (1989, 1993), Borromei & Quattrocchio (1990), Borromei (1992, 1995), Zavala et al. (1992), Gómez et al. (1992) Guerstein et al. (1992), Grill (1993, 1995, 1996), Grill & Guerstein (1995) and Grill & Quattrocchio (1996).

2 MATERIALS AND METHODS

The sediment profiles were subsampled for pollen analysis: 26 samples from Profile 1, 29 samples from Profile 2, 11 samples from Profile 3, and six samples from Profile 4.

The physical-chemical method preparation of Heusser & Stock (1984) was employed to extract the palynomorphs. Microplankton was treated with cold HCl, according to Dale (1976). Five, ten or twenty grams per sample (dry weight) were processed and two tablets of *Lycopodium* spores (11.267) were added to estimate the palynomorph concentration per gram of sediment (Stockmarr 1971). The marine samples, according to Stanley (1966), were stained with T-Safranine, to differentiate primary palynomorphs (homogeneously stained) from those reworked (not homogeneously stained, or unstained).

In order to guarantee that all taxa present are counting, we followed the method of Bianchi & D'Antoni (1986) of 'minimal area'.

In the marine samples, at least 150 palynomorphs were counted (pollen grains + spores + palaeomicroplankton). Percentages of palaeomicroplankton were calculated outside the total sum of palynomorphs.

Fossil pollen assemblages were compared with modern pollen assemblage recorded in the area (Borromei & Quattrocchio 1990) and with surface pollen samples (Prieto 1989, 1993). Likewise, the plant palaeocommunities were compared with present vegetation units studied by Verettoni & Aramayo (1976) for the area of Bahía Blanca, and by Cabrera (1976).

With respect to the marine palaeomicroplankton, the present day ecological characteristics of recent dinocysts (i.e. Wall et al. 1977, Harland 1983) provided the basis to infer palaeoecological conditions.

3 GENERAL FEATURES OF THE AREA

3.1 *Geomorphology and tectonics*

The Napostá Grande creek originates in the Sierras Australes (Napostá Hills) of Buenos Aires province and flows into the Bahía Blanca estuary 108 km to the SE. It is a permanent stream with two areas of different geomorphologic features: a hilly region and the lowland plains (Paoloni et al. 1987). Basement tectonics affect the strike of the creek, from E-W (coincident with a regional fault) to NE-SW (secondary fault of the basement) (Bonorino et al. 1986).

3.2 *Climate*

The Buenos Aires province occupies the central-eastern portion of Argentina, between lat. 33° to 43° S. Consequently, it lies within the temperate zone (Burgos 1968). Because of its position within the country and the South American continent, the oceanicity factor is significant (Burgos 1968), moderating the climate, especially near the coast. The continentality features intensify with distance from the coast (Verettoni & Aramayo 1976).

Both in warm and cold months, the prevailing direction of the winds is NE-SW, because of the activity of the South Atlantic and South Pacific anticyclones. Subantarctic cold air masses from the SW, and warm air masses from the north, caused by the occasional recession of the Pacific anticyclone (Burgos 1968), also affect the climate of the province.

3.3 *Phytogeography and vegetation*

According to Cabrera (1976), the Buenos Aires province belongs to the Neotropical Region, with a small area occupied by the Amazonic Dominion, and the rest by the Chaco Dominion. The latter is represented by the Monte, Espinal and Pampean plant geographic provinces.

114

The studied area comprises parts of the Austral (in Pampean province) and the Caldén (in Espinal province) districts.

According to the vegetation census made in the area of Bahía Blanca by Verettoni & Aramayo (1976), most of the region is used for grazing and farming and the natural vegetation that occupies the rest, has been modified. These authors recognized the following plant communities: grass steppe in the plains; herbaceous sandy steppe along the coast, on continental dunes and sandy soils (Verettoni 1965); halophytic steppe in the foreland of the coast and saline soils (Verettoni 1961), and shrubby and sub-shrubby forest in those places where the caliche outcrops or is near the surface. This is a modification of the original forest that extended along the coast.

4 STRATIGRAPHY AND AGE

The stratigraphic sections studied along the banks of the Napostá Grande basin comprise fluvial, aeolian and marine sediments. They are generally discontinuous, with a significant development of palaeosols. The stratigraphic units identified in this paper follow those defined by Rabassa (1989).

The Late Pleistocene is represented in the three profiles of the Napostá Grande basin. Although there are no dates for these sediments, their age is inferred on the base of the lithofacies and the fossil vertebrates content.

The Holocene sections have five radiocarbon dates: Profile 1: 2610 ± 60 yr BP (sample 8); profile 2: 1960 ± 100 yr BP (sample 25); profile 3: 5580 ± 130 yr BP (sample 8), 3850 ± 100 yr BP (sample 7) and 3560 ± 100 yr BP (sample 6).

In the layer with 'fossil footprints' at Monte Hermoso I region, sandy-silty deposits are discontinuos, outcropping on the marine abrasion platform. The stratigraphy follows Zavala et al. (1992). There are no dates for the oldest sediment; their age is based on vertebrate fossils. The Holocene sediments have two radiocarbon datings: 7125 ± 75 yr BP (sample 4) and 7030 ± 100 yr BP (sample 3).

The detailed study of the palynologic profiles with stratigraphic comments are given in Grill (1993).

5 PALYNOLOGICAL ZONATION

Pollen diagrams were divided according to the 'pollen zones' of Gordon & Birks (1972, in Birks & Birks 1980), and Interval Zones (Guía Estratigráfica Internacional 1980).

Relative pollen frequencies (%) and concentrations (grains per gram of sediment) of the identified taxa are shown as bar diagrams (Figs 2, 3, 4 and 5).

116

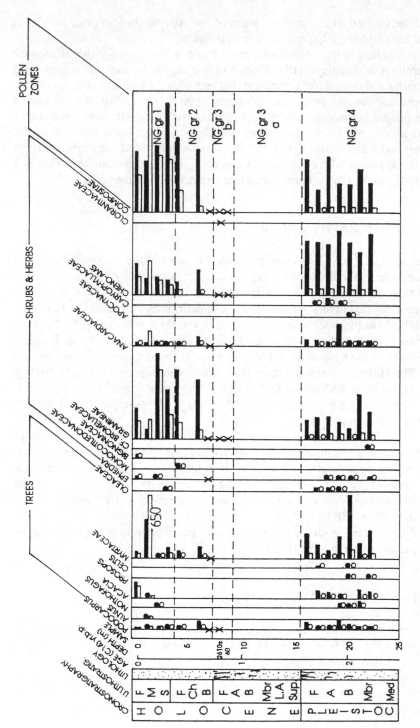

Figure 2. Pollen diagram. Profile 1 (García del Río).

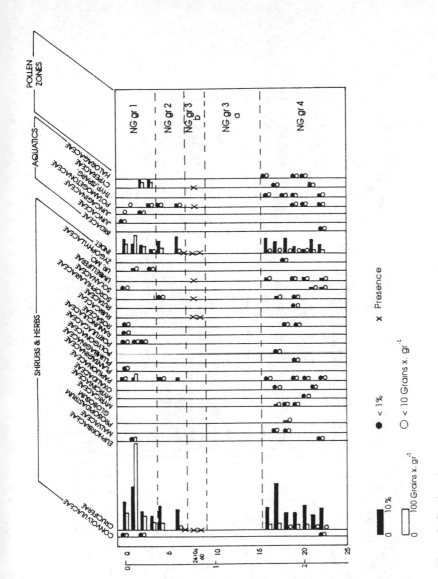

Figure 2. Continued.

117

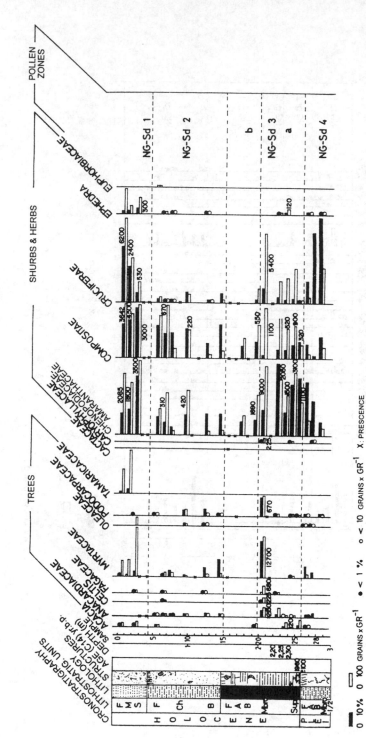

Figure 3. Pollen diagram. Profile 2 (Chacra Santo Domingo).

118

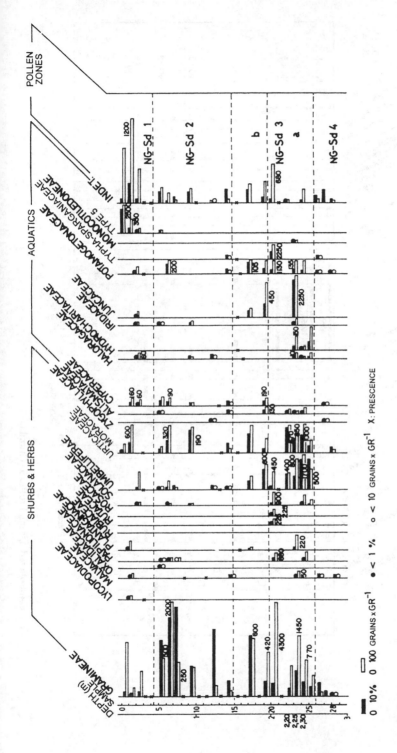

Figure 3. Continued.

119

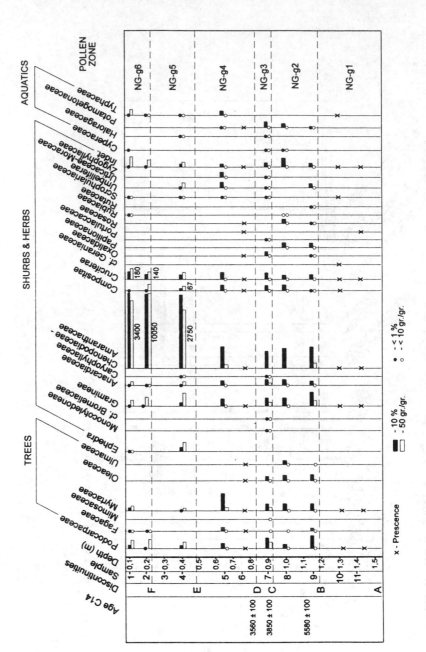

Figure 4. Pollen diagram. Profile 3 (Grumbein).

120

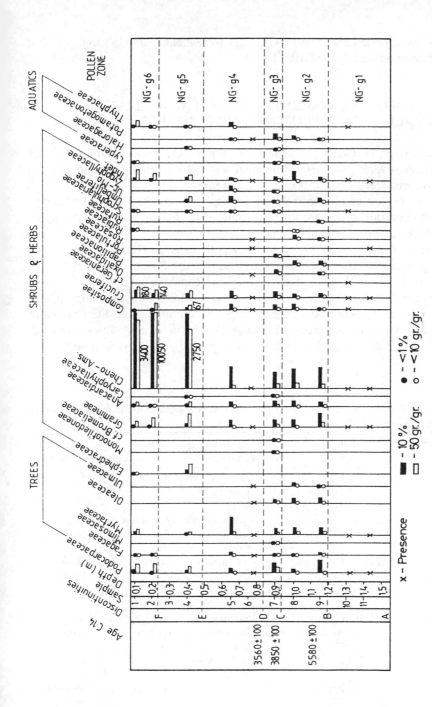

Figure 4. Continued.

121

Fifteen pollen zones and 4 subzones were identified in the four profiles. See detailed analysis in Grill (1993).

6 PALAEOENVIRONMENTAL AND PALAEOCLIMATIC EVALUATION. DISCUSSION AND CONCLUSIONS

Comparison between vegetation changes and associated lithofacies changes reveals palaeoenvironmental and palaeoclimatic changes for the Late Pleistocene and Holocene in this area (see Tables 1, 2, 3 and 4).

Chronostratigraphic correlation among the four profiles studied led to the elaboration of a chronostratigraphic (Fig. 6) and palynostratigraphic chart (Fig. 7) for the studied area.

This information is related to other pollen studies from the Buenos Aires province, including Borromei (1995) in the Sauce Grande river and Prieto (1989, 1993) in the Sauce Chico and Tapalqué creek.

Other data supplied by fossil vertebrates studied by Deschamps (Quattrocchio et al. 1988 and C. Deschamps, pers. comm.) and Deschamps & Tonni (1992) and ostracods studied by D. Martínez (in Quattrocchio et al. 1988 and Zavala et al. 1992) in the same profiles, are also discussed.

Table 1. Palaeoclimates, palaeoenvironments and vegetation communities registered in García del Río (Profile 1).

Epoch	Pollen zones		Vegetation communities	Palaeo-environments	Palaeo-climates
Holocene	NG-Gr1	Compositae, Gramineae and exotic pollen	Psammophytic Herbaceous steppe	Aeolian	Semiarid
	NG-Gr2	Compositae Gramineae	Psammophytic Herbaceous steppe	Aeolian with fluvial plains	Arid-semiarid
	NG-Gr3b	Gramineae Compositae	Gramineous steppe	Fluvial plains	Warm humid
	NG-Gr3a	Barren	?	?	
Late Pleisto-cene	NG-Gr4	Chenopodiaceae Amaranthaceae, Compositae	Psammophytic Herbaceous steppe with shrubby wood elements	Aeolian with ephemeral streams	Arid-semiarid
		Barren	?	Aeolian	Extremely arid

122

Table 2. Palaeoclimates, palaeoenvironments and vegetation communities registered in Chacra Santo Domingo (Profile 2).

Epoch	Pollen Zones		Vegetation communities	Palaeo-environments	Palaeo-climates
Late holocene	Ng-sd1	Chenopodiaceae-Amaranthaceae, Compositae, Cruciferae	Psammophytic Herbaceous steppe	Aeolian	Semiarid
	Ng-Sd2	Gramineae Compositae	Gramineous steppe	Aeolian with fluvial plains	Humid
	Ng-Sd3b	Gramineae Chenopodiaceae-Amaranthaceae	Gramineous steppe	Fluvial plains	Humid
	Ng-Sd3a	Chenopodiaceae-Amaranthaceae	Gramineous steppe with Hidrophytic communities	Lacustrine	Warm humid
Late pleisto-cene	Ng-sd4	Cruciferae	Psammophytic Herbaceous steppe with shrubby wood elements	Aeolian with ephemeral streams	Arid - semiarid

Table 3. Palaeoclimates, palaeoenvironments and vegetation communities registered in Grumbein (Profile 3).

Epoch	Pollen zones		Vegetation communities	Palaeo-environments	Palaeo-climates
Holocene	NG-g1	Chenopodiaceae Amaranthaceae	Halophytic steppe	Fluvial	Arid-Semiarid
	NG-g2	Chenopodiaceae Amaranthaceae	Halophytic steppe	Fluvial	Arid-Semiarid
	NG-g3	Chenopodiaceae-Amaranthaceae, Myrtaceae	Herbaceous Halophytic steppe/Gramineous	Fluvial plains marine	Humid
	NG-g4	Chenopodiaceae-Amaranthaceae, Podocarpaceae	Herbaceous Halophytic steppe/Gramineous	Marine	Humid
	NG-g5	Chenopodiaceae-Amaranthaceae, Gramineae	Herbaceous Halophytic steppe/Gramineous	Marine	Humid
	NG-g6	Chenopodiaceae-Amaranthaceae, Gramineae, Compositae (presence)	?	Marine	Humid

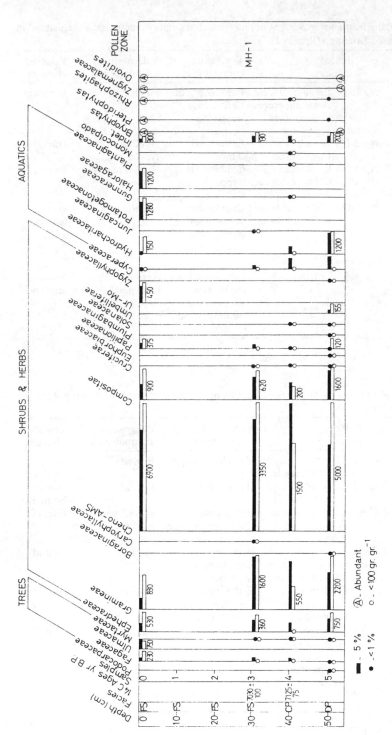

Figure 5. Pollen diagram. Profile 4 (Monte Hermoso I).

124

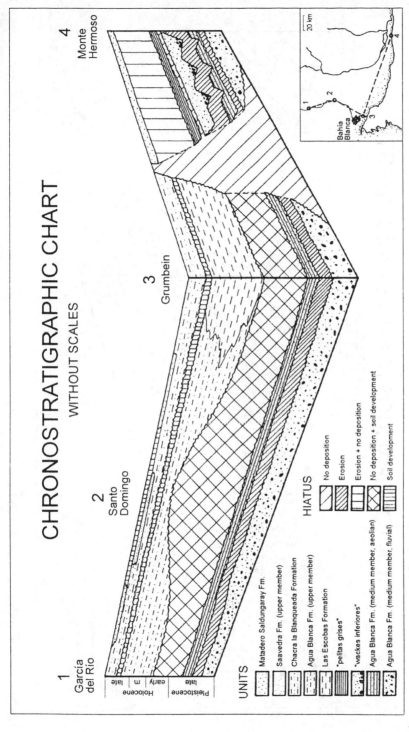

Figure 6. Chronostratigraphic chart of studied area.

125

126

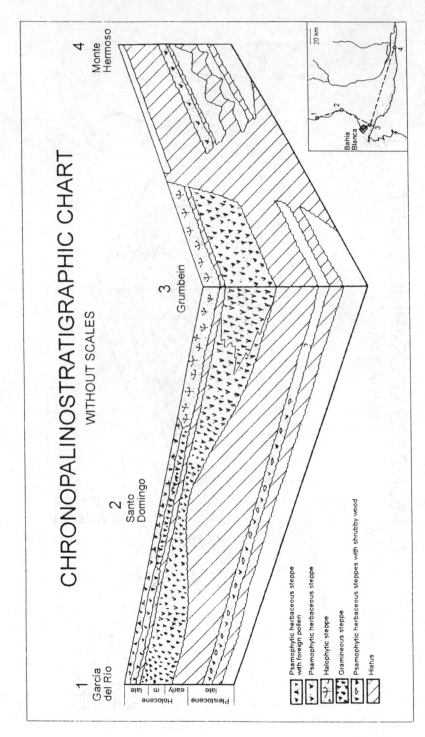

Figure 7. Palynostratigraphic chart of studied area.

Table 4. Palaeoclimates, palaeoenvironments and vegetation communities registered in Palaeoicnological Sites of Monte Hermoso I (Profile 4).

Epoch	Pollen zones		Vegetation communities	Palaeo-environments	Palaeo-climates
Holo-cene	MH-1	Chenopodiaceae-Amaranthaceae, Gramineae	Herbaceous Psamophytic steppe associated to Lacustrine body	Interdune	Humid (locality)

6.1 *Late Pleistocene*

The pollen assemblages recorded in sediments of Late Pleistocene age (the NG-gr4 and NG-sd4 Zones) are characteristic of vegetation communities of halophytic and herbaceous psammophytic steppes, with representatives of the shrubby forest. They suggest arid to semiarid palaeoclimatic conditions, similar to those today characteristic for the western Pampa, where precipitation ranges between 400 and 600 mm per year (Prieto 1989). Within these communities, Cruciferae dominate the assemblages in Profile 2. The development of these herbaceous taxa may imply an environmental disturbance, caused by aridity and strong aeolian activity, that may have caused also a decrease of grasslands (León & Anderson 1973).

Cool climatic conditions favor the development of quick growing plants such as the Cruciferae among others (Grime 1979). It can be inferred that during the Late Pleistocene the climate was cool-arid to extremely arid in the studied area.

The supposed extreme aridity could also be the cause of barren samples recorded in the bottom of Profile 1, upper middle basin of Napostá Grande creek.

The associated lithofacies suggest an aeolian environment with ephemeral water, reflecting extremely arid to semiarid conditions for the Late Pleistocene.

Another factor for the aridity may be related with a lower than present sea level, dated at 12.000 ± 100 yr BP ^{14}C by Aramayo & Bianco (1996) for a site approximately 75 km east from Napostá Grande and by Bayón & Zavala (1997) for Farola Monte Hermoso (Buenos Aires province).

The ostracods *Darwinula* sp. and *Larseypridopsis aculeata* (D. Martínez, pers. comm.), found in the Sandy Middle Member of the Agua Blanca Formation (Profiles 1 and 2 of Napostá Grande creek), suggest shallow and standing water bodies, fresh to slightly saline. The low species diversity and low frequency may imply alternating periods of temporary water and dryness (D. Martínez, pers. comm.).

The vertebrates found in the same profiles (*Lama guanicoe, Chaeto-phractus villosus* and *Lagostomus maximus*) suggest arid to semiarid open areas of grasslands and steppes (C. Deschamps, pers.comm.).

Similar conditions were recorded in the Sauce Grande river valley in the Sandy Middle Member, Upper submember, SG-4 Zone (Borromei 1995). Prieto (1989) in Tapalqué creek inferred, in sediments older than 10,000 yr BP similar conditions to those at present in western Buenos Aires province. The end of the EQ1 Zone is correlated with the NG-sd4 Zone, in the Sandy Middle Member of the Agua Blanca Formation in Naposta Grande creek (Profile 2). In both zones pollen associations are dominated by Cruciferae. Over the EQ1, SG-4 and NG-sd4 Zones there is a regional discontinuity.

The Late Pleistocene/Holocene contact is characterized by development of palaeosoils that can be correlated with Puesto Callejón Viejo Palaeosoil (Fidalgo et al. 1973, Fidalgo 1992). In some sectors this contact is transitional and in others there is a discontinuity. Because the palaeosoils are truncated, the pedogenetic event in the pollen record is inferred by presence of *Glomus* sp. and some fungal spores typical from palaeosoils. According to Dimbleby (1957), pollen grains percolate down to 30 cm from the soil. This may explain their absence in the pollen records within the development of the palaeosoils.

6.2 *Holocene*

The early Holocene pollen record of the MH-1 Zone in Profile 4 from Monte Hermoso I is dated at 7125 ± 75 and 7030 ± 100 ^{14}C yr BP and reflects the development of a vegetation community characteristic of coastal dunes (psammophytic herbaceous) and of interdune ponds with a slight marine influence. Sea level was still lower than present. This community suggests locally humid conditions.

The ostracods *Sarscypridopsis aculeata, Lymnocythere* sp. 1 and 2 and *Cyprinotus salinus* found in Profile 4, suggest the development of a standing water body probably poly-mesohalyne with abundant nutrients and good oxygenation (D. Martínez in Zavala et al. 1992).

Temperature and humidity reachs its maximum during the mid Holocene, recorded at the mouth of Naposta Grande creek in the NG-g6 Zone (Profile 3). Between 6000 and 5580 ± 100 ^{14}C yr BP, the high species diversity and abundance of marine dinocysts and acritarchs indicate the maximal marine transgression in this sector of the basin. These assemblages and the poor pollen preservation (NG-g5) reflect neritic/estuarine conditions enclosures during the deposition of the sediments (Grill & Quattrocchio 1996). This event may be correlated with the SG-3 Zone of Sauce Grande river (Borromei 1995).

128

Between 5600 and 3000 yr BP, (Profile 3; NG-g4 Zone) a higher diversity of the sporomorphs associated with representatives of the gramineous steppe approximately suggests more temperate conditions in this sector. This zone maybe correlated with the SCH-4 and 5 Zones of Sauce Chico creek (Prieto 1989).

This relative rise of sea level lead to flooding of the river-beds producing deposition of grey pelitic facies.

The warmer climatic conditions, associated with sea level rise, are partially inferred in the pollen records of the Napostá Grande basin, because the sediments of this age have been eroded or are palynologically barren. The destruction of the pollen grains may have been due to an intense activity of microorganisms under wet climate (Havinga 1970, Birks & Birks 1980, Dimbleby 1985, among others).

However, the expansion of the gramineous steppe during this interval is recorded in Buenos Aires province in the Sauce Grande river middle valley (the SG-3 Zone, with a top age of 5010 ± 120 ^{14}C yr BP, Borromei 1995). In the Sauce Chico creek, facies and hydrophytic associations recorded by Prieto (1989; SCH-3 and 2 Zones; SCH-2; 6170 ± 170 ^{14}C yr BP) may suggest the presence of humid grassland prairie and ponds within the valley. This Zone is equivalent to the EQ-2 (humid prairie and lacustrine) and EQ-3 (gramineous steppe) zones recorded in Tapalqué creek (Empalme Querandíes) by Prieto (1989).

Approximately at 3000 yr BP, the marine influence ended in the lower basin of the Napostá Grande creek (NG-g3 Zone, Profile 3).

After 2610 ± 100 ^{14}C yr BP (NG-gr2 Zone, Profile 1), psammophytic herbaceous steppe developed with representatives of the shrubby forest, suggesting arid to semiarid conditions. This Zone is equivalent to the SG-2 Zone recorded in the Sauce Grande river middle valley with a bottom age of 2830 ± 90 ^{14}C yr BP (Borromei 1995).

An interval of higher humidity is inferred at approximately 2000 yr BP (1960 ± 100 ^{14}C yr BP, NG-sd3 Zone, Profile 2), in the middle basin of the Napostá Grande creek (chacra Santo Domingo), based on the development of hydrophytic communities characteristic of lacustrine environments, associated with gramineous steppe. Likewise, a relative rise of temperature may be inferred by the southward expansion of the Brazilian fauna, i.e. *Holochilus brasiliensis* and *Cavia aperea* (Quattrocchio et al. 1988; Deschamps & Tonni 1992). In this profile, the Late Holocene lies discontinuedly over the Late Pleistocene. Neotectonic activities are inferred for this area (Quattrocchio et al. 1988).

The sympatry of Brazilian faunistic representatives with Patagonian elements (*Lestodelphys halli* and *Lama guanicoe*) may suggest vegetated water bodies that would locally modify the arid and semiarid regional conditions, providing pathways for Brazilian elements (Deschamps & Tonni 1992).

129

The increase in moisture can be recognized throughout the basin by development of palaeosoils, that can be correlated with the Puesto Berrondo Palaeosoil (Fidalgo et al. 1973, Fidalgo 1992).

Ostracods suggest shallow, isolated water bodies of low energy and salinity, and with rather dense aquatic vegetation. One of the species, *Clamidoteca incisa*, has its southern occurrence at this site. Brackish to fresh water conditions are inferred through the abundance of *Cyprideis salebrosa* (Bertels & Martínez 1990).

The persistence of gramineous steppe in the middle sector of the basin (NG-sd2 Zone) suggests the continuity of locally humid conditions, which are associated with fluvial overflows.

Influence of grazing and farming under semiarid conditions is registered in the upper middle and middle basin of Napostá Grande creek (Profiles 1, NG-gr1 Zone and 2, NG-sd1 Zone).

After 3000 yr BP, vegetation communities characteristic of halophytic steppe were established in the lower basin of Napostá Grande creek (Profile 3, NG-g2 and 1 Zone). These communities are similar to those of the present coastal environments.

ACKNOWLEDGEMENTS

The authors wish to acknowledge the aid of Lic. Cecilia Deschamps with the English version of this paper. We appreciate very much the comments of anonymous reviewers. Financial support was provided by CONICET and S.E.C.Y.T. (Universidad Nacional del Sur, Bahía Blanca).

REFERENCES

Aramayo S. & Bianco T. de 1996. Nuevos hallazgos en el yacimiento paleoicnológico de Pehuén-Có (Pleistoceno tardío), provincia de Buenos Aires, Argentina. *Primera Reunión Argentina de Icnología, Asociación Paleontológica Argentina*, Publicación especial N 4: 47-57. Buenos Aires.

Bayón, C. & C. Zavala 1997. Coastal sites in south Buenos Aires: a review of Piedras Quebradas. *Quaternary of South America & Antarctic Peninsula* 10 (1994): 229-254. Rotterdam: Balkema Publishers.

Bertels, A. & D. Martinez 1990. Quaternary ostracodes of continental and transitional littoral – shallow marine environments. *Cour. Forsch. Inst. Senckenberg* 123: 141-159. Frankfurt. Alemania.

Bianchi, M. & H. D'Antoni 1986. Depositación del polen actual en los alrededores de Sierra de los Padres (Provincia de Buenos Aires). *VI Congreso Argentino de Paleontología y Bioestratigrafía, Actas*: 16-27. Mendoza, Argentina.

Birks, H.J. & H.H. Birks 1980. *Quaternary Palaeoecology*. London: Arnold Publishers Limited, 289 pp.

130

Bonorino, G.A., R. Schillizzi & J. Kostadinoff 1986. Investigación geológica y geofísica en la región de Bahía Blanca. *III Jornadas Pampeanas de Ciencias*, Serv. Suplem. N 3: 55-63. Universidad Nacional de La Pampa. Santa Rosa, La Pampa, Argentina.

Borromei, A.M. 1992. Geología y Palinología de los depósitos cuaternarios en el valle del río Sauce Grande, provincia de Buenos Aires, Argentina. Ph.D. Dissertation, 200 pp. Universidad Nacional del Sur, Bahía Blanca. Unpublished.

Borromei, A.M. 1995. Palinología, estratigrafía y paleoambientes del Pleistoceno tardío-Holoceno en el valle del río Sauce Grande, provincia de Buenos Aires. Argentina. *Polen*, 7: 19-31. España.

Borromei, A.M. & M. Quattrocchio 1990. Dispersión del polen actual en el área de Bahía Blanca (Buenos Aires, Argentina). *Revista Anales de la Asociación de Palinólogos de Lengua Española*, 5: 39-52. España.

Burgos, J. 1968. El clima de la provincia de Buenos Aires en relación con la vegetación natural y el suelo. In: A. Cabrera, (ed.), *Flora de la provincia de Buenos Aires*, Tomo IV (Y): 33-39. INTA. Buenos Aires.

Cabrera, A. 1976. Regiones fitogeográficas argentinas. En: ACME (ed.) *Enciclopedia Argentina de Agronomía y Jardinería*, Tomo II, Fascículo 1: 1-85. Buenos Aires.

Dale, B. 1976. Cysts formation, sedimentation and preservation: factors affecting dinoflagellate assemblages in recent sediments from Trondheimsfjord, Norway. *Review of Palaeobotany and Palynology*, 22: 39-60. Amsterdam.

Deschamps, C. & E.P. Tonni 1992. Los vertebrados del Pleistoceno tardío – Holoceno del arroyo Napostá Grande, provincia de Buenos Aires; aspectos paleoambientales. *Ameghiniana* 29(3): 201-210. Buenos Aires.

Dimbleby, G.W. 1957. Pollen analysis of terrestrial soils. *New Phytology*, 56: 12-28.

Dimbleby, G.W. 1985. *The palynology of archaeological sites*. New York: Academic Press, Inc. 173 pp.

Fidalgo, F. 1992. Provincia de Buenos Aires. Continental. In: M. Iriondo (ed.), *El Holoceno en Argentina*. Vol. 1: 23-38. CADINQUA. Paraná, Argentina.

Fidalgo, F., F. De Francesco & U. Colado 1973. Geología superficial en las Hojas Castelli, J.M. Cobos y Monasterio (provincia de Buenos Aires). *Actas V Congreso Geológico Argentino*, 4: 27-39. Buenos Aires.

Gómez, E., D. Martínez, G. Cusminsky, M. Suárez, F. Vilanova & G.R. Guerstein 1992. Estudio del testigo Ps2, Cuaternario del estuario de Bahía Blanca, Provincia de Buenos Aires, Argentina. Parte Y: Sedimentología y Micropaleontología. *III Jornadas Geológicas Bonaerenses*, Actas: 39-46. La Plata.

González, M., H. Panarello, H. Marino & S. Valencio 1983. Depósitos marinos del Holoceno en el estuario de Bahía Blanca (Argentina). Isótopos estables y microfósiles calcáreos como indicadores paleoambientales. *Simposio Oscilaciones del nivel del mar durante el último Hemiciclo Deglacial en la Argentina*. Actas: 48-68. Mar del Plata.

Grill, S.C. 1993. Estratigrafía y paleoambientes del cuaternario en base a palinomorfos en la cuenca del arroyo Napostá Grande, provincia de Buenos Aires. Ph.D. Dissertation, 145 pages. Universidad Nacional del Sur. Bahía Blanca. Unpublished.

Grill, S.C. 1995. Análisis palinológico de un perfil cuaternario en la cuenca del arroyo Napostá Grande, Localidad García del Río, provincia de Buenos Aires. *IV Jornadas Geológicas Bonaerenses*, Actas: 99-107. Junín, Argentina.

Grill, S.C. 1996. Análisis palinológico de un perfil cuaternario en la cuenca media del arroyo Napostá Grande, provincia de Buenos Aires, Argentina. Implicancias paleoambientales. *Polen*, 8: 23-40. España.

Grill, S.C. & G.R. Guerstein 1995. Estudio palinológico de sedimentos superficiales en el estuario de Bahía Blanca, Buenos Aires. Argentina. *Polen*, Vol. 7: 41-49. España.

Grill, S. & M. Quattrocchio 1996. Fluctuaciones eustáticas durante el Holoceno a partir del registro del paleomicroplancton en la desembocadura del arroyo Napostá Grande, provincia de Buenos Aires. *Ameghiniana*, 33(4): 435-442. Buenos Aires.

Grime, J.P. 1979. *Plant strategies and vegetation processes*. New York: John Wiley & Sons, 222 pages.

Guerstein, G. R. & M. Quattrocchio 1984. Datos palinológicos de un perfil cuaternario ubicado en el estuario de Bahía Blanca. *IX Congreso Geológico Argentino*, Actas IV: 596-609. San Carlos de Bariloche. Río Negro, Argentina.

Guerstein, G. R., F. Vilanova, M. Suárez, G. Cusminsky, D. Martínez & E. Gómez 1992. Estudio del testigo Ps2, Cuaternario del estuario de Bahía Blanca, Provincia de Buenos Aires, Argentina. Parte II: Evaluación Paleoambiental. *III Jornadas Geológicas Bonaerenses*, Actas: 47-52, La Plata.

Guía Estratigráfica Internacional. 1980. Hollis D. Hedberg, eds.

Harland, R. 1983. Distribution maps of recent dinoflagellate cysts in bottom sediments from the north Atlantic Ocean and adjacent seas. *Paleontology*, 26: 81-98.

Havinga, A. J. 1970. An experimental investigation into decay of pollen and spores in various soil types. *Sporopollenin. Proceedings of a Symposium Held at the Geology Department, Imperial College*, pp. 446-479. London.

Heusser, L. & C. Stock 1984. Preparation techniques for concentrating pollen from marine sediments and other sediments with low pollen density. *Palynology* 8: 225-227. Dallas.

Horowitz, A. 1992. *Palynology of arid lands*. Amsterdam: Elsevier Science Publishers, 546 pp.

León, R. & D. Anderson 1973. El límite occidental del pastizal pampeano. *Tuexenia* 3: 67-82.

Paoloni, J., R. Vazquez & R. Rainieri 1987. Los parámetros morfométricos como una base de comparación en cuencas del sector sudoeste del Sistema de Ventania y su incidencia en los escurrimientos. *Congreso Nacional del Agua*, Actas: 1-11. Calafate. Argentina.

Prieto, A.R. 1989. Palinología de Empalme Querandíes, provincia de Buenos Aires. Un modelo paleoambiental para el Pleistoceno tardío – Holoceno. Ph.D. Dissertation, 207 pp. Universidad Nacional de Mar del Plata. Unpublished.

Prieto, A.R. 1993. Dispersión polínica actual en relación con la vegetación en la estepa pampeana: primeros resultados. *Asociación Paleontológica Argentina. Publicación especial* N 2: 91-95. Buenos Aires.

Quattrocchio, M., C. Deschamps, D. Martínez, S. Grill & C. Zavala 1988. Caracterización paleontológica y paleoambiental de sedimentos cuaternarios, arroyo Napostá Grande, provincia de Buenos Aires. *II Jornadas Geológicas Bonaerenses*. Actas: 37-46. Bahía Blanca.

Quattrocchio, M., C. Deschamps, C. Zavala, A. Borromei, S. Grill & R. Guerstein 1993. Cuaternario del sur de la provincia de Buenos Aires. Estratigrafía e inferencias paleoambientales. In: M. Iriondo (ed.) *El Holoceno en Argentina*, II: 22-34. CADINQUA. Paraná.

Quattrocchio, M., A. Borromei & S. Grill 1995. Cambios vegetacionales y fluctuaciones paleoclimáticas durante el Pleistoceno tardío – Holoceno en el sudoeste de la provincia de Buenos Aires (Argentina). *VI Congreso Argentino de Paleontología y Bioestratigrafía*. Actas: 221-229. Trelew, Argentina.

Rabassa, J. 1989. Geología de depósitos del Pleistoceno superior y Holoceno en las cabeceras del río Sauce Grande, provincia de Buenos Aires. *I Jornadas Geológicas Bonaerenses*. Actas: 765-790. Tandil, 1985. Argentina.

Rabassa, J., C. Heusser & N. Rutter 1990. Late-glacial and Holocene of Argentine Tierra del Fuego. *Quaternary of South America & Antarctic Peninsula* 7: 327-351. Rotterdam: Balkema Publishers.

Stanley, E.A. 1966. The problem of reworked pollen and spores in marine sediments. *Marine Geology*, 4: 397-408.

Stockmarr, J. 1971. Tablets with spores used in absolute pollen analysis. *Pollen spores*, 13: 615-621.

Verettoni, H.N. 1961. Las asociaciones halófilas del partido de Bahía Blanca. *Ed. Panzini, Patrocinada por la Comisión Ejecutiva del 150° aniversario de la Revolución de Mayo*. 105 pp. Bahía Blanca.

Verettoni, H.N. 1965. Contribución al conocimiento de la vegetación psamófila de la región de Bahía Blanca. *Diestra Producciones* 160 pp. Bahía Blanca.

Verettoni, H. & Aramayo, E. 1976. Las comunidades vegetales de la región de Bahía Blanca. Harris Eds. 175 pp., Bahía Blanca.

Wall, D., B. Dale, G. Lohmann & W. Smith 1977. The environmental and climatic distribution of dinoflagellate cysts in modern marine sediments from regions in the North and South Atlantic oceans and ayacents seas. *Marine Micropaleontology*, 2: 121-200.

Zavala, C., S. Grill, D. Martínez, H. Ortiz & R. González 1992. Análisis paleoambiental de depósitos cuaternarios. Sitio paleoicnológico Monte Hermoso I, provincia de Buenos Aires. *III Jornadas Geológicas Bonaerenses*, Actas: 31-37. La Plata. Argentina.

Palaeoenvironmental implications of a Holocene diatomite, Pampa Interserrana, Argentina

8

MARCELO A. ZÁRATE
Centro de Geología de Costas y del Cuaternario, Universidad Nacional de Mar del Plata & CONICET, Mar del Plata, Argentina

MARCELA A. ESPINOSA & LAURA FERRERO
Centro de Geología de Costas y del Cuaternario, Universidad Nacional de Mar del Plata, Mar del Plata, Argentina

ABSTRACT: Diatomites are important stratigraphic components of Holocene alluvial records of the Pampa Interserrana, Argentina. An Early to Middle Holocene palaeoenvironmental record has been obtained by diatom, ostracod and gastropod analysis of a 2.5 m thick section from the Río Quequén Grande, Pampa Interserrana, Argentina. The stratigraphic interval under study extends back to sometimes before 9000 ^{14}C yr BP when a soil forming interval prevailed in the fluvial environment. This palaeosol was alkaline and slighty brackish. Soon after 9000 ^{14}C yr BP, more humid-subhumid conditions started, evidenced by the flooding of floodplain depressions and the development of pond environments. Diatom, ostracods and gastropod asemblages along with the occurrence of soil features at some levels indicate a very shallow pond up to 2 m deep with seasonally fluctuating water levels. The high biogenic sedimentation that dominated in the pond between 9000 and circa 5000 ^{14}C yr BP is synchronic with the interval of soil formation in the catchment area. Therefore the diatomite deposition is recording an interval of biostasy. A shift to much drier conditions occurred sometime between 5000 and 4000 ^{14}C yr BP when desiccation and subaereal exposure took place giving rise to soil formation. This was followed by an interval of greater variability revealing environmental instability in the fluvial basin evidenced by the increased amount of siliciclastic deposition.

RESUMEN: Las diatomitas son importantes componentes estratigráficos de los registros aluviales holocenos de la Pampa Interserrana, Argentina. Con el propósito de analizar su significado paleoambiental se realizó el estudio de las asociaciones de diatomeas, ostrácodos y gasterópodos en una sección estratigráfica aluvial de 2,5 m de potencia, en el río Quequén Grande, Pampa Interserrana. El intervalo estratigráfico estudiado se extiende desde un momento anterior a los 9000 años ^{14}C AP cuando en el ambiente fluvial prevalecía la estabilidad indicada por la presencia de un paleosuelo alcalino y ligeramente salobre. Con posterioridad a los 9000 años ^{14}C AP, se produjo la inundación de las áreas deprimidas de la planicie de inundación y el consiguiente desarrollo de ambientes lagunares evidenciando el comienzo de condiciones húmedas-subhúmedas. Las asociaciones de diatomeas, ostrácodos y gasterópodos, además de la aparición de rasgos de suelo en algunos niveles indican una laguna somera de unos 2 m de profundidad con variaciones estacionales en los niveles de agua. La alta sedimentación biogénica que predominó en la laguna entre los 9000 y 5000 años ^{14}C AP es sincrónica con el intervalo de formación de suelos en la cuenca de drenaje, por lo cual la depositación de la diatomita está registrando un intervalo de biostasia. Entre los 5000 y 4000 años ^{14}C AP, se produjo un cambio a condiciones mucho más secas cuando tuvo lugar la desecación y exposición subaérea de los depósitos, dando lugar a la formación de un suelo. Este episodio fue seguido por un intervalo de condiciones variables caracterizado por la disminución de la estabilidad ambiental en la cuenca de drenaje, según lo evidencia el progresivo incremento de depósitos silicoclásticos.

1 INTRODUCTION

The Pampa Interserrana located between the Tandilia and Ventania ranges, occupies the southernmost part of the Buenos Aires pampas. The Late Pleistocene and Holocene stratigraphic record consists of a sandy loess to loessial sand apron that blankets the landscape (Zárate & Blasi 1993). Radiocarbon dates performed at archaeological sites located in the Tandilia range indicate that loess deposition ceased circa 10,000 ^{14}C yr BP followed by an interval of pedogenesis which gave rise to the present cultivated soils (Zárate & Flegenheimer 1991). In the plains of Pampa Interserrana, the Holocene is mostly documented by soil formation with very restricted episodes of deposition in the hillslopes of the Tandilia and Ventania ranges. Instead, during the Holocene sedimentation continued in fluvial environments which are characterized by well preserved valley fill sequences. Therefore, the alluvial sequences have a greater stratigraphic resolution than the aeolian sequences to interpret the environmental history of the region during the last 10,000 years.

To date, these alluvial records are only known in their general lithological composition and stratigraphic arrangement. Over the last 20 years valley fills have been the subject of study of several research teams providing new information on the alluvial stratigraphy and general environmental reconstructions (Fidalgo et al. 1991, Quattrocchio et al. 1988). The similarity of the stratigraphic records has been used for long range correlations of continental-scale without considering their particular geo-

136

morphological settings and the occurrence of time transgressive processes. This led to oversimplified models of climatic changes over the Pampas. Based on these previous results, research has been initiated to critically evaluate the temporal resolution and sensitivity of alluvial sequences to climatic changes.

The studied region is the fluvial system of Río Quequén Grande and the streams draining the area of Pampa Interserrana between Mar del Plata and Necochea (Fig. 1). Among several sections under study, our attention was addressed to a selected site named La Horqueta II with an excellent stratigraphic record of a diatomite. Owing to the usefulness of lacustrine organisms and particularly of diatoms as an important source of palaeo-climatic information (Forester 1987, Bradbury 1989) the purpose of this paper is to test the environmental significance of the diatomite. Changes in the environment are mostly inferred from the diatom assemblages and ostracods. Additional information on gastropods and other microfossils is also considered. Eventually, we discuss the local and regional palaeoenvironmental implications of this fossil record.

2 ENVIRONMENTAL AND GEOLOGICAL SETTING

La Horqueta II site is located in the mid basin segment of the Río Quequén Grande valley at the confluence between this master stream and a tributary (Arroyo Tamangueyú), both heading in the Tandilia range and flowing to the Atlantic Ocean (Fig. 1). The streams drain a relatively flat plain. The climate is humid temperate with a mean annual precipitation of around 800 mm and an average annual temperature of 14°C. The region is a grassland presently modified by intense agriculture; native vegetation is restricted to patches mostly composed of grasses, located along the margins of the main streams and in the mountain areas of Tandilia and Ventania.

The valley of the Río Quequén Grande is cut into Late Tertiary (?) fine sandstones cropping out on the extensive interfluvial surfaces. These are capped by thick calcrete crusts, in turn overlain by an apron of Late Pleistocene deposits ranging from sandy loess to loessial sands. Isolated outcrops of Middle to Late Pleistocene (?) fine sandstones are exposed along the stream bankcuts. Ameghino (1889) divided the Late Pleistocene and Holocene alluvial record of the Pampas into two stratigraphic units: 'Lujanense' and 'Platense'. This stratigraphic subdivision has been traditionally followed by other authors (i.e. Frenguelli 1928). More recently, Fidalgo et al. (1973) grouped these units into the lithostratigraphic Luján Formation which comprises two members: Guerrero ('Lujanense') and Río Salado ('Platense').

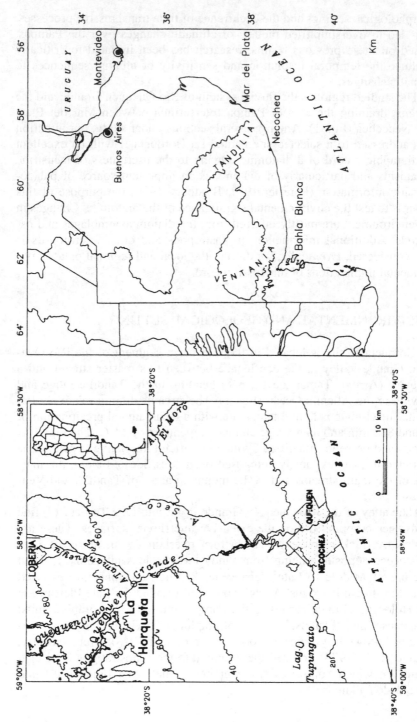

Figure 1. General geographical setting and location of the studied site.

The 'Lujanense' consists of fluvial (sandstones and siltstones) and aeolian facies (sandy loess to loessial sands) with extinct fossil megafauna. The deposition of the 'Lujanense' began prior to around 29,000 ^{14}C yr BP (Carbonari et al. 1992) and ceased sometime between 11,000 and 10,000 ^{14}C yr BP. A relatively well developed soil is formed on the top of the 'Lujanense' and it is named Suelo Puesto Callejón Viejo (Puesto Callejón Viejo Soil) (Fidalgo et al. 1973). The 'Platense', in which the stratigraphic section of La Horqueta II is included, is made up of diatomaceous deposits, traditionally interpreted as lacustrine and paludal facies of Holocene age (Frenguelli 1928).

La Horqueta II section (Fig. 2) is illustrative of the Late Pleistocene-Holocene alluvial record of the Pampas. Similar stratigraphic sequences, although not as well preserved as this one, are present not only along the valley of Río Quequén Grande but in other Pampean fluvial basins.

3 MATERIALS AND METHODS

The micropalaeontological analysis was performed on a stratigraphic interval bounded by radiocarbon dates, whereas the uppermost part of the diatomite between 175 and 70 cm was characterized on the basis of its lithological features. Prior to the sampling the outermost 10 cm were removed.

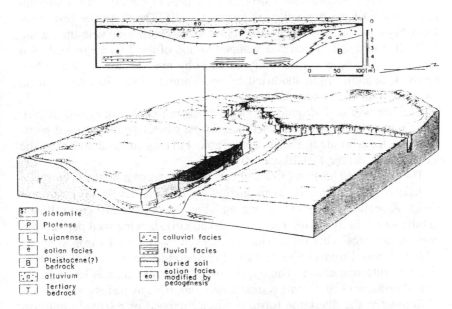

Figure 2. Geometry and geomorphological setting of La Horqueta II section.

139

Diatom samples were treated with hydrochloric acid to remove carbonates, and then oxidized with hydrogen peroxide to remove organic matter. After washing with distilled water, samples were mounted in Canada Balsam. Between 500 to 1000 diatom frustules were counted in each sample under a magnification of 1000×. Species assemblages were plotted in a diagram showing taxa percentages for each sample. Those taxa higher than 1% are grouped alphabetically.

Ostracode samples of 50 grams were wet-sieved to remove the sediment fraction finer than 65 µm. The remaining fraction was dried and, depending on the samples, split for counting. The specimens were picked and retained on micropalaeontological slides. Adults and juveniles were differentiated and when possible adult sex was also determined. To approximate the total number of adult individuals within each sample, only complete carapaces and the most abundant valves (right or left) were recorded for counting. The percentage abundance of the taxa in each sample and the total abundance per 50 g of unwashed sample were determined and plotted. No percentages were calculated when the number of individuals was lower than 50. Gastropods and characean gyrogonites were recognized in the same samples and their presence was plotted in the diagram.

4 CHARACTERISTICS OF THE DIATOMITE

Stratigraphic relationships: Horizontal surfaces of both the lower and upper contacts of the diatomite are sharp (Fig. 2). The diatomite rests conformably on a buried soil (350-370 cm interval) correlated with the Puesto Callejón Viejo soil, which is developed on top of the 'Lujanense' beds. On top, the diatomite is conformably overlain by massive coarse silts to fine sands of aeolian origin modified by soil formation (20-70 cm), in turn, buried by recent alluvium (20-0 cm).

Composition and structures: The deposit is mainly composed of biogenic remains mostly of diatoms including a siliciclastic fraction of particles which varies along the section (Table 1) being more abundant at the 210-180 cm and the 120-70 cm levels.

The lowermost section (350-270 cm) is characterized by the occurrence of horizontally stratified layers including discrete levels of gastropod shells. A section composed of two massive layers (270-210 cm) follows, in turn, overlain by a darker layer (210-180 cm) showing well defined pedogenic features. This layer was buried by a layer with gastropod shells (180-175 cm). From 175 to 120 cm, the diatomite consists of alternating layers of different colors. The uppermost section which includes archaeological remains (120-70 cm) is moderately modified by pedogenesis.

Geometry: the diatomite forms a tabular deposit of relatively uniform thickness (around 2 m), elongated along the valley axis and located in a

140

Table 1. Internal features and composition of the diatomite.

Depth (cm)	Description
70-120	Dark grayish-brown (10YR 5/2) diatomaceous sediments grading to light gray (10YR 7/2); massive; very common invertebrate bioturbations, very frequent root traces with $CaCO_3$; abundant siliciclastic grains; archaeological remains (fragments of bones, charcoal and chalcedony microflakes), gradual to diffuse and smooth lower contact.
120-175	Light brownish gray (10YR 6/2) diatomaceous layers alternating with darker layers; gastropod shells distributed throughout the matrix; root traces and bioturbations; abundant siliciclastic grains, clear and smooth lower contact.
175-180	Discrete layer of whole gastropod shells in a dark grayish-brown (10YR 5/2) diatomaceous matrix; sharp and smooth lower contact.
180-210	Very dark gray (5Y 3/1); irregular prismatic to subangular blocky moderate structure; abundant root traces; whole gastropod shells concentrated in discontinuous levels of 2-3 cm; abundant siliciclastic content and phytoliths; clear and smooth lower contact.
210-270	Light gray (10YR 7/2); predominantly massive, megascopically no lamination is observed; weak defined layering, two main beds are recognized with upper boundaries defined by darker colors (grayish brown 10YR 5/2 grading to light gray 10YR 7/2) and low degree of bioturbations; dispersed gastropod shells; occasional bone fragments of cricetid rodents at the lowermost bed; abundant volcanic shards; root traces; clear and smooth lower contact.
270-350	White (10YR 8/1) to light gray (10YR 7/1), horizontally stratified; layering determined by slight changes in colour from white to light gray; freshwater gastropod shells are present in discrete levels and also chaotically dispersed in the matrix; massive or internally laminated layers of variable thickness (1 to 10 cm); no siliciclastic content is present; internal contacts between layers are very abrupt and smooth.

marginal setting within the floodplain of Río Quequén Grande, resting much closer to the western valley margin. The deposit is 300 m long and between 70 to 100 m wide according to the coring made in the floodplain. It pinches out both upstream and downstream (Fig. 3).

Upstream, the diatomite grades into slope wash and colluvial facies that rest unconformably on the Pleistocene (?) bedrock. Downstream, it grades into diatomaceous siltstones highly cemented by $CaCO_3$ which unconformably overlie the Late Pleistocene aeolian facies (Zárate et al. 1996).

The water ponding, and therefore the deposition of the diatomite, occurred in a depression limited by the bedrock outcrop upstream and the aeolian facies downstream, which formed a 2 m high hill (Fig. 3).

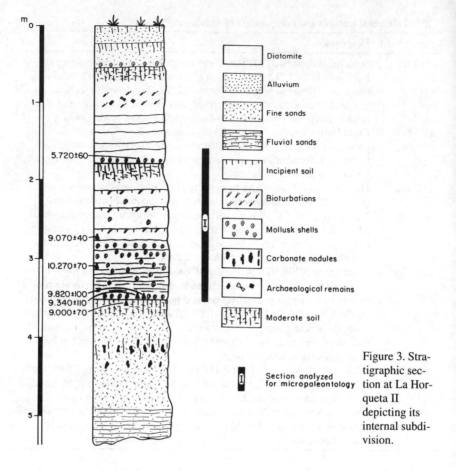

Figure 3. Stratigraphic section at La Horqueta II depicting its internal subdivision.

5 RADIOCARBON CHRONOLOGY AND ABSOLUTE AGE

The chronology of the section is based on conventional radiocarbon dates of gastropods and organic matter from the palaeosol (Puesto Callejón Viejo Soil) (Table 2). The datings of gastropod shells collected from the lowermost 50 cm of the diatomite have yielded ages appreciably older than the dating of organic matter from the uppermost 5 cm of the buried soil. The discrepancy of ^{14}C dating between gastropod shells and organic matter samples has been also reported by Figini et al. (1996) from an alluvial stratigraphic section of similar age in Arroyo Tapalqué, 200 km northwest of the studied site. These authors interpreted the older ^{14}C ages of gastropod shells to be the result of an estimated reservoir effect of circa 1100 ± 300 ^{14}C yr. The upper fluvial systems of both streams drain areas of Precambrian calcareous rocks, which may account for the hard water effect of calcareous shells on the radiocarbon ages. Instead, the organic

Table 2. Radiocarbon dates on gastropods and organic matter.

Depth (cm)	^{14}C Age (years BP)	Lab number	Material dated
360-355	9000 ± 70	Beta 79439	organic matter
355-350	9340 ± 110	Beta 84180	gastropods
355-350	9820 ± 100	1972 AECV	gastropods
305-300	10270 ± 70	Beta 79440	gastropods
275-270	9070 ± 140	Beta 84182	gastropods
180-175	5720 ± 60	2007 AECV	gastropods

matter samples from the buried soils were not subject to this age offset because the carbon uptake of terrestrial vegetation appeared to be dominated by photosynthesis (Figini et al. 1996). Although we were unable to calculate the reservoir effect at La Horqueta II, the proximity of the neighboring basin of Arroyo Tapalqué and the similar geological setting allow us to assume a magnitude of the reservoir effect comparable to that obtained by Figini et al. (1996). Therefore, under these circumstances, the radiocarbon dates of gastropods would be at least in the order of 1100 yr younger than the conventional radiocarbon dates recorded.

6 MICROFOSSIL ASSEMBLAGES

6.1 Diatoms

The stratigraphic section under analysis was subdivided into twelve levels in which sixty-eight diatom species were identified (Fig. 4).

The assemblage found at the uppermost part of the A horizon (375-350 cm interval) of the Puesto Callejón Viejo Soil is mainly composed of the epiphytic and freshwater taxa *Synedra ulna* and *Denticula tenuis* var. *crassula,* along with the brackish and epiphytic diatom *Rhopalodia gibberula.* Also important in this level are *Diploneis smithii,* a benthic and euryhaline taxon that occurs in brackish and marine environments (Germain 1981), and the benthic taxa *Synedra platensis* and *Hyalodiscus subtilis* (= *H. schmidtii),* living in slightly brackish environments (Frenguelli 1941).

The diatom assemblage of the stratified diatomite (350-270 cm) is dominated by the freshwater taxa *Fragilaria pinnata, F. construens* var. *venter* and *Synedra ulna.* The first two species are tychoplanktonic, frequently present in the water column but also related to either benthic or epiphytic habitats (Vos & de Wolf 1993), whereas *Synedra ulna* is epiphytic. The diatom assemblage between 350-300 cm points to a deeper water body and more freshwater with a high nutrient availability. At the upper part between 300-270 cm, the increase of *Fragilaria construens* var.

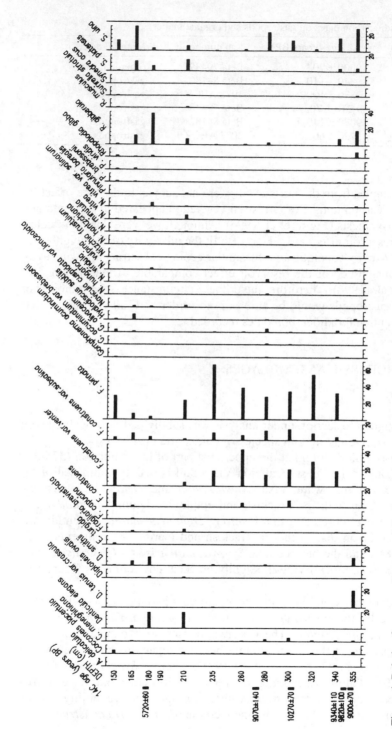

Figure 4. Relative frequency diagram of diatom species composition.

144

subsalina may indicate a slightly higher salt content. Also, the growth of epiphytic and benthic taxa suggests an increment of aquatic vegetation (Fontes et al. 1985) and a shallower water body.

The diatom assemblage of the massive diatomite (270-210 cm) is dominated by the tychoplanktonic taxa *Fragilaria pinnata, F. construens* var. *venter* and *F. construens* var. *subsalina.* The depth would have increased again with an important growth in nutrient availability. The high abundance of diatom frustules in this level gradually decreases towards the top.

The dominant species at level 210 cm are the epiphytic and benthic taxa, *Denticula elegans, Synedra platensis* and *Rhopalodia gibberula,* and the tychoplanktonic *Fragilaria pinnata.* The assemblage suggests a depth decrease and a gradual increment of salinity. The next level analyzed was sterile. At 180 cm, the brackish and epiphytic taxa *Rhopalodia gibberula* and *Diploneis smithii* and the freshwater taxa *Denticula elegans* show the highest proportion recording a brackish and shallower environment or even a saturated groundsurface.

Synedra ulna, S. platensis and *Rhopalodia gibberula* dominate in the massive layer at 165 cm, recording a brackish and very shallow swampy environment. The next stratified level (150 cm) is mainly composed of the freshwater taxa *F. pinnata, F. construens* and *F. construens* var. *venter* and *Synedra ulna,* which show a decrease of salinity.

6.2 *Ostracodes*

Sixteen ostracoda species were recorded in the twelve levels studied along the stratigraphic interval under analysis (Fig. 5). Microfossils are generally well preserved, except for some ostracoda mainly juvenile with delicate valves that have suffered breakage.

The dominant ostracoda species are *Darwinula* sp., *Sarscypridopsis aculeata* and *Cyprideis salebrosa hartmanni.* In present environments, the fresh water genus *Darwinula* has no particular ecologic requirements (Würdig 1984), although it is normally found in organic rich environments (Kotzian 1974). *Sarscypridopsis aculeata* is a brackish water form apparently found in salinities not exceeding 1%, but it has also been recorded in fresh water (Bronshtein 1947). This species commonly occurs in slightly salty coastal and inland waters and is only rarely found in pure fresh water (Meisch & Broodbakker 1993). *Cyprideis* is considered a brackish water genus, *C. salebrosa hartmanni* is a mixohaline ostracode (Ornellas & Würdig 1983) common in estuaries, but is also found in lakes and ponds (Lister 1976).

The dominance of ostracode species varies along the stratigraphic section under analysis. *Darwinula* sp. prevails in the A horizon of the palaeosol underlying the diatomite, indicating an organic rich setting.

145

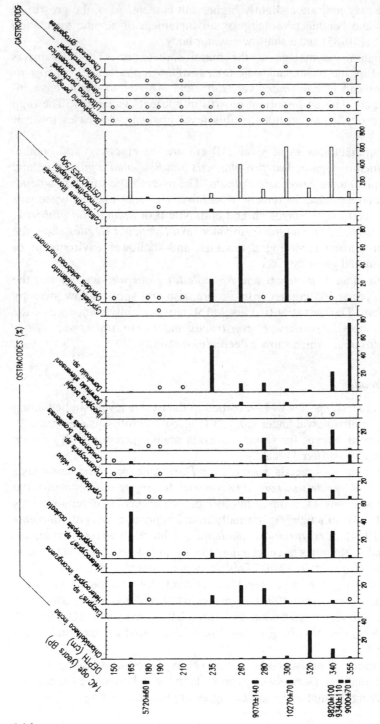

Figure 5. Relative frequency diagram of ostracode species and presence of gastropod species and charophytes. (°) means 'Presence'.

The ostracode assemblage of the stratified diatomite (350-270 cm) is mostly composed of *Sarscypridopsis aculeata, Cyprideis salebrosa hartmanni, Chlamidotheca incisa* and *Candonopsis* sp. This assemblage consists of complete ostracode populations, which along with the presence of carapaces, suggest calm water. Important percentages of *S. aculeata* in some levels and the dominant presence of *C. salebrosa hartmanni* in others, may indicate a slightly saline environment.

The lower level of the massive diatomite (at 260 cm) is again dominated by *C. salebrosa hartmanni,* recording similar conditions to the underlying levels. The next level (235 cm) shows a low ostracode and gastropod abundance, and is dominated by *Darwinula* sp. followed by *C. salebrosa hartmanni*. The change of the ostracode association may be related to shallowing of the water pond.

The studied levels between 210 and 180 cm were almost sterile which is in agreement with the shallowing and subaerial exposure recorded by the soil features present at this stratigraphic interval. A restoration of appropriate conditions for ostracode living is recorded by the assemblage in the overlying level. Here, at 165 cm, *S. aculeata* is dominant followed by *Heterocypris incongruens* and *Cypridopsis* cf. *vidua,* representing a slightly saline pond with aquatic vegetation.

6.3 *Gastropods and other microfossils*

Four taxa of gastropods were recorded along the studied sequence. *Littoridina parchappei* is abundant and dominant. This taxon is commonly found in ponds from the eastern Pampas of Buenos Aires province, living on submerged aquatics, such as *Potamogeton striatus, Ceratophyllum demersum* and on filamentous algae *Oedogonium* sp., *Spyrogyra* sp. and Characeae (Gaillard & Castellanos 1976).

Biomphalaria peregrina is the second most abundant taxon. It is euryoic, sometimes an invader of high reproductive potential and considered as an r-strategist. *Biomphalaria* lives preferably in environments with predominant submerged vegetation (Rumi 1991).

Gundlachia concentrica lives attached to leaves of aquatic macrophytes (Fernández 1981). The last species of gastropod identified is *Chilina parchappei*, which is poorly represented.

Gastropods are present throughout the studied section, occurring either dispersed or concentrated in discrete layers (Fig. 3).

Among other microfossils, characean gyrogonites appeared at 300 cm, where they are very abundant, and at 260 cm. Charophytes occur generally in freshwater, but some are known from saline waters, living anchored to different kinds of substrates.

7 DISCUSSION AND CONCLUSIONS

The diatom assemblage, the occurrence of *Darwinula* sp and the presence of phytoliths in the uppermost 25 cm of the buried A horizon indicate a vegetated surface of a slightly brackish and alkaline soil under very moist or saturated conditions.

The radiocarbon dates of the uppermost part of Puesto Callejón Viejo Soil indicate that deposition of the diatomite started soon after 9000 ^{14}C yr BP. A comparable radiocarbon dates of organic matter was reported from a swampy deposit at the same stratigraphic position of the Puesto Callejón Viejo Soil at the locality of Salto de Piedra in Arroyo Tapalqué (Bonadonna et al. 1995). The distinctness and topography of the contact between the diatomite and the palaeosol suggest a rapid event (flood episode) that inundated the depressions along the valley floodplain. The presence of well preserved microfossils and gastropods shells concentrated in a discrete layer at the very bottom of the diatomite indicates low energetic conditions. The radiocarbon dates obtained between 350-270 cm reveal a relative high sedimentation rate (80 cm in circa 1000 years).

According to Bradbury (1989), the diatom assemblage of *Fragilaria construens* var. *venter*, *F. construens* var. *subsalina* and *Fragilaria pinnata* indicates a shallow freshwater pond 1 to 3 meters deep. This is in agreement with the diatomite geometry and the lateral stratigraphic relationships that reveal a gentle palaeotopography of up to 2 meters of relative relief. At present these diatoms live under a pH range between 7.6 and 7.8 (Bradbury 1971) which together with the presence of carbonate fixing organisms, such as charophytes, gastropods, and calcareous small stems, give evidences of alkaline waters. The composition of the ostracode association, the abundance of gastropods like *Littoridina parchappei*, and also the presence of numerous calcareous stems together with characean gyrogonites (300 cm level) indicate a shallower pond with abundant submerged vegetation.

The stratigraphic section between 270-210 cm and the interval of soil formation (210-180 cm) encompasses a longer time interval bracketed between 9070 ^{14}C and 5720 ^{14}C yr BP, recording a much lower sedimentation rate than the underlying sediments. Brief episodes of desiccation and subaerial exposure seem to have followed the deposition of each of the massive layers between 270-210 cm, judging from the invertebrate bioturbations, root traces, and slight darker colors present at their uppermost contacts.

Probably the end of the deposition of the stratified diatomite (270 cm) is related to a notorious decrease in the preservation potential of sedimentary structures due to a reduction of the sediment storage capacity of the pond. This reduction may have increased the possibility of slight reworking of the bottom, preventing the preservation of sedimentary structures

148

during the deposition of the massive diatomite (270-210 cm). The changes found in the relative abundance of the taxa point to variations in the water depth of the pond connected with desiccation events. Accordingly, the presence of brackish diatoms and peaks of ostracoda revealing increments of salinity must have been related to higher salt concentration due to evapotranspiration under shallower conditions.

The interval between 210 and 180 cm reveals a more pronounced episode of water depth reduction. The diatom assemblage indicates a brackish and very shallow environment which eventually underwent subaerial exposure and moderate soil formation. The very low abundance of ostracodes in this interval is consistent with the occurrence of a desiccation event and inappropriate condition for their existence. The duration of the subaerial exposure of the pond is not known, although in view of the moderate degree of development of the soil profile, it may have lasted for a relatively short interval (200-300 years), tentatively placed between 5000 and 4000 ^{14}C yr BP.

This interval of subaerial exposure is interrupted by the deposition of a discrete layer of gastropod shells at circa 5700 ^{14}C yr BP, which is interpreted as a flood episode. The massive and stratified layers of increased siliciclastic content evidence a spell of pronounced environmental variability with alternation of marsh and pond conditions. Finally, the uppermost part of the diatomite between 120-70 cm, which shows pedogenetic features, marks the beginning of the definitive desiccation of the pond followed by the accumulation of aeolian sediments.

8 REGIONAL ENVIRONMENTAL IMPLICATION OF THE DIATOMITE

The palaeosol underlying the diatomite (Puesto Callejón Viejo Soil) is traced for several kilometers along the valley of Río Quequén Grande and is also present in other fluvial basins of the Pampa Interserrana. This interval of floodplain stability is correlated with the soil formation interval initiated sometime between 10000 and 11000 ^{14}C yr BP (Zárate & Flegenheimer 1991) which follows the deposition of the Late Pleistocene aeolian mantle (Zárate & Blasi 1993). The interval gave rise to the development of most of the present cultivated soils. However, in fluvial environments such as the Río Quequén Grande valley, soil formation was interrupted around 9000 ^{14}C yr BP by diatomite deposition whereas pedogenesis continued in the interfluvial areas. Therefore, the evolution of pond environments in the floodplains is contemporaneous with the regional episode of soil formation. Pedogenesis may have released significant amounts of nutrients, which may explain the high and almost exclusive biogenic sedimentation for most of the Early and Middle Holocene. Within this framework, while the dominant biogenic sedimentation is reflecting general landscape stability in the fluvial basin, the diatomite indicates an interval of biostasy (Erhardt 1967). Erosion was inhibited and consequently the supply of siliciclastic sediments from the basin was reduced, whereas occasional shards from volcanic eruptions, occur in the diatomite.

At present these pond environments are found in northeastern Buenos Aires province where climate is characterized by average annual rainfall of around 900 mm and by a dry summer season. Here, ponds usually undergo water level variations and even complete desiccation during very dry years. Considering this present analog, the Early to Middle Holocene pond was subject to more humid conditions than those prevailing in the Pampas after 5000 ^{14}C yr BP, supporting the results obtained from pollen analysis (Prieto 1996). Water level may have seasonally fluctuated with several spells of much drier conditions. These changes may account for the shallower and more brackish waters alternatively recorded by the biological assemblages and the brief episodes of subaerial exposure recorded in the stratigraphic section.

The desiccation interval, and the concomitant soil formation recorded by the 210-180 cm section, documents a relatively short interval of drier conditions in the floodplain, followed by a return to alternating shallow ponds and marshes under an increased environmental instability in the fluvial basin. Considering the hard effect of the ^{14}C ages from gastropods, this desiccation interval may correspond with the episode of moderate and localized soil truncation that occurred in the upper catchment areas of the Tandilia range after 5000 ^{14}C yr BP (Zárate & Flegenheimer 1991). This climatic fluctuation, also recorded by pollen assemblages (Páez & Prieto 1993) is interpreted as a result of a climatic shift to much drier conditions, which seems to have prevailed during the Late Holocene.

The small size of the pond made it very sensitive to changes of both internal and external factors. The microfossil assemblages found between 350-210 cm suggest a rather similar environment under a seasonally dry, humid-subhumid climate. Therefore, the different internal structure of these stratigraphic intervals is interpreted as representing internal changes of the water body: depth reduction due to sedimentation, which increased its sensitivity to fluctuations of the water level. The fluctuations of the water level and the subaereal exposures recorded by the 210-70 cm interval apparently reflect changes in the dynamic of the fluvial basin.

ACKNOWLEDGEMENTS

This research was supported by the Universidad Nacional de Mar del Plata grant to project EXA/10. Special thanks are extended to Virginia Bernasconi and Juliana Bo for their assistance in figure preparation. Jordan

Sharon, Cristian Favier and Walter Brotz assisted during field work. Susana Serra and Griselda Golfieri treated the samples.

REFERENCES

Ameghino, F. 1889. Contribución al conocimiento de los mamíferos fósiles de la República Argentina. *Actas Academia Nacional de Ciencias,* 6: 1-1027. Buenos Aires, Argentina.

Bonadonna, F.P., G. Leone & G. Zanchetta 1995. Composición isotópica de los fósiles de gasterópodos continentales de la provincia de Buenos Aires. Indicaciones paleoclimáticas. In: M.T. Alberdi, G. Leone & E. Tonni (eds), *Evolución biológica y climática de la región pampeana durante los últimos cinco millones de años. Un ensayo de correlación con el Mediterráneo occidental:* 77-104. Monografías. Museo Nacional de Ciencias Naturales. Madrid.

Bradbury, J.P. 1971. Paleolimnology of Lake Texcoco, Mexico. Evidence from diatoms. *Limnology and Oceanography* 16(2): 180-200. Baltimore.

Bradbury, J.P. 1989. Late Quaternary lacustrine paleoenvironments in the Cuenca de Mexico. *Quaternary Science Reviews* 8: 75-100. Oxford.

Brohnstein, Z.S. 1947. *Fresh-water ostracoda.* United States Department of the Interior and the National Science Foundation, Washington. New Delhi: Amerind Publishing Co. 470 pages.

Carbonari, E.J., R.A. Huarte & A.J. Figini 1992. Miembro Guerrero, Formación Luján (Pleistoceno, Pcia de Buenos Aires): edades [14]C. *3ras Jornadas Geológicas Bonaerenses, Actas:* 245-247. La Plata.

Erhardt, H. 1967. *La genèse des sols en tant que phénomène géologique.* Paris: Masson et Cie, Éditeurs.

Fernández, D. 1981. Mollusca Gasteropoda Ancylidae. In R.A. Ringuelet (dir.) *Fauna de Agua Dulce de la República Argentina* 15(7): 99-114. La Plata.

Fidalgo, F., O. De Francesco & U. Colado 1973. Geología superficial en las hojas Castelli, J.M. Cobo y Monasterio (Provincia de Buenos Aires). *V Congreso Geológico Argentino,* Actas IV: 27-39. Buenos Aires.

Fidalgo, F., J.R. Riggi, H. Correa & N. Porro 1991. Los sedimentos postpampeanos continentales en el ámbito sur bonaerense. *Revista de la Asociación Geológica Argentina* 45: 239-256. Buenos Aires.

Figini, A.J., F. Fidalgo, R. Huarte, J. Carbonari & R. Gentile 1996. Cronología radiocarbónica de los sedimentos de la Fm Luján en arroyo Tapalqué, Provincia de Buenos Aires. *IV Jornadas Geológicas Bonaerenses,* Actas 1: 119-126. Junín. Argentina.

Fontes, J.Ch., F. Gasse, Y. Callot, J.C. Plaziat, P. Carbonelli, P.A. Dupeuble & Y. Kaczmarska 1985. Freshwater to marine-like environments from Holocene lakes in Northern Sahara. *Nature* 317: 608-610. London.

Forester, R.M. 1987. Late Quaternary palaeoclimate records from lacustrine ostracodes. In: W.F. Ruddiman & H.E. Wright Jr. (eds), *North America and adjacent oceans during the last deglaciation,* The Geology of North America V. K-3, Chapter 12: 261-276. Geological Society of America. Boulder.

Frenguelli, J. 1928. Observaciones geológicas de la región costanera sur de la provincia de Buenos Aires. *Anales de la Facultad de Ciencias de la Educación, Universidad Nacional de Litoral* 2: 1-145. Santa Fe, Argentina.

Frenguelli, J. 1941. Diatomeas del Río de la Plata. *Revista del Museo de La Plata* III: 213-334. La Plata, Argentina.

Gaillard, C. & Z.A. de Castellanos 1976. Mollusca Gasteropoda, Hydrobiidae. In: R.A. Ringuelet (dir.) *Fauna de Agua Dulce de la República Argentina* 15(2): 1-40. La Plata.

Germain, H. 1981. Flore des diatomées. Diatomophycées. *Société Nouvelle des Editions Boubée*, Paris.

Kotzian, S. 1974. New fresh-water ostracodes of the genus *Chlamydotheca* from Brazil. Ecology, Geographic and Stratigraphical position. *Annais Academia Brasileira Ciências* 46: 423-67. Rio de Janeiro.

Lister, K.F. 1976. Temporal changes in a Pleistocene Lacustrine Ostracode Association; Salt Lake Basin, Utah. In: R.W. Scott & R.R. West (eds), *Structure and Classification of Paleocommunities:* 193-213. Dowden: Hutchinson & Ross.

Meisch, C. & N.W. Broodbakker 1993. Freshwater Ostracoda (Crustacea) collected by Prof. J.H. Stock on the Canary and Cape Verde islands. With an annotated checklist of the freshwater Ostracoda of the Azores, Madeira, the canary, the Selvagens and Cape Verde islands. *Travaux scientifiques du Musée National d'Histoire Naturelle de Luxembourg* 19: 3-47. Luxembourg.

Ornellas, L. P. de & N.L. Würdig 1983. *Cyprideis salebrosa hartmanni* Ramírez, 1967 a new subspecies from Brasil and Argentina. *Pesquisas* 15: 94-112. Brasil. Porto Alegre.

Páez, M. & A. Prieto 1993. Palaeoenvironmental reconstruction by pollen analysis from loess sequences of the southeast Buenos Aires (Argentina). *Quaternary International* 17: 21-26. Oxford.

Prieto, A. 1996. Late Quaternary Vegetational and Climatic Changes in the Pampa Grassland of Argentina. *Quaternary Research* 45: 73-88. Seattle.

Quattrocchio, M.E., C. Deschamps, D. Martínez, S. Grill & C. Zavala 1988. Caracterización paleontológica y paleoambiental de sedimentos cuaternarios, Arroyo Napostá Grande, provincia de Buenos Aires. *II Jornadas Geológicas Bonaerenses, Actas:* 37-46. La Plata.

Rumi, A. 1991. La Familia Planorbidae Rafinisque, 1815 en la República Argentina. In: R.A. Ringuelet (dir.), *Fauna de Agua Dulce de la República Argentina* 15(8): 1-51. La Plata.

Vos, P.C. & H. de Wolf 1993. Diatoms as a tool for reconstructing sedimentary environments in coastal wetlands; methodological aspects. In: H. van Dam (ed.), *Twelfth International Diatom Symposium*: 285-296. Dordrecht: Kluwer Academic Publishers, The Netherlands.

Würdig, N.L. 1984. Ostracodes do sistema lagunar de Tramandaí, RS, Brasil. Sistemática, ecología e subsidios a paleoecologia. Ph.D. Dissertation, Universidad Federal do Rio Grande do Sul, Brasil. Porto Alegre.

Zárate, M. & A. Blasi 1993. Late Pleistocene-Holocene aeolian deposits of the southern Buenos Aires province, Argentina: A preliminary model. *Quaternary International* 17:15-20. Oxford.

Zárate, M. & N. Flegenheimer 1991. Geoarchaeology of Cerro La China locality: site 2 and site 3. *Geoarchaeology, An international Journal* 6: 273-294. John Wiley & Sons.

Zárate, M., M. Espinosa & L. Ferrero 1996. La Horqueta II, Rio Quequén Grande: Ambientes sedimentarios de la transición Pleistoceno-Holoceno. *4tas Jornadas Geológicas y Geofísicas Bonaerenses, Actas* 1: 195-204. Junín, Argentina.

Southernmost South America climate and glaciers in the 16th century through the observations of Spanish navigators

9

MARÍA DEL ROSARIO PRIETO
Unidad de Historia Ambiental, IANIGLA, CRICYT, Mendoza, Argentina

ROBERTO G. HERRERA
IADIZA, CRICYT, Mendoza, Argentina

ABSTRACT: The objective of this work is to achieve an approximation to the climate and to the phenomena of glaciation occurred in the Magellan Strait during the 16th and part of the 17th centuries through historical records. The data presented here complement the results obtained by scientists from other palaeoclimatic disciplines in the area, verifying if the historical sources confirm the conclusions of those authors, who postulate a very cold interval between 1520 and 1670 AD.

To accomplish this investigation, the 'Navigation journals' of the first Spanish sailors that went through the Strait were used obtained from Spanish Archives.

By means of the technique of the content analysis, the stability in the vocabulary referring to the climate and the glaciers throughout these centuries was verified and the events were qualified according to their magnitude. Some categories were determined, and snow and rain, temperature, winds and storms occurrence have been selected to be analyzed in this work. The icebergs presence in the Magellan Strait and the descriptions of the Fuegian glaciers as low-temperature indicators were specially studied.

It is concluded that direct references exist concerning iceberg presence originating from glaciers in the Magellan Strait, only in the second half of the 16th century. The advance of the ice would suppose an accentuation of colder conditions during that century, which can be confirmed by the data of the ships, that show cold and very cold summer temperatures in the Pacific Ocean slope. These data would coincide with dendrochronological results suggesting that the interval from 1520 to 1670 AD was one of the coldest of the Little Ice Age.

RESUMEN: El objetivo de este trabajo es lograr una aproximación al clima y a los fenómenos de glaciación ocurridos en el Estrecho de Magallanes durante el siglo XVI y parte del XVII a través de registros históricos. Se complementan los resultados obtenidos por investigadores de otras disciplinas paleoclimáticas en el área. Se verifica si las fuentes históricas corroboran las conclusiones de esos autores, quienes postulan un intervalo muy frío entre los años 1520 y 1670 de nuestra era.

Para realizar esta investigación se utilizaron los 'diarios o relaciones de navegación' de los primeros marinos españoles que pasaron por el Estrecho, recopilados en su mayor parte en archivos históricos españoles.

A través de la técnica del análisis de contenido se verificó la estabilidad en el vocabulario referido al clima y los glaciares a lo largo de esas centurias y se calificó los eventos de acuerdo a su magnitud. Se determinaron diversas categorías de las cuales se han seleccionado precipitaciones sólidas y líquidas, temperatura, dirección del viento y ocurrencia de tormentas para analizar en este trabajo. Se estudió especialmente la presencia de témpanos en el Estrecho y las descripciones de los glaciares fueguinos y del sur de la Patagonia chilena como indicadores de bajas temperaturas.

Se concluye que existen referencias directas a la presencia de témpanos provenientes de glaciares en el Estrecho, sólo en la segunda mitad del siglo XVI. El avance de los hielos supondría una acentuación de las condiciones de frío durante esa centuria, lo que puede ser corroborado por los datos de los navíos que consignan temperaturas estivales frías y muy frías en la vertiente pacífica. Estos datos coincidirían con los resultados dendrocronológicos en cuanto a que el intervalo entre los años 1520 a 1670 de nuestra era fue uno de los más fríos de la Pequeña Edad del Hielo.

1 INTRODUCTION

The literature about palaeoclimatic studies in the southern area of South America (Fig. 1) using information proceeding from Spanish historical sources is scarce. Martinic (1972, 1973) in his historical-geographical studies about the region has dealt with the subject, but without a systematic study of the past climatic events. Other authors have used the descriptions of the Spanish and English navigators about the Patagonian glaciers in the 18th and 19th centuries to reinforce or to illustrate the results of glaciological studies in the terrain (Mercer 1982).

The object of the present paper is to achieve an approach to climate and glaciation phenomena occurred in the region comprised between lat. 50°S and the Magellan Straits shores, lat. 54°S, during the 16th and part of the 17th centuries, through historical data. This paper complements the results obtained by scientists from other palaeoclimatic disciplines, in southern South America.

It is generally accepted that the 16th century, the subject of our present analysis, is included in the Little Ice Age (LIA) time period where extreme climatic conditions, developed in the last centuries, were enhanced and

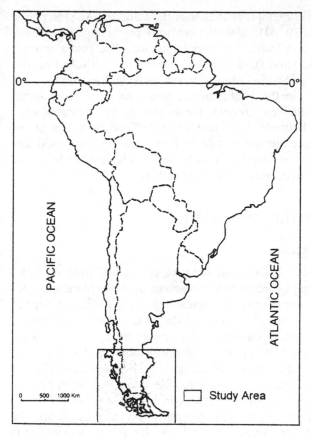

Figure 1. Study Area.

culminated in the middle of the 19th century. This phenomenon is considered as one of the most significant climatic events of the last millennium. Even though evidence is stronger in the Northern Hemisphere and the Northern Atlantic (Lamb 1977, Le Roy Ladurie 1990, Pfister 1985, Jones & Bradley 1992), its existence in South America, with some chronological unadjustments has also been verified through glaciological studies (Hastenrath 1981, Mercer 1965, Grove 1988). Rothlisberger (1986) according to a compilation of radiocarbon data in the glaciers of the southern Patagonia, points at lapses without glacial advances between 1440-1550, 1700-1800 and 1850-1900 AD. From this it can be inferred that the 1550-1700 and 1800-1850 AD intervals could be characterized as periods of advance or at least, as periods without receding glaciers.

The studies on tree-ring carried out by Villalba (1994) with *Fitzroya cupressoides* at the Valley of the Rio Alerce (Province of Rio Negro, Argentina), allowed him to reconstruct the summer temperatures of the last 1000 years, finding three episodes where high and low temperatures alter-

155

nate. He points out as contemporaneous with the Little Ice Age the period between 1270 and 1670 AD, where conditions prevailing were lower summer temperature and lower winter precipitation. The colder interval happened between 1520 and 1670 AD, at the same time with the first stories about the passage of Spanish ships through the Magellan Strait.

Climatic conditions in the Southernmost South American end during the 16th century would have coincided with one of the episodes when conditions worsened between 1520 and 1670 AD and with one of the moments when glaciers did not recede in Patagonia between 1550 and 1700 AD. The purpose of this paper is to verify whether the historical sources confirm the conclusions of the cited authors.

2 DESCRIPTION OF THE INFORMATION

2.1 *Historical sources used*

The climatic information from South America proceeding from the colonial period has time gaps, linked to the moment of the exploration, conquest and colonization from each particular region, that did not happen simultaneously in all areas. At the end of the 15th century, the Spanish conquerors set foot in Antilles and in half a century of rapid and constant advance, most of the New World was under their domain. The southernmost South American region is one of the areas with the earliest information, obtained by Magellan when crossing the Strait that separates the continent from the island of Tierra del Fuego in 1520 AD.

Once the conquest was on its way, the traffic between Spain and the colonies happened through convoys of galleons that arrived to Cartagena de Indias and to Portobelo. From these cities the inland communication was through the Panama Isthmus with the Pacific coast. Starting at the second half of the 16th century the annual traffic of the Galeón de Manila is established, commercially linking Mexico with the Philippines (Morales Padrón 1973).

This would not have been possible without the first expeditions that risked the crossing of the Magellan Straits searching for the land of the species. These expeditions were armed by the Spanish Crown or by individuals interested in the commerce with the Far East. The ships were mainly the ones generically named at the time as 'oarless and heavy'. According to Alonso de Chaves in his work 'Quatri partibus: Espejo de Navegantes' (mirror of navigators) (1537, in Castañeda Delgado et al. 1983: 210), the most common were the 240 ton ships 'as they were the most used for navigation'.

Fleets were composed by several ships, whose leader was put by the Crown, armers, merchants or bankers, a 'Capitán General' in charge of the destiny of the expedition (Randier 1960). Each ship was commanded by a captain. The pilot governed the ship navigation and lead the routes to be followed, through the use of navigation charts, mariner's needles, sand watches, compasses, probes, astrolabes and wood quadrants (Castañeda Delgado et al. 1983).

2.2 *The navigation journals*

The sources used in this paper are mainly the journals and reports of the first Spanish navigators that went through the Magellan Straits, some of which are published and some are still unpublished. The last ones were mostly compiled in the Archivo del Museo de la Marina (Madrid) and in the Archivo General de Indias (Sevilla). They were named 'Navigation Journals' as they were a second and more carefully and complete writting of the logbooks.

The navigation journals selected were those of the seven more representative expeditions from Spain, that crossed the strait between 1520 and 1619 AD. Other were discarded as the climatic data were scarce or too general, like the one written by Juan de Mori, from the expedition of Ximón de Alcazaba (1535), and that by Juan de Ladrillero (1558). In some cases the journals corresponding to two or three ships from the same fleet have been useful to compare and verify the information. Francisco de Albo and Antonio Pigafetta (1520) wrote about the Magellan's expedition; in the army of Fray García Jofré de Loaysa (1526-1527) the daily journal was written by Fray García himself, by Andrés de Urdaneta and by Juan de Areyzaga; the expedition of Francisco de Ulloa (1553-1554) was recorded by the Captain Hernando Gallego and Gerónimo de Vivar; the expedition of Juan Ladrillero (1557-1558) was written by Captain Cortés de Ojea; in the expedition of Pedro Sarmiento de Gamboa (1578-1580) besides the General Captain, the pilots Antón Pablos and Hernando Lameros kept a record, and the daily journal of the Nodal brothers' Army (1618-1619) was registered by the Captain G. Nodal and by the cosmograph Diego Ramírez de Arellano.

As for each trip of the Obispo de Plasencia's Army (1539-1540) and for those trips of Sarmiento de Gamboa (1581-1582, 1583-1584) there is only one journal respectively (Table 1).

Even though most of the fleets sailed from Spain, the fleets of Ulloa and Ladrillero left from Puerto de Valdivia in the Chilean coast (lat. 39°45'S) and the first expedition of Sarmiento de Gamboa from the Puerto del Callao, Peru. Their mission was to '...discover and explore the Strait of Magellan'.

2.2.1 *Characteristics of the data afforded by the journals*
The climatic database of Tierra del Fuego for the 1520-1618 AD period

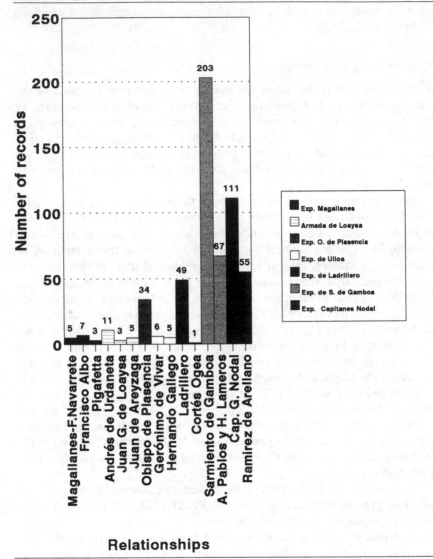

Relationships

has 753 records corresponding to navigation journals and stories by Spaniards, Englishmen and Dutchmen.

A previous selection and the need to apply linguistic techniques such as content analysis, lead to consider only the daily entries written in Spanish and this restricted the data number. Data about the Atlantic Ocean zone south of Isla de los Estados were also discarded.

The information used was that corresponding only to the Southern Hemisphere spring-summer period from October to March; the scarce winter records were disregarded owing to their minimum representativity. Therefore the base was reduced to 513 records, as it can be seen in Table 1.

The first expeditions provided little meteorological information. The noted events were mainly those that would endanger the ship and its crew, and eventually they would also write down some other interesting information related to sailing. Towards the middle of the 16th century the daily entries are more complete, but not written at fixed hours. In the expeditions of Sarmiento de Gamboa (1579) and of the Nodal brothers (1618) a routine already exists: observations are registered thrice a day. The great amount of information produced at the end of the 16th century provokes a marked lack of equilibrium between it and that provided by the first records. This has been solved analyzing the information into percentages.

The close dependence of the sailing ships to the intensity and direction of the wind, to the marine currents and, in a lesser way, to other climatic events, is clearly seen in the priority given to those phenomena in the navigation journals. Consequently, the sea status and the weather are the events more frequently consigned. The description of the landscape at both sides of the strait had a great importance, as well as the presence of wood, ice, snow and rocks.

2.3 *Localization of the sites*

The localization of the observations carried out by the ships, especially those that have to do with glaciers and ice, presents some difficulties derived from the little accuracy of the navigation instruments used at that period. The astrolabe and the sun declination tables allowed to determine quite approximately the latitude at midday calculating the sun height (Castañeda Delgado et al. 1983). Knowing the latitude, the distance in leagues or in degrees travelled from one point to another and the name of the localities that are still existing, there is a relatively certainty in locating the mentioned sites in the navigation records.

The relationship marine league-degrees is defined by Alonso de Chaves (in Castañeda Delgado et al. 1983). He states that 5 degrees and five sevenths make one hundred leagues according to the Spanish use, therefore one degree corresponds to seventeen and a half leagues. A Spanish marine league was equivalent in 1587 to four miles, or nowadays to 5,572 2/3 meters (Nagy 1991).

The zeal to carefully point the latitude and the distances corresponded to the necessity of making detailed 'pathes of the Strait', where to locate natural ports, bays and other geographical features of both coasts along the passage. Every ship that crossed the strait made its own path and accompanying charts. This poses a problem since each expedition would ascribe the discovery to itself, and would name the geographical features all over again. This has provoked confusions at the moment of localizing descriptions and phenomena.

These inconveniences have been greatly overcome partly owing to the access to the geographical charts drawn during the 16th century and above all owing to the Mapa Marítimo del Estrecho de Magallanes (Straits of Magellan maritime chart) draw by the geographer Juan de la Cruz Cano y Olmedilla in 1769 (Archivo General de Indias, M. y P. Buenos Aires 239). To accomplish it, Cano y Olmedilla used all the existing information about the pass through the Strait derived from journals and travel memories, mostly those by Sarmiento de Gamboa. It is very useful because it registers all the successive names given by the Spanish navigators and by navigators from other countries to each geographical feature they 'discovered' (Fig. 2).

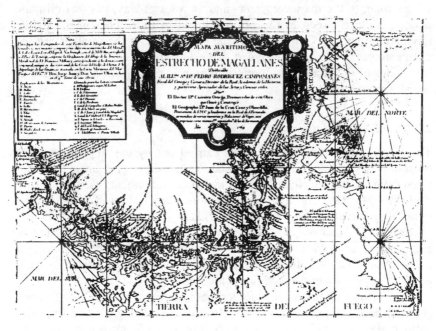

Figure 2. Maritime map of the Magellan Straits accomplished by the geographer Juan de la Cruz Cano y Olmedilla in 1769.

3 PROCESSING OF THE INFORMATION

3.1 *Obtaining climatic categories*

Ten climatic categories were derived from the texts under consideration. To obtain these variables the journals were thoroughly examined so as to determine a complete range of climatic information (Table 2).

The content analysis technique was used to verify the stability of the vocabulary that referred to the climate and to the glaciers along the century. In this regard it is remarkable the terminological consistency shown by the set of records (Baron 1982). This consistency can be related to the original regional provenance of the Spanish sailors of that century. According to Morales Padrón (1973) most of them came from northern Spain: the Basque country and Galicia. This also has to do with the sailor's perception of the meteorological phenomena. Since they were used to cold weather and snow, due to their geographical origins, their qualitative appreciation of the phenomena can be surely trusted.

At the same time, the following climatic categories were developed from the contemporaneous meteorological perspective, using the corresponding technical vocabulary:

Precipitation: two major divisions were established: rain and snow.

Table 2. Straits of Magellan: considered variables from the available data.

Position	Degrees	
	Site	
Source		
Day Month Year		
Rain	Qualitive appreciation	
	Quantitive appreciation	
Storms	Presence	
	Qualitive appreciation	
Snow	Qualitive appreciation	Intermittent-moderate-abundant
	Quantitive appreciation	
Weather		Without clouds-semicloudy -fog-stormy without precipitation
Temperature	Qualitive appreciation	Very warm-warm-mild-cold-very cold
	Degrees	
Status of the sea		Calm-choppy-rough
Currents	Presence	
	Direction	
Icebergs		
Wind	Qualitive appreciation	Without wind-regular-intense
	Direction	
Hail	Qualitive appreciation	Intermittent-moderate-abundant
Observations		

The qualitative appreciation of each phenomenon was established as: intermittent – moderate – abundant.

Present weather: four/five categories were included: without clouds, semi-cloudy, cloudy fog, stormy without precipitation.

Storms: the occurrence was recorded to determine their frequency, and whenever possible, the qualitative appreciation of the event.

Non-instrumental temperature: four categories were developed with subjective references: very warm, warm, temperate, cold, very cold.

Sea status: calm, choppy, rough.

Direction of the wind: it was registered through the compass rose with its thirty-two compass lines. It was usually reduced to sixteen quadrants.

Intensity and type of wind: generally there is not a net separation between these two variables in the documentation. The appreciation of the event was used to characterize the intensity: without wind, soft, regular, intense. Types of wind are 'side' or 'crosswind' and 'brisk wind'.

The ranges 'abundant precipitation' and 'intense wind' are those with the highest quantity of significant equivalents: 22 and 24 respectively, that could indicate a predominance of these events in the analyzed journeys. The sighting of glaciers and ice and the geographical position were also taken into account.

3.2 *Assimilation of the climatic expressions in the 16th century to the present categories*

The meaning of the past expressions and references were established and calibrated so as to adequate the climatic categories of the 16th century to the present language. For this task, and to determine the equivalencies in meaning of the climatic language of the 16th century with the language used nowadays, the tools used were the 'Tesoro de la Lengua Castellana o Española' written in 1611 (Cobarrubias Orozco 1994), the already mentioned 'Espejo de Navegantes of 1537' (Castañeda Delgado et al. 1983) and the 'Diccionario Marítimo Español' (1831).

The content analysis technique also permitted to qualify the events according to their magnitude, establishing a range between phenomena of the same nature (Baron 1982).

In Table 3, the present climatic categories and the words, sentences or expressions equivalent used in the 16th century by the Spaniards can be observed. This work also permitted to ensure that each expression or climatic reference would always be written down within the same category.

Solid and liquid precipitations, temperature, direction of the wind and storm occurrence were selected from the determined variables to thoroughly analyze this job. The presence of icebergs in the Strait is specially studied, as well as the description of the Fuegian glaciers and those from southern Chilean Patagonia as indicators of low temperatures.

162

Table 3. Significant equivalences.

Category: Rain
Descriptor

Intermitent	Moderate	Abundant
aguacerillos de agua y nieve	llovió todo el día	Temporal de viento y agua
Tiempo inconstante	lluvia	Muchos aguaceros
	chubasco de viento y lluvia	Mucha agua
	tiempo oscuro y lluvioso	Mucha agua y granizo
	chubascos violentos	Mucha agua granizo y nieve
	agua del cielo	Tormenta
	agua menuda, espesa	Duros temporales
	mal tiempo	Llueve mucho
	nos llovió poco	Fue creciendo la tormenta
	surco de agua y cerrazón	Agua con tanta fuerza
		Recia tormenta
		Tormentas fuertes y lluvias
		Grande viento y tormenta
		Tormentas fuertes y lluvias
		Fuerte temporal y varios aguaceros
		Temporal de rayos, truenos, relámpagos y agua
		Grandes temporales
		Mayores tormentas y peligros
		Recios aguaceros
		Días malos y aguaceros
		Recios tiempos de nieves y aguaceros
		Frescachón con tiempo muy nebuloso y abundantes lluvias

Category: Snow and graupel
Descriptor

Moderate	Abundant
Aguaceros de agua, envueltos en nieve y granizo	Mucha agua y granizo y nieve
Bonanza con lluvia, nieve y mucho frío	Nevó muchísimo
Cargó mucho viento, nieve y mal tiempo	Grandes aguaceros de nieve
Frío y nieve	Grandes refriegas de viento, nieve y granizo
Granizó y llovió muy bien	Aguaceros de granizo
Mucha mar, aguaceros y granizo	Granizó muchísimo
	Temporal de granizo tan grueso
	Granizo en tanta copia que cubrió el combés

Table 3. Continued.

Category: Wind			
Descriptor			
Without wind	Soft	Regular	Intense
Calma	Bonancible	Vientos naturales	muchas tempestades
extraordinaria	viento bonanza	ventó	viento tempestuosísimo
calmo	vientos calmosos	vientos variables	gran refriega
cesó el viento	poco viento	ventando	tempestad de viento y
noche serena	viento escaso	buen viento	agua
falta de viento	no hacía tiempo	las refriegas de viento	gran tempestad de agua y
	para partir	no nos fatigaron	viento
	vientos blandos	vientos más favorables	viento a refriegas
	abonanzó	vientos moderados	vientos fuertes
	bonanza especial	el tiempo moderó	viento furioso
		muy ventoso	aire recio
		próspero viento	borrascas
			tanto viento y mar y re-
			friegas y travesía
			tormentas de viento
			ventó tan recio
			viento violento
			tiempo recio con tanta
			pujanza
			grande viento
			recreció el viento
			terribilidad y tempestuo-
			sidad de los tiempos
			fortísimo viento
			rebolones de viento
			vientos duros
			vientos muy soberbios
			viento de ruin semblante
			refriegas infernales
			recias refriegas

Category: Temperature			
Descriptor			
Very warm-warm	Mild	Cold	Very cold
sol caluroso	Templado	poco verano	mucho frío
mucho calor	buen temple	frío seco	frío intenso
días de calor	tierra templada		gran frío
tanto calor como en			muchas heladas
Lima por cuaresma			tiempo muy frío
caluroso			tierra fría y nevada
calma y calor			padeciendo frío muy gran frío
tiempos calurosos			

Nowadays, the eastern and western Patagonia precipitation patterns are driven by the circulation of the westerlies. Nevertheless, the Andes cordillera exerts a strong orographic control in the amount of the precipitation (Endlicher 1995) situation that reaches our study area (Fig. 3).

South of lat. 50°S, the western slope of the Andean Cordillera presents abundant and well distributed precipitations along the year (between 4000 mm/yr in Bahía Félix and 2028 mm/yr in Cabo Rapel). The climate is maritime, with mean temperatures in the spring-summer season of 7.5°C

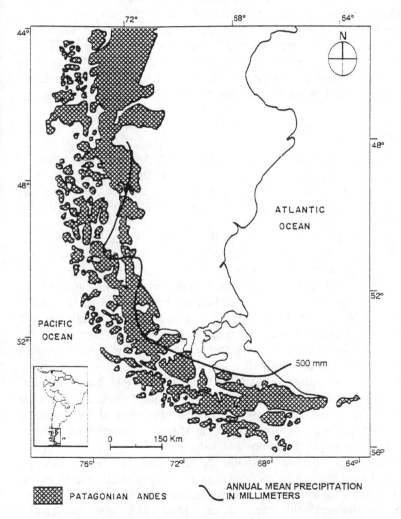

Figure 3. Eastern and western slope of the Magellan's Straits and isohyet of 500 mm.

165

in Evangelistas (lat. 52°24'S). At this same site the NW winds prevail throughout the year, exception made of the months of June and July, when those from the SW are more frequent (Miller 1976).

East of the Andes the climate is continental, and the spring-summer temperatures reach an average of 9.5°C in Punta Arenas. The precipitations diminish to almost only 200 mm per year in the plain lands. According to data proceeding from the Argentine and Chilean Meteorological Services (Endlicher 1995), an important local variation can be observed in the annual rainfall (for example, 447 mm/yr in Punta Arenas and 226 mm/yr in Río Gallegos). The westerlies dominate in this sector. Endlicher (1995) points out that in Kampenaike (lat. 52°41'S, long. 70° 54'W) the westerlies prevail in more than 50%.

On the base of the evident climatic differences between both slopes, the historical data compiled were included according to their geographical localization, in the maritime climate region or in the continental climate region. We have taken as a boundary the dividing line between the forest and the steppe that lies south of the Brunswick Peninsula.

4.1 *Temperatures*

In the navigation journals between 1520 and 1620 AD analyzed in this paper, there is a prevalence of low summer temperatures in the zone with maritime climate as well as in the zone of continental climate. Low and very low temperatures are reported in the years 1520/1521, 1539/1540, 1553/1554 and 1557/1558 (Fig. 4).

During the 1579/80 summer and in those years that followed (1582/1583 and 1583/1584), there were changes in the thermal conditions. In the first cited season there is an appreciable percentage of warm and very warm days in the slope with Atlantic influence: 69.5%. Sarmiento de Gamboa writes on February 15, 1580 that on the Straits at lat. 53° and a half (long. 71°W, nearby the present city of Punta Arenas) '... this day and the day before were as hot as in Lima during Lent or as Spain in July' (Sarmiento de Gamboa 1988: 121). In the western zone, there were also recorded warm days but in a smaller percentage: 10.6%.

In the following seasons (1582-1583 and 1583-1584), there is no information about the area corresponding to the Pacific Ocean, as Sarmiento de Gamboa dedicated his efforts to the reconnaissance of the adequate sites for the foundation of two forts in the Magellan Straits on the continental climate slope. In the summer of 1583-1584 the new inhabitants settle in the cities just founded, Nombre de Jesús and Rey Felipe, the first one situated at '... thirty leagues from the Cabo Vírgenes ...' (Sarmiento de Gamboa 1988: 270) and the second at the site where the present Puerto Hambre is located. In the season 1582/1583, the percentage

166

ORIENTAL SLOPE   OCCIDENTAL SLOPE

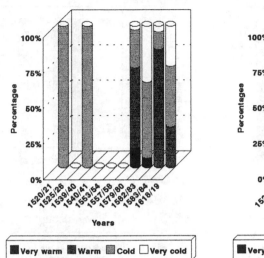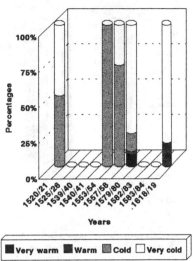

Figure 4. Qualitative temperature in the zone of the Magellan Straits in the 16th century (in percentages).

of warm days diminishes to 6.7% and there are no warm days consigned. On the second season there is an absolute prevalence of warm days: 83%.

During the 1618/1619 season in the Atlantic Ocean cold and very cold days prevailed, although with an interesting percentage of warm days, 28.6%. On the other hand, in the Pacific Ocean slope the very cold days were dominant: 83.3%.

4.2 Precipitation

Until the years 1618/1619, the days prevailing in the slope with Atlantic Ocean influence were those without precipitation in a range oscillating between 85 and 100%. The scarce precipitations recorded happened as rainfall (Fig. 5). During that last season the days without precipitation diminished 75%.

This trend is also observed in the maritime slope but, in coincidence with the present climatic pattern, there is a higher percentage of days with precipitations in relation to the Atlantic Ocean, in all of the analyzed seasons. In the years 1557/1558, there is a 48% of days with precipitation, diminishing in 1579/1580 to 37%.

The summer season of 1618/1619 shows a 63.5% with abundant rainfall, that coincides with the increase recorded at the eastern sector.

ORIENTAL SLOPE   **OCCIDENTAL SLOPE**

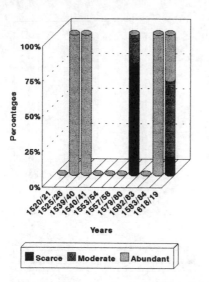
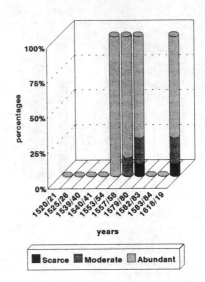

Figure 5. Summer rains in the Magellan's Strait zone in the 16th century by types (in percentages).

4.3 *Winds*

In the regional general descriptions, the authors of the journals make an unconscious synthesis of the climatic observations done during the journey, with a generalization of the areal climate.

Generally the navigation journals register a predominance of winds from the SSW, SW and WSW in the summer season, approximately like in the present times.

Only two of the navigators points out the importance of the northern winds in the western area of the Magellan Straits. Ladrillero states in 1557 (Gay 1852) that in the western mouth of the Strait, until approximately 30 leagues (140 km) towards the east, the northern winds prevailed in the summer and winds from the east and west during the winter.

The same is said by Sarmiento de Gamboa in 1579: 'a very tempestuous wind from the N, NW and W and when the northern tempest wants to finish and the crosswind hails greatly and the cold is intense and with the north it is more temperate' (Sarmiento de Gamboa 1579: 58).

Going back to the analysis of the selected years in both slopes, until the 1553/1554 summer season the winds from the southern quadrant prevail. It must be said that these data have to be taken cautiously, as there are very few records of the wind direction, not enough to generalize this information.

In the data available since 1557/58 and following years (1579/1580; 1582/1583 and 1583/1584) winds from the N and NE quadrant are more frequent, and their percentage oscillates around the 40% in the western sector, reaching in the eastern part a 47.6% in the 1583/1584 summer (Fig. 6).

In 1618/1619, the winds again prevail from the S and SW in both sectors: 50% in the Pacific Ocean zone and 60.9% in the Atlantic Ocean side.

4.4 *Storms*

We analyze this variable because the changes in its frequency may be direct pointers of changes in the atmosphere stability (Ball 1992).

It must be pointed out that the storm occurrence in the Atlantic Ocean slope during the years 1520/1521 and 1525/1526 although with scarce precipitations. In the Pacific Ocean slope, the seasons of 1557/1558 and 1579/1580 were characterized by a significant percentage of storms: 42% and 24% respectively, accompanied in this case by rain and snowfall (Fig. 7).

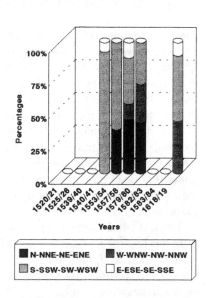

Figure 6. Direction of the winds in the zone of the Magellan's Strait in the 16th Century (in percentages).

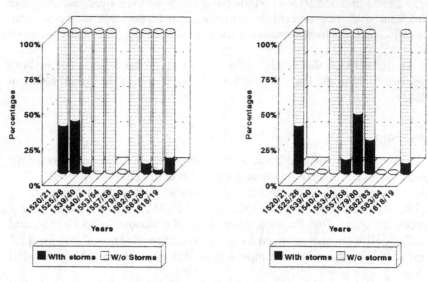

Figure 7. Days with and without storms in the zone of the Magellan's Strait in the 16th Century (in percentages).

5 GLACIERS AND ICEBERGS IN THE 16TH CENTURY

The anomalous phenomena observed in the years 1557/1558 and 1579/1580 were the following: a prevalence of winds from the northern quadrant, presence of days extremely warm for that region and abundant precipitations. These conditions may have favored the calving of the ice blocks that the sailors found in the western area of the Magellan Straits in the second half of the 16th century.

In spite of the detailed descriptions recorded since 1520 until 1557 AD, the analyzed navigation journals do not mention the presence of icebergs when crossing the Magellan Straits. This phenomenon would have not gone unnoticed by sailors at the beginning of the 16th century owing to their potential danger for navigation, and also because of the novelty that those huge masses of floating ice at sea would have been for them.

To confirm the above, a revision was made of other navigation journals from navigators that had crossed the Straits of Magellan during the studied period, such as James Weddell in 1523 (Randier 1960) and Juan Mori from the Expedition of Ximón de Alcazaba in 1535 (Pastells 1910) with the same negative results.

In 1557 the exploratory expedition commanded by Captain Juan de Ladrillero was sent by Governor García Hurtado de Mendoza from Valdivia,

170

recording the presence of large glacierized areas in the Chilean fjords from lat. 51°S to the Magellan Straits. The information about big icebergs floating in the channels and bays that had been passed by suggests changes in the glacier conditions, in relation to the descriptions from the Ulloa expedition, that does not record that phenomenon.

The first mention about the presence of 'snow banks' comes from the captain of one of the ships, Cortés Ojea, who thinking that he was entering the strait introduced the thirty leagues (150 km) through the Channel of the Conception (lat. 50°50'S) (Pastells 1910):

'... en este paraje hallamos muchos pedazos e islillas de nieve que iban nadando sobre el agua las cuales pareció salían de un abra e valle nevado que está al SE deste dho. Puerto Bonifacio e surtos que fuimos bien cerca de tierra en treinta brazas dimos proa en tierra en la cual estaba zabordada una isla de nieve tan dura como peña, que los remos no la podían romper' (Gay 1852: 62); ('...in this place we found many snow pieces and islets that went swimming over the water, and they seemed to have come from a snowed cove and valley that is SE of this said Puerto Bonifacio and when we were provided very close to the land at thirty fathoms we had the bow to land, where it was stranded a snow island as hard as rock, that the oars could not break it').

Looking for the source of the ice, they sailed between twenty and twenty-five leagues towards the inside of the fjord until they found a great bay from where 'much snow came swimming', finding the impressive front of the glacier reaching the sea: '...el camino le halló cerrado de nieve e llegándonos más cerca lo vimos desde el navío estar cerrado de nieve de cerro a cerro; esta nieve era tan alta que enchía hasta la mitad de los cerros... hasta que vimos se remataba en unos tres balcones o cerros altísimos e cuajados de nieve hasta la lengua del agua de los cuales descendía mucha nieve que cuajaba la dicha bahía...' (Gay 1852: 64). ('... the route was closed with snow and getting nearer we saw it from the ship that it was closed with snow from hill to hill; this snow was so high that it filled until the half of the hills... until we saw that it finished in three very high hills and filled with snow that reached the water tongue, and from these hills much snow descended that filled up the said bay...').

There is no doubt, considering the latitude, that Cortés Ojea went into the Strait of the Conception and then through the Canal Ancho until reaching the tongue of one of the glaciers that form the present Dr. Brüggen icefield (Fig. 8).

Ladrillero, commanding another ship, observed the same phenomenon where the Eyre Cove ends (lat. 49°S):

'... donde se acaba entre unas islas nevadas donde hallamos tantas islas de nieve que había algunas que tenían siete estados de alto y del tamaño de un solar y otras menores y más pequeñas que no podíamos pasar aunque el brazo tenía legua y media de ancho y hallándole cerrado dimos la

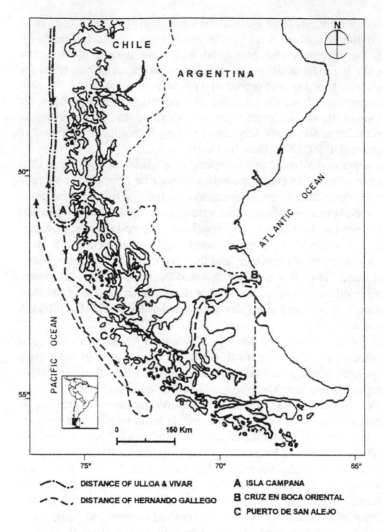

DISTANCE OF ULLOA & VIVAR   **A** ISLA CAMPANA

DISTANCE OF HERNANDO GALLEGO   **B** CRUZ EN BOCA ORIENTAL

C PUERTO DE SAN ALEJO

Figure 8. Course of the expedition of the Captain Juan Ladrillero in 1557.

vuelta'. (1 estado or furlong = 200 m) (Pastells 1910: 345). ('...
where it finishes among some snowed islands where we found so many
snow islands that some of them were as high as seven furlongs and the
size of a manor and others were less big and smaller that we could not
pass although the arm was a league and a half wide, and finding it closed
we turned around').

Mercer (1967) confirms that over the west side of the Brüggen (or Pius
XI) icefield chunks of ice calved from the glaciers within the Eyre Cove,

172

entering from the side at short distance from the fjords head (lat. 49°10'S and long. 73°58'W).

At the southern side, mostly in the Magellan Straits, Ladrillero (1558) also found icebergs, although it is more difficult to locate the sites he describes. He generally alerts navigators about the enormous ice masses calving from the glaciers of the cordillera at the western portion of the strait from the western entrance to lat. 54°S, a phenomenon that he describes thoroughly:

'... si vieren sierras nevadas que caygan sobre el canal por donde fueren que se aparten dellas porque hay en muchas partes donde cae tanta nieve que las tierras tienen sobre sí cinco y seis y siete y ocho y diez líneas de nieve... y según parece debe estar recogida de muchos tiempos y cuando la sierra está muy cargada della quiebra la nieve y viene rodando haciéndose pedazos a cien estados y doscientos y trezientos y mill...y viene con gran ruido a manera de truenos por la sierra abajo y da en el brazo o canal gran multitud della en pedazos como naos como casas y casi tamaños como solares y menores y de seis y de siete y de ocho estados della en el agua y son tan duros como una peña... y en legua y media de brazo no podiamos pasar con un bergantín sin topar con aquellos pedazos andaban encima del agua como islas que algunas tenían tres y cuatro estados debajo del agua y otros tantos encima della y csto cs apartados dcl mar y dc las bahías por los canales que dellas se apartan la tierra adentro hacia la cordillera...' (Pastells 1910: 366). ('... were you to see snowed hills falling in the channels wherever you went, get away from them as there are many parts where it falls so much snow that the lands have above them five and six and seven and eight and ten lines of snow... and it seems it must have been gathered there since a long time and when the hill is too loaded with snow the snow breaks and it comes rolling down breaking into many pieces at one hundred furlongs and two hundred and three hundred and at a thousand... and it comes with a great noise like thunder down the hill and it hits in the branch or channel a great crowd of chunks like ships, like houses and almost the size of manors and smaller and of six and seven and eight furlongs in the water and they are as hard as rocks... and in a league and a half of branch we could not go through with a brig without hitting those pieces over the water like islands that some of them had three and four furlongs underneath the water and the same amount over the water and this is apart from the sea and from the bays through the channels that from them go apart from the inland toward the cordillera...').

This phenomenon is observed by Ladrillero (1558) in the northern coast, between the mouth of the strait and the Otway Cove (lat. 53°S), distance that he estimated at of 35 leagues (175 km). The glaciers he mentioned could be those located at the west half of Riesco Island, that presents many peaks and mountainous groups covered by ice. In the map by

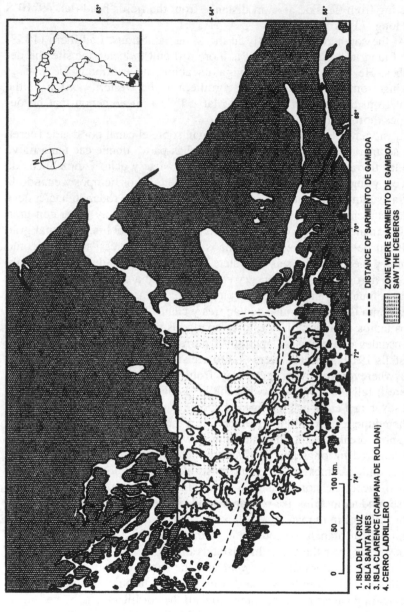

Figure 9. Zone in which Sarmiento de Gamboa saw snow bankings and icebergs during the expedition of 1579-1580.

1. ISLA DE LA CRUZ
2. ISLA SANTA INES
3. ISLA CLARENCE (CAMPANA DE ROLDAN)
4. CERRO LADRILLERO

- - - - DISTANCE OF SARMIENTO DE GAMBOA

ZONE WERE SARMIENTO DE GAMBOA
SAW THE ICEBERGS

0   50   100 km.

174

Cano y Olmedilla that zone is localized as 'Puerto de la Playa Parda', were 'a lot of snow' is registered.

Presently, a small icefield is found over the Ladrillero Hill (1665 m) but the mouth of the glaciers has slightly receded in recent times (Mercer 1967). Besides it can be also seen an icefield near lat. 52°50'S, long. 73°10'W that extends over a surface of 20 by 10 km, that may have been one of the glaciers described by the navigator (Mercer 1967).

In 1579-1580 Sarmiento de Gamboa (1988) observed on February 8th. the same phenomenon as Ladrillero did in this area, when passing by the Isla de la Cruz (Fig. 9), situated SE of the Península de Córdoba according to the map by Cano y Olmedilla (it could be the present Carlos III Island):

'Allí se vieron grandes pedazos de nieve andar sobreaguados por la mar, que salen de las islas nevadas que están al sur de esta Isla de la Cruz [a] tres leguas, y las tormentas del viento despedazan la nieve y la echan y sacan a la mar' ('There, great chunks of snow were seen moving floating on the sea water, that come out from these snowed islands that are south of this Isla de la Cruz at three leagues, and the windstorms tear the snow and they throw it and take it out to the sea').

The icebergs he mentioned could come from the glaciers of the Isla Santa Inés, situated as Sarmiento states, approximately at 15 km or at three leagues from the Isla de la Cruz to the south. According to Mercer (1967), the glaciers from the Isla Santa Inés are the ones less known in South America. Nevertheless, it is known that an icefield covers the greatest part towards the southeast of the island. Mount Wharton (1317 m), in the northern side has glaciers gliding on both slopes, that could be those seen by Sarmiento de Gamboa. Navigating towards the southeast:

'... está un monte muy alto agudo delante de unas sierras nevadas. Este monte es el que llaman las relaciones antiguas la Campana de Roldán. Toda esta bahía de la Campana es cercada de sierras altas y nevadas ...Aquí son las islas nevadas que dicen las relaciones viejas... a la hora que esto escribo, hace calor, estío y calma, sabe muy bien la agua fría con estar cercados de sierras nevadas y balsas de nieve por la mar en 53 grados y dos tercios, donde en muchos meses no suele verse el sol...' (p. 112). ('... there is a very sharp high hill in front of some snowed hills . This mount is that one named as Campana de Roldán in the ancient reports. All this Campana bay is surrounded by high and snowed hills... Here are the snowed islands as the old reports say... at the time I write this, it is hot, summer and calm, the cold water tastes very good being surrounded by snowed hills and snow rafts by the sea at 53 degrees and two thirds, where in many a month the sun is not seen...').

We think that Sarmiento de Gamboa was here referring to Isla Clarence, which highest part, where the Campana Hill comes out, bears many glacier cirques that presently seem to have moderately receded (Mercer 1967). Cano y Olmedilla exactly marks both 'Snowed islands' and the hill Campana de Roldán, distinguishing them clearly from the 'Cordillera with much snow', that undoubtedly is the Darwin Cordillera.

What attracts our attention in these observations from the 16th century is the magnitude of the glaciers of these 'snowed islands', Santa Inés and Clarence Islands, as the glacier snouts reached the sea, with the calving of great chunks of ice.

After all these sightings there are no more references recorded of icebergs proceeding from the glaciers west of the Magellan Straits until the end of the period under study. Ramírez de Arellano (1619) author of this last analyzed report, only writes down snow mountains in the northern coast, at the Puerto de San Joseph (Brunswick Peninsula?) '... towards the north it has some very high hills filled with trees and snow and a tree reef between the port and the cove...'.

Also, no reference was found in the journals of Thomas Cavendish that went through the Magellan Straits in 1587, or those from Jacob Maher (Randier 1960) or in those from J. Le Maire & Guillermo Scouten in 1619 (Manuscript AMNM).

6 DISCUSSION AND SOME CONCLUSIONS

The sighting of icebergs calving from glaciers in the Magellan Straits and in the southern tip of the Chilean Patagonia occurred, in the studied period, exclusively in the second half of the 16th century in the years 1557-1558 and 1579-1580. These unusual event could have been originated owing to a considerable increase in the ice volume in the interval previous to those years, and because of the consequent advance of the glaciers, whose fronts reached the sea. The ice calving from the Brüggen Glacier is common to its behavior even presently, so its condition as a climatic indicator is relative.

It is more important the presence of icebergs in the Strait of Magellan that were not originated in the glacierized area of the Darwin Cordillera, specially on sites where such phenomenon is not presently seen, in the Riesco, Clarence and Santa Inés islands, 'Snowed islands' for the Spanish navigators.

The advance of the ice would suppose an accentuation of the colder conditions during that century, this can be confirmed by data from ships that registered cold and very cold summer temperatures in the Pacific Ocean slope.

These data would coincide with Villalba (1994) results, regarding to the 1520-1670 AD interval as being one of the coldest of the Little Ice Age.

The 1579/1580 season escapes to the general characterization of this cold interval because, even though days with very low temperatures prevail there also warm and very warm days (10.6%). Besides during that

176

spring-summer the more important winds were those from the first quadrant, from the N and NE. This fact may have contributed to the calving of great ice masses from the glaciers of the Magellan Straits, after an advance and a considerable volume increase in the cold years before. Although in the summer period 1557/1558 there was not a temperature rise, the percentage of winds from the northern quadrant increased. These are warmer winds that could have contributed together with the rains of that year, to the calving of glacier ice blocks.

According to Quinn et al. (1992) the first El Niño-Southern Oscillation (ENSO) event of extreme severity is registered in the written records in the years of 1578 and 1579. A link between both phenomena should not be discarded.

REFERENCES

Albo, F. 1837. Diario o derrotero del viaje de Magallanes desde el Cabo de San Agustín hasta el regreso a España de la Nao Victoria, escrito por (1520). In: L. Fernández de Navarrete, (ed.) *Colección de los viajes y descubrimientos que hicieron por mar los españoles*. T. IV. Madrid.

Ball, T.F. 1992. Historical and instrumental evidence of climate: western Hudson Bay, Canadá, 1714-1850. In: R. Bradley & P. Jones (eds), *Climate since A.D. 1500*: 40-73. London and New York: Routledge.

Baron, W. 1982. The reconstruction of eighteenth century temperature records through the use of content analysis. *Climatic Change*, 4.

Barros, J.M. 1981. Expedición al Estrecho de Magallanes en 1553. *Anales del Instituto de la Patagonia* 12: 31-40. Punta Arenas. Chile.

Castañeda Delgado, P., M. Cuesta & P. Hernández 1983. *Alonso de Chaves (1537) Quatri Partitu en Cosmografia practica y por otro nombre llamado Espejo de Navegantes*. Madrid: Instituto de Historia y Cultura Naval.

Cobarrubias Orozco, S. 1611 (1994). *Tesoro de la lengua castellana o española*. Felipe R. Maldonado (ed.). Madrid: Editorial Castalia.

Cortés de Ojea, F. 1852. Relación diaria del Capitán Francisco Cortés de Ojea, Capitán del Bergantín San Salvador, escrita por el escribano Miguel de Goyzueta (1557/1558). In: Gay, C. (ed.) *Historia Física y Política de Chile. T. II, Documentos sobre la historia, la estadística, la geografía*: 55-97. Imprenta de E. Thunot.

Diccionario Marítimo Español 1831. Madrid: Imprenta Real

Endlicher, W. 1995. Climatological Aspects of Landscape Degradation in Patagonia. *Climatology and Air Pollution*. Universidad Nacional de Cuyo, Facultad de Filosofía y Letras: 251-254. Mendoza, Argentina.

Fernández de Navarrete, L. 1837. *Colección de los viajes y descubrimientos que hicieron por mar los españoles*. Madrid.

Jofré de Loaysa, G. 1910. Expedición de Fray García Jofré de Loaysa (1526). In: Pastells, P. (ed.) *El descubrimiento del Estrecho de Magallanes en conmemoración del IV Centenario del Descubrimiento de América*. Madrid.

Jones, P.D. & R.S. Bradley. 1992. Climatic variations over the last 500 years. In: R. Bradley & P. Jones (eds), *Climate since A.D. 1500*: 649-665. London and New York: Routledge.

Gallego, H. 1981. Declaración del Estrecho de Magallanes. Expedición de Francisco de Ulloa. Relación de Hernando Gallego (1553). In: Barros, J.M., Expedición al Estrecho de Magallanes en 1553. *Anales del Instituto de la Patagonia*, 12: 31-40. Punta Arenas, Chile.

Gay, C. 1852. *Historia Física y Política de Chile. Documentos sobre la historia, la estadística, la geografía.* Volume II: 55-97. E. Thunot, París.

Grove, J. 1988. *The Little Ice Age.* London and New York: Routledge.

Hastenrath, S. 1981. The glaciation of the Ecuadorian Andes. Rotterdam: Balkema Publishers.

Ladrillero, J. 1910. Descripción de la costa del Mar Océano desde el sur de Valdivia hasta el estrecho de Magallanes inclusive (1558). Manuscrito en AGI. In: Pastells, P. *El descubrimiento del Estrecho de Magallanes en conmemoración del IV Centenario del Descubrimiento de América.* Document 15: 338-367. Madrid.

Lamb, H.H. 1977. *Climate, Present, Past and Future* Vol. 2. London: Methuen.

Le Roy Ladurie, E. 1990. *Historia del clima desde el año mil.* México: Fondo de Cultura Económica.

Martinic, M. 1972. *Magallanes. Síntesis de Tierra y Gentes.* Edit. Fco. de Aguirre. Buenos Aires.

Martinic, M. 1973. Crónica de las Tierras del Sur del Canal Beagle. Edit. Fco. de Aguirre. Buenos Aires.

Mercer, J. 1965. Glacier variations in southern Patagonia. *Geographical Review* 55: 390-413.

Mercer. J. 1967. *Southern Hemisphere Glacier Atlas.* New York: American Geographical Society.

Mercer, J. 1982. Holocene glacier variations in Southern South America. In: W.Karlén (ed.), *Holocene Glaciers,* Striae, 18: 35-40. Uppsala.

Miller, A. 1976. El clima de Chile. In: W. Schwerdtfeger (ed.), *Climates of Central and South America* 12: 113-145. World Survey of Climatology. Amsterdam-Oxford-New York: Elsevier Sc. Publishing Co.

Morales Padrón, F. 1973. *Historia del Descubrimiento y Conquista de América.* Edit. Nacional. Madrid.

Mori, J. 1910. Relación de la expedición de Ximón de Alcazaba (1535). In: Pastells, P. (1910) *El Descubrimiento del Estrecho de Magallanes en conmemoración del IV Centenario del Descubrimiento de América:* 263. Madrid.

Nagy, A.S. 1991. La legua y la milla de Colón. *Cuadernos Colombinos* 17: 22-41. Universidad de Valladolid, Spain.

Pastells, P. 1910. *El Descubrimiento del Estrecho de Magallanes en conmemoración del IV Centenario del Descubrimiento de América.* Madrid.

Pfister, Ch. 1985. Snow cover, snow-lines and glaciers in Central Europe since the 16th century. Reprinted from: Tooley, M.J. & G.M. Sheail (eds) *The climatic Scenes:* 154-174. Allen & Unwin, London.

Pigafetta, A. 1954. *Primer viaje en torno del globo* (1520). Colección Austral, Espasa-Calpe Argentina, Buenos Aires.

Prieto, María del R. 1993. Reconstrucción del clima de América del Sur mediante fuentes históricas. Estado de la cuestión. Preprint 1º Workshop of Project 341 IGCP/IUGS/UNESCO Southern Hemisphere Paleo- and Neoclimates. A review of the State of Art, Mendoza.

Quinn, W.H. & V.T. Neal 1992. The historical record of El Niño events. In: R. Bradley & P. Jones (eds), *Climate since A.D. 1500*: 623-648. London and New York: Routledge.

Randier, J. 1960. *Hommes et navires au Cap Horn*. Hachette, Paris.

Rosales, D. de 1889. *Crónica General del Reyno de Chile (1666)*. Colección de Historiadores de Chile. Volume XIV. Santiago, Chile: Ercilla.

Rothlisberger, F. 1986. *10000 Jahre Gletchergeschichte der Erde*. Sauerlande, Salzburg.

Sarmiento de Gamboa, P. 1988. *Viajes al Estrecho de Magallanes*. Introducción, transcripción y notas de F. Saravia Viejo. Madrid: Alianza Editorial.

Villalba, R. 1994. Tree-rings and glacial evidence for the Medieval Warm Epoch and the Little Ice Age in Southern South America. *Climatic Change* 30: 1-15.

Vivar, Gerónimo de 1966. *Crónica y Relación copiosa y verdadera de los Reynos de Chile (1553-58)*. Edición facsimilar del Fondo Histórico y Bibliográfico J. T. Medina, Santiago de Chile.

Williams, L.D. & T.M. Wigley 1983. Comparison of evidence for Late Holocene summer temperature variations in the Northern Hemisphere. *Quaternary Research* 20: 286-307.

Manuscripts and Unpublished reports

Anónimo. Estrecho de Magallanes y Mayre. 1618. Manuscrito 25v, Chile. Sección Manuscritos, Biblioteca Nacional, Madrid.

Discurso y derrotero del viaje a los Estrechos de Magallanes y Mayre por el Cosmógrafo Diego Ramírez de Arellano, 1618/1619. Biblioteca Nacional, Madrid. Manuscrito 3019.

Reconocimiento de los Estrechos de Magallanes y San Vicente mandado hacer por S.M. en el Real Consejo de Indias. Partieron de Lisboa en 27 de setiembre de 1618 y llegaron de vuelta a San Lúcar a 9 de julio de 1619. Cabo de dos carabelas Bartolomé García de Nodal y Capitán Gonzalo de Nodal: Cosmógrafo Diego Ramírez: Piloto Juan Manço. Manuscrito Nº 19873, Sección Manuscritos. Biblioteca Nacional, Madrid.

Relación diaria que dio Juan de Areyzaga, clérigo natural de Guipuzcoa, sobre la navegación que hizo la Armada de S.M. de que iba por Capitán el Comendador Loaysa hasta la desembocadura del Estrecho de Magallanes el año 1525. Archivo del Museo Naval de Madrid. 197 Ms, folio 9.

Relación de la Armada de Loaysa por Andrés de Urdaneta. 1525. Archivo del Museo Naval de Madrid, Manuscrito 197, folio 67.

Relación de la navegación del Estrecho de Magallanes de la banda del norte del Navío que volvió a España de los del Obispo de Plasencia, 1539/1541. Archivo del Museo Naval de Madrid. Manuscrito 197, folio 96.

Relación del primer descubrimiento que hizo el General y los Pilotos Antón Pablos y Hernando Lameros por el Golfo de la Santísima Trinidad. Biblioteca Nacional de Madrid, Manuscrito 3102.

Relación del suceso de la Armada del Obispo de Plasencia que salió de España en el año de 1539 compuesta de cuatro navíos para la especiería por el Estrecho de Magallanes... sacada de una carta escrita a Lázaro Aleman por Cristobal Rayzer, 12 de julio de 1541. Archivo del Museo Naval de Madrid, Manuscrito 141, doc. 12.

Viaje que hicieron J. Le Maire y Guillermo Schouten...a la parte austral del Estrecho de Magallanes. Año 1619. Relación diaria que imprimió en Amsterdam Pedro Xeno en 1622. Archivo del Museo Naval de Madrid, Manuscrito 142: 324.

Magnetic measurements used for palaeoenvironmental reconstruction in Pampean loess and soils

10

PAULINA NABEL
CONICET, Museo Argentino de Ciencias Naturales, Buenos Aires, Argentina

ABSTRACT: Magnetostratigraphic and magnetic susceptibility studies have contributed to the understanding of palaeoclimatic and palaeoenvironmental changes of the Pampean Region.

The dating of the Hisisa palaeosol at the end of the Matuyama Chron, and fossil remains and a marine transgression of the Brunhes Chron contribute to the establishment of a chronological framework for past environmental changes.

Most of the magnetic susceptibility measurements carried out in Pampean loess-palaeosols sequences have shown an opposite trend to that observed in the majority of the studied regions of the world. In the studied Pampean Region, the magnetic susceptibility measurements show lower values for palaeosols levels.

To understand this disparity and establish a regional comparative model for the behavior of Pampean palaeosols, the study of magnetic susceptibility from present Pampean soils has been started. The first results of this study suggest that the loessic parent material with frequent volcanic windblown ash contributions is the most important source of magnetic input in the studied area. Topography, which controls the effect of erosion and transport by water, and the presence of sodium alkalinity are other modifing factors of the magnetic susceptibility values.

These studies point out that magnetic susceptibility behavior has to be considered in a regional context before it can be used as a tool in palaeoclimatic studies.

RESUMEN: Los estudios magnetoestratigráficos y de susceptibilidad magnética contribuyen a la comprensión y reconstrucción de los cambios paleoclimáticos y paleoambientales de la Región Pampeana.

La datación del Geosol Hisisa al final del Cron de Matuyama, de restos fósiles y de ingresiones marinas en el Cron Brunhes, contribuyen a establecer una estructura cronológica para los cambios ambientales del pasado en la región.

Por otra parte, la mayor parte de las mediciones de susceptibilidad magnética realizadas hasta el presente en las secuencias de loess y paleosuelos pampeanos, han evidenciado una tendencia opuesta a la registrada mayoritariamente en secuencias equivalentes de otras partes del mundo. En la Región Pampeana estudiada, los niveles de paleosuelos evidencian valores menores de susceptibilidad magnética que los sedimentos loéssicos intercalados.

Con el fin de analizar los motivos de esta disparidad, como así también establecer un modelo regional comparativo para el comportamiento de los paleosuelos pampeanos, se inició el estudio de la susceptibilidad magnética de los suelos actuales en la región. Los primeros resultados de este estudio sugieren que el material loéssico parental con frecuentes contribuciones de cenizas volcánicas es la fuente más importante en la señal magnética. La topografía que controla la erosión y circulación de agua y la alcalinidad sódica son factores, entre otros, que modifican los valores de susceptibilidad magnética.

Surge de este estudio la necesidad de considerar el comportamiento de la susceptibilidad magnética en un contexto regional antes de utilizar estos valores como herramienta en los estudios paleoclimáticos.

1 INTRODUCTION

The purpose of this paper is to summarize results of magnetic measurements in the study of the Plio-Pleistocene records of the Buenos Aires Province, Argentina, emphasizing both the contribution to the palaeoclimatic and palaeoenvironmental knowledge of the region, and the difficulties of carrying forward such studies.

Buenos Aires Province is located at the eastern central region of Argentina. It is an area of some 300,000 km^2 most of which belongs to the Pampean region, characterized by an extensive flat plain composed of loess and loesslike sediments (Fig. 1).

The predominant mineralogical composition of the loessic sediments is of volcaniclastic nature. Its main source is the Andean Cordillera, developed as a belt along the western boundary of the country (Teruggi 1957). Loess and loesslike sediments extend to the eastern and northeastern regions of the country, along more than 1,200,000 km^2 supplied by southwestern winds, mainly by dust storms or by direct volcanic ashfalls (Zárate & Blasi 1993).

Magnetostratigraphic studies have given a chronological framework to the loess-palaeosols sequences and the environmental changes recorded in them, i.e. sea level changes and extinct mammal remains as well as paly-

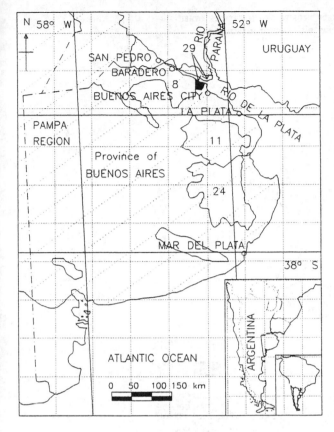

Figure 1. Location map.

nological records. Magnetic susceptibility studies of the loess-palaeosol sequences reflect past climatic changes.

The areas of loess deposition, like the Pampean region, provide one of the best deposits to study the records of climatic changes in continental environments. Nevertheless, the scarcity of natural exposures that characterize this region makes their study difficult.

2 MAGNETOSTRATIGRAPHY

The first aim of this magnetic study of Pampean Quaternary sequences was to determine the age and time span of their deposition.

In the early 1980's, magnetostratigraphy was used to assign an absolute age to the Pampean sediments (Nabel & Valencio 1981) (Fig. 2). At that time, radiocarbon was the only absolute dating tool used to establish a chronologic framework for these sequences. The unselective use of this

183

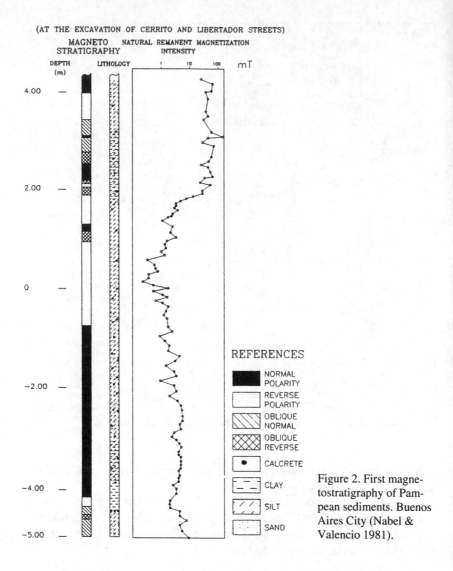

Figure 2. First magnetostratigraphy of Pampean sediments. Buenos Aires City (Nabel & Valencio 1981).

methodology together with its inherent limitations suggested an absolute age of the Pampean sediments of no more than a few tens of thousand of years.

The first magnetostratigraphic information moved back the basal age of these deposits to more than two million years, to the Plio-Pleistocene, as had been originally proposed by Ameghino as early as 1889.

This experience is a good example of the risk of the unselective use of dating methods which although they can provide excellent results, should not be extended beyond their useful limit and can only be used in con-

184

junction with geological and stratigraphic criteria. An interdisciplinary approach and careful field work are key points to produce accurate and useful magnetostratigraphies.

The recognition and proper interpretation of geological structures such as palaeovalleys, caves, palaeosols and calcretes are not only useful for the palaeoclimatic and environmental interpretation of the studied profiles and the possibility of dating these events, but their recognition is necessary for adequate sampling and the production of reliable results.

The Hisisa Geosol from the Ensenada Formation, located at the end of the Matuyama Chron (Nabel et al. 1993), has been recognized as a result of this interdisciplinary approach.

The time setting of this palaeosol suggests an environmental change in the Pampean region at the end of the Matuyama Chron, shortly before the beginning of the Brunhes Chron. The top of the Hisisa Geosol is characterized by the absence of an A Horizon and the presence of an erosional surface, suggesting a deflation period at this time.

This palaeosol has been recognized from several studies in Buenos Aires City (Nabel & Spiegelman 1988), in the Paraná River cliffs from the northeastern section of the province (Nabel et al. 1990, 1993) and in the surroundings of the city of La Plata (in progress), as well as a synthesis of sedimentological and magnetostratigraphic information (Riggi et al. 1986, Bobbio et al. 1986, Bidegain 1991) from the regions of La Plata and Mar del Plata (Orgeira 1987, Ruocco 1989) (Fig. 3).

Loess deposits characterize the Brunhes onset suggesting a change to dryer and probably cooler environmental conditions (Nabel 1993). There are currently in progress magnetostratigraphic studies of the Quaternary sea transgressions interbedded in Pampean sediments. Up to now, the studied marine sediments are younger than the Brunhes-Matuyama boundary, ca. 780 ka ago.

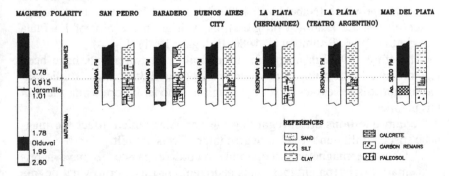

Figure 3. Hisisa Geosol at the end of Matuyama Chron in the eastern Buenos Aires Province.

3 MAGNETIC SUSCEPTIBILITY MEASUREMENTS OF PALAEO-SOLS

The palaeosols interbedded in Pampean loess are indicators of warmer environmental conditions during the Quaternary. Although the distinction between loess (drier conditions) and palaeosols (moister conditions) is an important tool for deciphering continental climatic history, the scarcity of natural exposures that characterizes the flat Pampean Region, makes their identification particularly difficult. The recognition of palaeosols sometimes requires the application of many and complicated methods.

As a result of this interdisciplinary approach, it was shown that the presence of a palaeosol is always associated with changes in magnetic susceptibility values (Nabel et al. 1990, 1993). This behavior means that magnetic susceptibility measurements can be used as an indicator of palaeosols, as observed by many researchers in previous works (Liu 1985a, Heller et al. 1985, Heller & Liu 1986, Kukla et al. 1988, Zhou et al. 1990, Maher & Thompson 1991, 1992, Liu et al. 1992, Evans & Heller 1994, among others).

Although the magnetic susceptibility measurements provide a sensitive method of discriminating between loess and palaeosols in common with European and Chinese deposits, most of the measurements carried out on the palaeosols of the Pampean region have shown the opposite correlation to those reported from European and Chinese palaeosols. In the latter the susceptibility values are commonly higher in the palaeosols than in the interbedded loess, while in the Pampean region the magnetic susceptibility measurements recorded so far show lower values for palaeosol levels. Recently a similar behavior has been reported in loess-palaeosols sequences from Alaska (Beget & Hawkins 1989) and from Siberia (Rutter et al. 1995).

This trend has been recorded in various studies from Buenos Aires City (Nabel & Spiegelman 1988), from the Paraná River cliffs from the NE part of the province (Nabel et al. 1990, 1993), and from the surroundings of the city of La Plata (in progress), as well as in the city of La Plata (Riggi et al. 1986, Bobbio et al. 1986, Bidegain 1991) (Fig. 4).

Therefore, although magnetic susceptibility measurements have been recommended as a proxy climatic indicator (Liu 1985b), these studies indicate that it is necessary to set up a regional model before using them in this manner.

Certain questions arise, regarding the variables which affect the magnetic susceptibility and to what degree their effects are felt.

1. Does the magnetic susceptibility record of palaeosols have a primary origin (parent loess plus pedogenic process) or to what degree does it reflect a superimposed history of different effects related to the climatic history of the region?

186

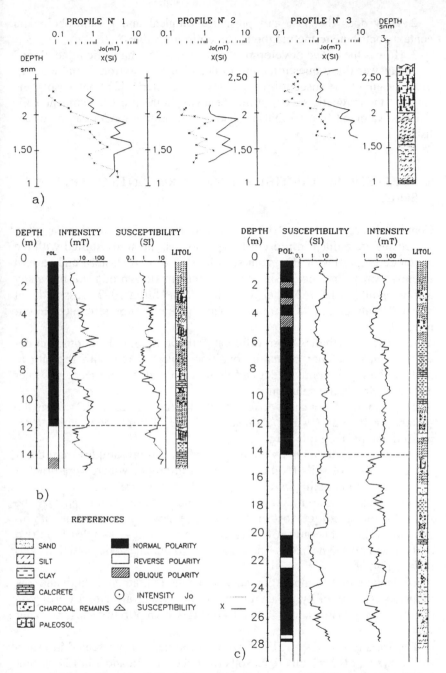

Figure 4. Magnetic susceptibility of Pampean palaeosols. a) Cazadores profiles (Nabel & Spiegelman 1988), b) Baradero section (Nabel et al. 1993), c) La Plata, adapted from Riggi et al. 1986 and Bobbio et al. 1986).

187

2. How does the past and present geological and geomorphological context contribute to the magnetic susceptibility values?

3. How is the above development reflected in the magnetic mineralogy?

To answer these questions and to decipher the environmental and climatic origin of the susceptibility values of the interbedded palaeosols, the study of the magnetic susceptibility behavior in the present Pampean soils will be examined first so that a reliable comparative model can be established.

4 MAGNETIC SUSCEPTIBILITY MEASUREMENTS OF PRESENT SOILS

Convinced that the clues to the magnetic signature of the studied Pampean palaeosols are hidden in present Pampean soils, the study started with the magnetic properties of Pampean soils located in northeastern and eastern Buenos Aires Province. Significative information produced by many authors (Mullins 1977, Maher & Taylor 1988, Hilton 1987, Fassbinder et al. 1990, Valy et al. 1987, among others) have been key references to our research.

At first, soil samples from different geomorphologic locations, but developed under similar climatic conditions, were taken with the aim to analyze the topographic effect and its relation with the presence of water and other circulating solutions.

The samples belong to different edaphic dominiums (ED) from the 'Pampa ondulada' ('wavy Pampa') and the 'Pampa deprimida' ('depressed Pampa') regions (Fig. 1).

Different measurements were performed on natural samples from the A and B horizons of soils (Nabel & Petersen in press), which were dried at room-temperature taking care to preserve their structure.

Directional measurements were carried out with the aim to analyze the soil magnetic stability. Most of them have been well grouped, but with inclination values generally low (Fig. 5a). Only in a few samples from two of the ten analysed horizons, the dispersion of the magnetic directions was higher (Fig. 5b).

To analyze the magnetic domain and magnetic grain size, high and low frequency susceptibility measurements were carried out with a Bartington dual frequency instrument (470 and 4700 Hz) susceptibility sensor (Maher 1986).

The variation of susceptibility with profile depth is presented in Fig. 6 and shows that B horizons from soils from the 8a, 11a and 24a ED present increased susceptibility values in relation to their A horizons, whereas soils from the 29a and 8c ED show the opposite trend.

The soils from the well drained 'Pampa ondulada' region, the 8a ED,

188

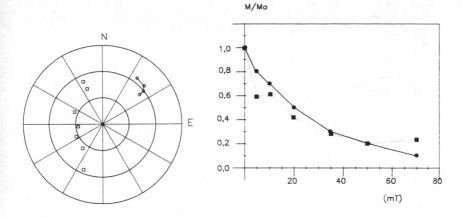

Figure 5. Directional behavior after AF treatement.

have the maximum values of susceptibility, which increases with depth (Fig. 6a).

Soils located on a flat plain and on lower topographic levels (the 11a ED) show susceptibility values about half that of the above soils and they also show an increase in susceptibility with depth (Fig 6c).

Soils located on a flat-concave plain (the 24a ED) show similar susceptibility values for both A and B horizons.

Soils from the 8c ED are waterlogged. They have the lowest susceptibility values recorded (Fig. 6b) but a relative increase of susceptibility in the top soil. This type of soil is located in the depressed areas of the 'Pampa ondulada' region. Soils from the 29a ED belong to the delta region. Susceptibility values also show an increase in the top soil (Fig. 6e).

In all cases, the magnetic intensity values show the same behavior as susceptibility, suggesting that both are influenced by the same sedimentological processes (Fig. 7).

It is possible that the increase of magnetic susceptibility values with depth in the well drained soils could be related with its volcanoclastic origin (Teruggi 1957). The first results of this study (Nabel 1996) suggest that the main magnetic mineral input in this region came from the loessic parent material with frequent wind-blown volcanic ash contribution.

In studies in progress (Nabel & Petersen in press, Nabel 1996, Nabel & Morrás 1996) in the same climatic region, different factors modify the magnetic susceptibility values. In addition to the edaphic processes and composition of the parent material, topography which influence the degree of soil drainage, erosion and transport by water, and sodium alkalinity have great importance on them.

189

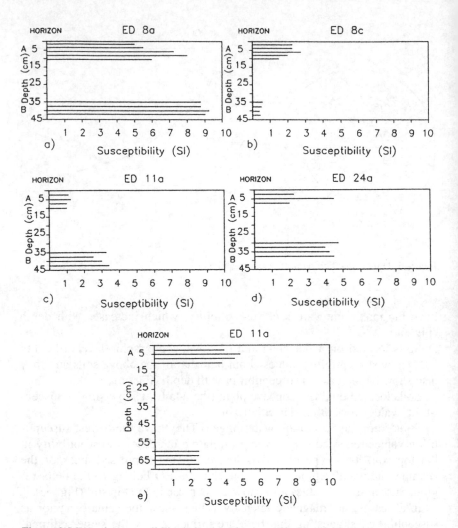

Figure 6. Magnetic susceptibility behavior of present Pampean soils.

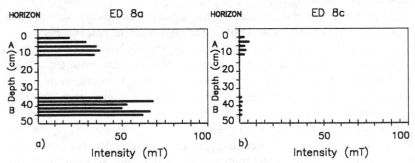

Figure 7. Magnetic intensity behavior of present Pampean soils.

190

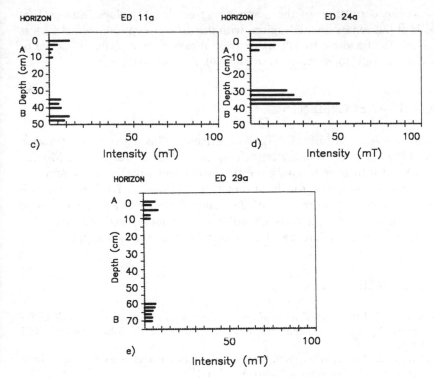

Figure 7. Continued.

Local and regional characteristics related to sedimentological, hydrological, geomorphological, climatic and biological features, as well as anthropogenic effects, give rise to particular correlations that have to be recognized.

5 CONCLUSIONS

The contribution of the magnetic measurements to the palaeoclimatic and palaeoenvironmental knowledge of the Pampean Region can be considered in three different ways:

1. Dating of environmental change indicators such as palaeosols, fossil remains and marine transgressions on basis of magnetostratigraphic studies. The magnetic remanence dating of these records contributes to the establishment of a chronological framework necessary in order to understand the frequency and trend of past environmental changes.

2. The magnetic susceptibility studies of the loess/palaeosol sequences contribute to the knowledge of the climatic changes of this region.

191

3. The correlation of the magnetic susceptibility values from present soils of the same climatic region, with the different environmental factors which affect and modify them, provide results related to the influence of sedimentology, topography, water and alkali solutions.

ACKNOWLEDGEMENTS

I am very grateful to Prof. Nikolai Petersen for his generous cooperation and his kind offer of analysis techniques, to Daniel Vargas and Monika Hanesch for their help during the field work and with the measurements. To Munich University for the use of their equipment. To Prof. Friedrich Heller for the critical reading of the manuscript. To CONICET (Consejo Nacional de Investigaciones Científicas y Técnicas) and the Argentine Museum of Natural Sciences, I am grateful to both for their support.

REFERENCES

Ameghino, F. 1889. *Contribución al conocimiento de los mamíferos fósiles de la República Argentina*. Academia Nacional de Ciencias de Córdoba. Córdoba, 1027 pages.

Beget, J.E. & D.B. Hawkins 1989. Influence of orbital parameters on Pleistocene loess deposition in Central Alaska. *Nature* 337: 151-153.

Bidegain, J.C. 1991. Sedimentary development, magnetostratigraphy and sequence events of the Late Cenozoic in Entre Ríos and surrounding areas in Argentina. Ph.D. Dissertation, Stockholm University, Sweden. 128 pages. Unpublished.

Bobbio, M.L., M. Devicenzi, M.J. Orgeira & D.A. Valencio 1986. La magnetoestratigrafía del 'Ensenadense' y 'Bonaerense' de la Ciudad de La Plata: su significado geológico. *Revista de la Asociación Geológica Argentina* 41(1-2): 7-21. Buenos Aires.

Evans, M.E. & F. Heller 1994. Magnetic enhancement and paleoclimate: study of a loess/palaeosol couplet across the loess plateau of China. *Geophysical Journal International* 117: 257-264.

Fassbinder, J.W., H. Stanjek & H. Val 1990. Occurrence of magnetic bacteria in soil. *Nature* 343: 161-163.

Heller, F. & T. Liu 1986. Paleoclimatic and sedimentary history from magnetic susceptibility of loess in China. *Geophysical Research Letters* 13(11): 1169-1172.

Heller F., B. Meili, W. Junda, L. Huamei & T. Liu 1985. Magnetization and Sedimentation History of Loess in Central Loess Plateau of China. In: Liu Tungsheng (ed.), *Aspects of Loess Research*: 147-163. China Ocean Press.

Hilton, J. 1987. A simple model for the interpretation of magnetic records in lacustrine and ocean sediments. *Quaternary Research* 27: 160-166.

Kukla G., F. Heller, Liu Xiu Ming, Xu Tong Chun, Liu Tungsheng & A.Z. Sheng 1988. Pleistocene climates in China dated by magnetic susceptibility. *Geology* 16: 811-814.

Kukla G., Z.S. An, J.L. Melice, J. Gavin & J.L.Xiao 1990. Magnetic susceptibility record of Chinese Loess. *Trans. of the Royal Soc. of Edinburgh: Earth Sciences* 81: 263-288.

Liu, T. 1985a. *Loess and the environment*. China Ocean Press: 1-251.

Liu, T. 1985b. The loess-paleosol sequence in China and climatic history. *Episodes* 8 (1): 21-28.

Liu X.M., J. Shaw, T. Liu, F. Heller & B. Yuan 1992. Magnetic mineralogy of Chinese loess and its significance. *Geophysical Journal International* 108: 301-308.

Maher, B. 1986. Characterization of soils by mineral magnetic measurement. *Physics of the Earth and Planetary Interiors* 42: 76-92.

Maher, B. & R. Taylor 1988. Formation of ultrafine-grained magnetite in soils. *Nature* 336: 368-370.

Maher, B. & R. Thompson 1991. Mineral magnetic record of Chinese loess and palaeosols. *Geology* 19: 3-6.

Maher, B.A. & R. Thompson 1992. Paleoclimatic significance of the mineral magnetic record of the Chinese loess and palaeosols. *Quaternary Research* 37: 155-170.

Mullins, C. 1977. Magnetic susceptibility of the soil and its significance in Soil Science. *Journal of Soil Science* 28: 223-246.

Nabel, P. 1993. The Brunhes-Matuyama boundary in Pleistocene sediments of Buenos Aires Province, Argentina. *Quaternary International* 17: 79-85.

Nabel, P. 1996. Aspectos ambientales registrados en suelos de la región pampeana, identificados por sus parámetros magnéticos. *Revista de la Asociación Geológica Argentina* 51(2): 147-155. Buenos Aires.

Nabel, P. & D.A.Valencio 1981. La magnetoestratigrafía del Ensenadense y Bonaerense de la Ciudad de Buenos Aires: su significado geológico. *Revista de la Asociación Geológica Argentina* 36: 7-18. Buenos Aires.

Nabel, P. & A. Spiegelman 1988. Caracterización sedimentológica y paleomagnética de una sección del Pampeano, en el subsuelo de la Ciudad de Buenos Aires. *Revista de la Asociación Geológica Argentina* 43: 224-230. Buenos Aires.

Nabel, P. & H.J. Morrás 1996. Influencia de la litología y la pedogénesis en la susceptibilidad magnética de suelos de la pampa ondulada, Argentina. *Congreso Latinoamericano de Suelos*. CD Rom publication.

Nabel, P. & N. Petersen in press. Magnetic signatures of Pampean soils, a case study. *Southern Hemisphere Paleo and Neoclimates*. Project 341/IGCP/IUGS/UNESCO (1995).

Nabel, P., G. Machado & A. Luna 1990. Criterios diagnósticos en la estratigrafía de los 'Sedimentos Pampeanos' del NE de la Provincia de Buenos Aires. *XI Congreso Geológico Argentino*, Actas II: 121-124.

Nabel P., C. Camilión, G. Machado, A. Spiegelman y L. Mormeneo 1993. Magneto y litoestratigrafía de los sedimentos pampeanos en los alrededores de la Ciudad de Baradero Provincia de Buenos Aires. *Revista de la Asociación Geológica Argentina* 48(3-4): 193-206. Buenos Aires.

Orgeira, M.J. 1987. Estudio paleomagnético de sedimentos del Cenozoico tardío de la costa atlántica bonaerense. *Revista de la Asociación Geológica Argentina* 42: 362-376. Buenos Aires.

Petersen, N., T. von Dobeneck & H. Vali 1986. Fossil bacterial magnetite in deep-sea sediments from the South Atlantic Ocean. *Nature* 320: 611-615.

Riggi, J.C., F. Fidalgo, O.R. Martínez & N.E. Porro 1986. Geología de los 'Sedimentos Pampeanos' en el partido de La Plata. *Revista de la Asociación Geológica Argentina* 41(3-4): 316-333. Buenos Aires.

Ruocco, M. 1989. A 3 Ma paleomagnetic record of coastal continental deposits in Argentina. *Paleogeography, Paleoclimatology, Paleoecology* 72: 105-113. Elsevier.

Rutter, N.W., J. Chlachula & M.E. Evans 1995. Magnetic susceptibility and remanence record of the Kurtak loess, southern Siberia, Russia. *Terra Nostra*: 235. Abstracts XIII INQUA Congress, Berlin 3-10 August 1995.

Teruggi, M.E. 1957. The nature and origin of Argentine loess. *Journal of Sedimentary Petrology* 27(3): 322-332.

Thompson, R. & F. Oldfield 1986. *Environmental Magnetism*. Allen & Unwin, London. 219 pages.

Vali, H., O. Forster, G. Amarantidis & N. Petersen 1987. Magnetotactic bacteria and their magnetofossils in sediments. *Earth and Planetary Science Letters* 86: 389-400.

Zárate, M. & A. Blasi 1993. Late Pleistocene-Holocene eolian deposits of the Southern Buenos Aires Province, Argentina: a preliminary model. *Quaternary International* 17: 15-20.

Zhou, L.P., F. Oldfield, A.G. Wintle, S.G. Robinson & J.T. Wang 1990. Partly pedogenic origin of magnetic variations in Chinese loess. *Nature* 346: 737-739.

Magnetic properties and environmental conditions: study of a palaeosol of the Chaco-Pampean plain (Argentina)

11

CARLOS A. VÁSQUEZ
Depto. de Ciencias Geológicas, Facultad de Ciencias Exactas y Naturales, Universidad de Buenos Aires; CONICET, Argentina.

ANA MARIA WALTHER, MARIA JULIA ORGEIRA & INES DI TOMASSO
Consejo Nacional de Investigaciones Científicas y Técnicas (CONICET). Laboratorio de Paleomagnetismo, Universidad de Buenos Aires, Argentina

MARIA SUSANA ALONSO & HORACIO F. LIPPAI
CONICET, Laboratorio de Química Geológica y Edafológica (LAQUIGE) y Centro de Investigaciones de Recursos Geológicos (CIRGEO). Buenos Aires, Argentina.

JUAN FRANCISCO VILAS
Depto. de Ciencias Geológicas, Facultad de Ciencias Exactas y Naturales, Universidad de Buenos Aires, CONICET. Argentina

ABSTRACT: The objective of this contribution is to study the magnetic properties in a palaeosol from the Chaco-Pampean plain (Buenos Aires, Argentina), in order to identify changes in the magnetic carriers due to pedogenetic processes. In addition, non magnetic analysis were carried out: chemical analysis of ferrous and ferric iron, and X-ray diffraction. The results are compared with those published from China region.

The analized magnetic properties of the Chaco-Pampean plain palaeosol are different from the Chinese palaeosols. Due to parental dissimilarity in both areas, the pedogenetic minerals originated under mild conditions are different. While in the Chinese loess the temperate conditions generate SP magnetite, in the Argentine loess an antiferromagnetic fraction is detected.

RESUMEN: El objetivo de esta contribución es el estudio de las propiedades magnéticas de sedimentos pertenecientes a un paleosuelo de la Llanura Chaco-Pampeana (Buenos Aires, Argentina), con el fin de identificar fluctuaciones en los portadores magnéticos originadas por los procesos pedogenéticos. Por otra parte, determinaciones no magnéticas fueron llevadas a cabo: análisis químicos de óxido férrico y óxido ferroso y espectrometría por difracción de rayos X. Los resultados obtenidos fueron comparados con los obtenidos por otros autores en loess y paleo-suelos de China.

Las propiedades magnéticas analizadas de los paleosuelos de la llanura Chaco-Pampeana son diferentes a aquellas observadas en los paleosuelos del loess chino. Las diferencias en el material parental de los paleosuelos de ambas áreas podrían ser la causa de la existencia de distintos minerales pedogenéticos hallados en ambas regiones. Condiciones ambientales benignas originan en el loess chino paleosuelos portadores de magnetita SP, en tanto en el loess argentino se detecta una fracción antiferromagnética.

1 INTRODUCTION

Even though the mineral magnetic carriers are a minor fraction in rocks and sediments (O'Reilly 1984), their behavior under chemical changes make them excellent tracers for environmental changes (Verosub et al. 1993).

The study of soils and palaeosols by mineral magnetic properties began in the 1970s (Mullins 1977). It is not until in the 1990s when techniques and equipment reached a suitable degree of development. Research carried out in sequences of loess and palaeosols in northern China (Xining, Xifeng region; Hunt et al. 1995, Banerjee et al. 1993, Xiuming et al. 1993, among others) showed that it is possible to infer palaeoclimatic changes by measuring different magnetic properties.

The Chaco-Pampean plain is one of the most extensive loess regions in the world. There are many loess and palaeosols sequences well exposed all over the region. The parental material of these sediments differs notably from the Chinese loess. The most important differences are the following (Frenguelli 1955):

a) volcanic material is not present in the Chinese loess, but it is abundant in the Chaco-Pampean loess;

b) the quartz content in the Chinese loess is much higher than in the Chaco-Pampean loess;

c) the detritic fraction in the Chinese loess derives from a strong mechanic weathering, whereas the Chaco-Pampean loess was produced by a strong hydrolic decomposition. Besides, the Chaco-Pampean loess shows high solubility in acidic media.

Therefore, the expected physico-chemical response driven by environmental changes on both sediments could be different as well as the palaeosol magnetic signal. As a matter of fact, the magnetic susceptibility variation follows different patterns: it decreases in the Chaco-Pampean loess (i.e. Nabel 1996, Orgeira et al. 1997), and increases in the Chinese loess (i.e. Heller et al. 1991, Verosub & Roberts 1995).

The objective of this work is to study the magnetic properties in a palaeosol from the Chaco-Pampean region in order to identify changes in the magnetic carriers due to pedogenetic processes. Results are compared with those from China.

196

Changes in the magnetic carriers can be recognized by different methods. In the present contribution the following techniques were used:

1. Magnetic methods: magnetic susceptibility, isothermal remanent magnetization, anhysteric remanent magnetization, low temperature demagnetization and the Lowrie test (Lowrie 1990);

2. Chemical analysis: by this method it is possible to quantify fluctuations in bulk iron and ferric fractions along the profile;

3. X-ray diffraction (XRD): magnetic carriers can be identified by this technique. However, if a pre-concentration process is not carried out, a direct identification of very scarce components is very difficult.

The studied section is located in a building excavation in City of Buenos Aires (lat. 34°30'S, long. 58°20'W), in the Fray Justo Santa María de Oro street, where a loess-palaeosol sequence outcrops. The present results include the detailed study of a 2 m-thick palaeosol (1 m over sea level; 4 m under street level). The exposed sedimentary sequence is correlated with the Buenos Aires Formation (Middle to Late Pleistocene; Brunhes magnetic age <0.7 Ma). No anthropogenic modifications were found within the sedimentary sequence.

Twelve levels were sampled. The lower three ones belong to a lower loess, from level 4 to level 10 to a palaeosol, and the top levels 11 and 12 to an upper loess. Bag samples (approximately 100 g) were taken at 15 cm intervals from a smoothed fresh face. In the laboratory, subsamples from the bags were weighed and packed into plastic boxes for magnetic measurements.

2 MAGNETIC METHODS

2.1 Magnetic susceptibility

Double frequency (470Hz and 4700Hz) susceptibility was measured using a Bartington susceptibilimeter MS2 (Fig. 1a). The susceptibility values range in both frequencies between 1.5E-7 m^3/Kg to 5E-7 m^3/Kg. The double frequency measures do not show any significant difference; by these means, the superparamagnetic magnetite fraction (d<30E-9m) is not detected.

On the other hand, the decreasing trend of susceptibility (see Fig. 1a) in the palaeosol is very different from that observed in the Chinese palaeosols, in which a conspicuous increase has been observed (Hunt et al. 1995, among others).

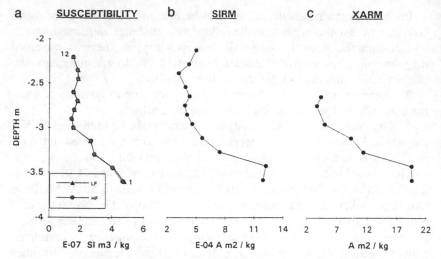

a SUSCEPTIBILITY   **b** SIRM   **c** XARM

Figure 1. Profiles of a) magnetic susceptibility at two frequencies (470 Hz and 4700 Hz), b) Saturation of isothermal remanent magnetization (SIRM), and c) anhysteric susceptibility (X_{ARM}) in a loess-palaeosol section of the Buenos Aires Fm (Argentina).

2.2 *Isothermal Remanent Magnetization (IRM) and Anhysteric Remanent Magnetization (ARM)*

The IRM is the magnetization acquired at constant temperature in a magnetic field. When this field is varied, a curve of acquired remanence vs. applied magnetic field is obtained. The determination of the magnetization as a function of the applied field is characterized by the following parameters: Saturation remanent magnetization (SIRM) and Coercivity of remanence (Bcr).

IRM curves were obtained at room temperature and the higher applied field was 2,3T. In addition, back field measurements were done in all sampled levels.

Anhysteric remanent magnetization (ARM) is the magnetization acquired when a dc bias field was applied to the sample in presence of a decreasing alternating field.

The applied dc bias field in this sequence was 0.1mT and the alternating field peak 100mT. The results are presented in terms of its susceptibility (X_{ARM}).

The SIRM and X_{ARM} from different levels show a similar behavior (Figs 1b and 1c). The dropping of both parameters in the palaeosol (similarly to susceptibility behavior) allow to infer a lower proportion of ferrimagnetic minerals in this level.

The isothermal remanent magnetization (IRM) acquisition curves (Fig. 2), show a behavior lead by magnetite and/or titanomagnetite (Dankers 1978).

198

On the other hand, the S ratio (IRM-300 mT/ SIRM) (Fig. 3a) was calculated. Using this parameter, the relationship ferri/antiferromagnetic is inferred. If it is far from 1, an antiferromagnetic fraction is present, but if it is close to 1, the magnetic mineralogy is mainly ferrimagnetic. Figure 3a shows that in the palaeosol levels the value keeps nearly to 0.8, whereas in

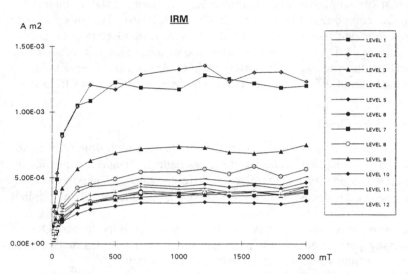

Figure 2. Isothermal remanent magnetization curves of some representatives samples of loess and palaeosol levels in a loess-palaeosol section of the Buenos Aires Formation (Argentina).

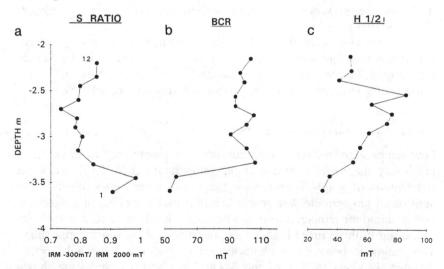

Figure 3. Profiles of a) S ratio (IRM-300mT/SIRM), b) remanence coercivity (Bcr), c) mean destructive field (H1/2$_I$) in a loess-palaeosol section of the Buenos Aires Fm (Argentina).

199

the loess levels it becomes close to 1. On analyzing S ratio fluctuations it is possible to infer that an antiferromagnetic fraction associated with a ferrimagnetic one could be present in the palaeosol section.

Finally, the SIRM/X ratio was calculated for the 12 sampled levels. Resulting values lay between 1.7 to 3.5 KA/m. This range is in agreement with that of samples bearing multidomain magnetite (MD) as their main magnetic component (Thompson & Oldfield 1986). The inferred MD magnetite is consistent with the results presented by Orgeira et al. (1997) who detected MD magnetite in a sequence of loess and palaeosols, close in location and age to the one described here. Those results were obtained upon the basis of different experiences, i.e. detection of Verwey transition.

2.3 *Coercive Fields*

When an alternating field which amplitude changes with time (AF) is applied to a ferromagnetic material, it suffers demagnetization. A previously magnetized sample is demagnetized at different values of maximum field. A magnetization decaying is then observed as a function of the applied field peak.

Ferromagnetic minerals have characteristic coercivity forces. One of the driving parameters of this phenomenon is the mean destructive field ($H1/2_I$): the field value AF which reduces IRM by half.

$H1/2_I$ fluctuations (Fig. 3c) show an increase in the palaeosol levels. This increment, together with the magnitude of the parameter, indicates the presence of a mixture of low and high coercivity components. In short, it shows the presence of a ferri/antiferromagnetic magnetic mixture (Dankers 1978).

Similar results were obtained for the Bcr parameter (Fig. 3b). Values are around 60mT in loess whereas they increase to 100mT in the palaeosol. Again, these values are indicative of a mixture of ferri/antiferromagnetic minerals in the palaeosol layers (Dankers 1978).

2.4 *Low temperature measurements*

Low temperature measurements consisted in submerging in liquid air a previously magnetized sample (room temperature IRM 2.3T). Then the fluctuations of magnetization were measured under null magnetic field until room temperature was reached. Initial measurements of magnetization at liquid air temperature in the palaeosol levels showed a notable increase, up to three times higher than measurements at room temperature. This magnetic behavior is characteristic of goethite, α-FeOOH, orthorhombic (Dekkers 1988, Ozdemir & Dunlop 1996). Examples are shown in Figure 4.

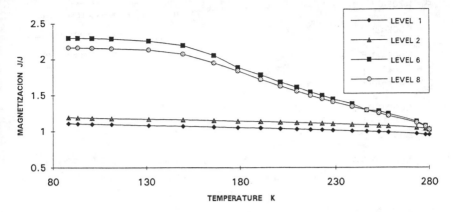

Figure 4. Low temperature behavior of loess levels 1 and 2, and palaeosol levels 6 and 8, of a loess-palaeosol section of the Buenos Aires Fm (Argentina).

This initial magnetization increment is negligible in the loess samples. On entering the palaeosol, going from bottom to top, the low temperature magnetization values increase up to three times the initial value. This behavior could indicate the presence of goethite in the palaeosol with its concentration growing up from bottom to top. According to this experience, goethite was not detected in the loess layers.

2.5 *Lowrie Test (Lowrie 1990)*

In samples previously magnetized, the high temperature experience produces a demagnetization depending on the blocking temperature of each of the magnetic minerals present. On this basis, the coercivities affected by temperature allow the characterization of the analyzed material.

Three different magnetic fields along mutually orthogonal axis were applied to the sample. Then a step by step thermal demagnetization is carried out in atmospheric conditions.

In this study a modification of the original test was used (Lowrie 1990); the intensities of the applied fields were 0.3T, 2.3T and 4T in order to separate low, middle and high coercivity components. Then the thermal demagnetization was progressively carried out when possible from 100°C to 700°C.

During the warming up, mineralogical transformations were inferred for most of the specimens. This was shown by a noteworthy increase of the residual remanent magnetization.

Despite these transformations, some observations could be done. Firstly, minerals of low, middle and high coercivity were found throughout the whole sequence. The higher intensities of remanent magnetization

201

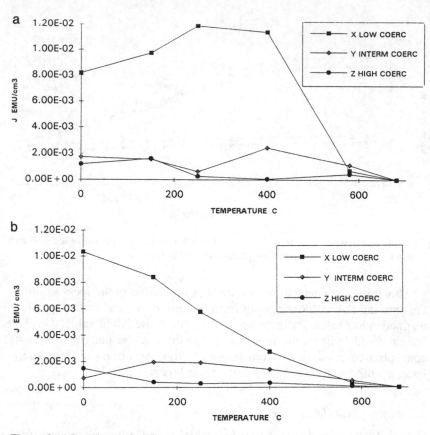

Figure 5. a) Lowrie test in a loess sample of level 1 and b) Lowrie test in a palaeosol sample level 10 of a loess-palaeosol section of the Buenos Aires Formation (Argentina).

are always related to the axis of low coercivity minerals. The low coercivity thermal demagnetization curves are consistent with magnetite and/or titanomagnetite. As an example, Figure 5a shows the behavior of a loess (level 1) sample, in which the blocking temperature spectrum matches that of magnetite. In Figure 5b (level 10 of the palaeosol), a blocking temperature spectrum consistent with the presence of titanomagnetite is displayed.

3 CHEMICAL ANALYSIS

Ferric and ferrous iron content were determined in several levels of loess and palaeosol (Figs 6a, 6b). All the samples were milled in an agate mor-

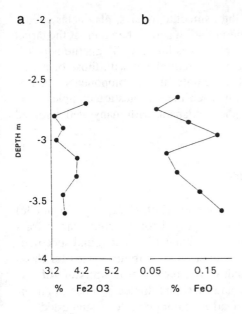

Figure 6. Chemical results a) Ferric iron and b) Ferrous iron of some samples of a loess-palaeosol section of the Buenos Aires Formation (Argentina).

tar to #150 and dried at 100-105°C. The used methods were the following:

a) ferric iron: Acidic treatment and atomic absorption spectrometry (Buck spectrometer). Measurements were carried out vs. andesite AGV-1, granite G-2 and W1 US Geological Survey reference materials. Method uncertainty: 0.2‰.

b) ferrous iron: Volumetric method as recommended by the US Geological Survey. Uncertainty: 0.5‰.

As shown in Figures 6a and 6b, the amount of iron does not show significant changes throughout the sampled section. No relevant migration processes could thus be inferred for iron and no accumulation zones were observed either.

4 X-RAYS RESULTS

Samples from levels 2 and 9 were analyzed by X-rays diffraction. A Phillips 1130 diffractometer, operated with 40KV and 20mA, range of 1×10(3), time constant 2 and scanning and paper speed 1°2Θ and 1cm/min, respectively. The measurements have been repeated with Cu radiation (filtered by Ni), Fe (filtered by Mn) and Co (filtered by Fe) in order to enhance the signal and eliminate the secondary fluorescent radiation. One sample was analyzed also with an automatic equipment, Siemens 5000, yielding similar results.

The identified minerals were: illite, smectite, quartz, plagioclase. Considering the intensity of background radiation and its behavior at the target exchange, the presence of volcanic glass was inferred. Magnetite is possibly present and some reflections could be correlated with those of hematite and goethite. Identification of extremely minor components, such as ferromagnetics here, is very difficult when the samples are complex mixtures containing plagioclases and other feldspars, with many well resolved diffraction lines.

5 DISCUSSION OF THE RESULTS

The values of the coercivity forces parameters (Bcr, H1/2$_I$; Figs 3b and 3c) in the palaeosol are higher than those expected for a ferromagnetic fraction represented by magnetite and/or titanomagnetite. The found spectra of coercivity forces lay beyond the characteristic maximum values for those minerals. It is explained as the result of the presence of a mixture of low and high coercivity force minerals, and therefore, the occurrence of an antiferromagnetic mineral, such as hematite and/or goethite, is suggested.

The low temperature tests (Fig. 4) allow to infer the presence of goethite (Dekkers 1988); at low temperature the remanent magnetization progressively increases in samples towards the palaeosol, which could mean an increment in the goethite amount. In agreement with these results the antiferromagnetic behavior detected could be explained by the presence of goethite. However, hematite can not be discarded because a magnetic fraction with middle coercivity forces was detected by using the Lowrie test.

The X and SIRM diminishing in the palaeosol indicates a decrease of the ferrimagnetic fraction present. In addition, the appearance of an antiferromagnetic fraction in the same levels, could suggest partial alteration of magnetite or titanomagnetite and the consequent genesis of goethite. This is in full agreement with current ideas about the genesis of this mineral. The goethite generation, a frequent mineral in palaeosols, occurs as the result of different processes: weathering of iron silicates, oxidation of magnetite, iron sulphates and carbonates and hematite hydration (Zitzmann 1978, Dekkers 1988).

Finally, from the results obtained it is possible to conclude that the palaeosol bears magnetite-titanomagnetite, as a main magnetic component, which signal is the strongest. Due to the magnitude of the SIRM/X ratio the magnetic grain size is MD. The MD behavior is characteristic for particles coarser than 8×10^{-9} m diameter (Hunt et al. 1995). This grain size is equivalent to particles around fine silt. In agreement with this the magnetite should have a detrital origin.

An antiferromagnetic signal overprinted to the main magnetic one is observed. It could be originated by a goethite signature.

From the analyzed data it can be concluded that, the difference of susceptibility behavior in the Chaco-Pampean and Chinese palaeosols is due to the presence of different magnetic components. While in the Chinese palaeosols, where susceptibility increases, the conspicuous magnetic signal is lead by SP magnetite superparamagnetic (Hunt et al. 1995, Banerjee et al. 1993, Xiuming et al. 1993), in the Chaco-Pampean palaeosol the susceptibility could drop as a consequence of a decrease of MD magnetite which is partially replaced by an antiferromagnetic fraction.

The different pedogenic minerals originated in northern China (Xining and Xifeng) and South America (Buenos Aires City) could be due to differences in the parental materials.

6 CONCLUSIONS

The analyzed magnetic properties of the Chaco-Pampean plain palaeosol are different from the Chinese palaeosols. Consequently the magnetic fraction in both areas is different.

Due to the parental dissimilarity in both areas, the pedogenic minerals which originated under mild conditions are different. While in the Chinese loess the temperate conditions generate SP magnetite, in the Chaco-Pampean loess an antiferromagnetic fraction, probably represented by goethite, is formed.

The magnetic method has turned to be the most sensible tool for the determination of the magnetic fraction as well as to detect changes in palaeoenvironmental conditions.

ACKNOWLEDGEMENTS

The authors thank the Universidad de Buenos Aires (UBA) (UBACYT ex 012); the Laboratorio de Paleomagnetismo 'Daniel Valencio', Departamento de Ciencias Geológicas, Facultad de Ciencias Exactas y Naturales, Universidad de Buenos Aires; CIRGEO (Consejo Nacional de Investigaciones Científicas y Técnicas, CONICET) and CONICET for their support.

REFERENCES

Banerjee, S.K., C.P. Hunt & L. Xiu-Ming 1993. Separation of local signal from the regional paleosoon recorded in the Chinese plateau: a rock magnetic approach. *Geophysical Research Letters* 20(9): 843-846.

Dankers, P.H.M. 1978. Magnetic properties of dispersed natural iron-oxides of known grain-size. Ph.D. Dissertation. University of Utrecht. 143 p. Utrecht. Unpublished.

Dekkers, M.J. 1988. Some rock magnetic parameters for natural goethite, pyrrhotite and fine grained hematite. Geologica Ultraiectina, N 51. Ph.D. Dissertation. University of Utrecht. 231 p. Utrecht. Unpublished.

Frenguelli, J. 1955. Loess y limos pampeanos. *Serie Técnica y Didáctica 7*. Universidad Nacional de La Plata; 88p. La Plata.

Heller, F., L. Xiuming, L. Tungsheng & X. Tongchun 1991. Magnetic susceptibility of loess in China. *Earth and Planetary Science Letters* 103: 301-310.

Hunt, P.C., S.K. Banerjee, J. Han, P.A. Solheid, E. Oches, W. Sun & L. Tungsheng 1995. Rock magnetic proxies of climate changes in the loess-palaeosol sequences of western Loess Plateau of China. *Geophysical Journal International* 123: 232-244.

Lowrie, W. 1990. Identification of ferromagnetic minerals in a rock by coercitivity and unblocking temperatures properties. *Geophysical Research Letters* 17: 159-162.

Mullins, P. 1977. Magnetic susceptibility of the soil and its significance in soil sciences. *Journal of Science* 14: 179-187.

Nabel, P. 1996. Aspectos ambientales registrados en los suelos de la región pampeana. *Revista de la Asociación Geológica Argentina* 51(2): 147-155. Buenos Aires.

Orgeira, M.J., A.M. Walther, C.A. Vásquez, I.M. Di Tommaso, S. Alonso, G. Sherwood, Yuang Hu & J.F.A. Vilas 1997. Magnetismo ambiental II: loess y paleosuelos de la Fm Buenos Aires (Pcia de Buenos Aires). *Revista de la Asociación Geológica Argentina*. In press. Buenos Aires.

O'Reilly, W. 1984. *Rock and mineral magnetism*. New York: Chapman & Hall, 220 p.

Ozdemir, O. & D.J. Dunlop 1996. Thermoremanence and Néel temperature of goethite. *Geophysical Research Letters* 23(9): 921-924.

Thompson, R. & F. Oldfield 1986. *Environmental magnetism*. London: Allen & Unwin, 227p.

Verosub, K.L. & A.P. Roberts 1995. Environmental magnetism: past, present and future. *Journal of Geophysical Research* 100(B2): 2175-2192.

Verosub, K.L., P. Fine, M.J. Singer & J. TenPas 1993. Pedogenesis and paleoclimate: interpretation of the magnetic susceptibility record of Chinese loess-palaeosol sequences. *Geology* 21: 1011-1014.

Xiuming, L., J. Shaw, L. Tungsheng & F. Heller 1993. Magnetic susceptibility of the Chinese loess-palaeosol sequence: environmental change and pedogenesis. *Journal of the Geological Society* 150: 583-588.

Zitzmann, A. 1978. *The iron ore deposits of Europe and adjacent areas*. Bund. Geowiss. Rohst., Volume 2, 386 p. Hannover: Schwizepart.

New evidence of the Brunhes/Matuyama polarity boundary in the Hernández-Gorina Quarries, north-west of the city of La Plata, Buenos Aires Province, Argentina

12

JUAN CARLOS BIDEGAIN

CIC-LEMIT & Departamento de Geomagnetismo, Facultad de Ciencias Astronómicas y Geofísicas, Universidad Nacional de La Plata

ABSTRACT: Palaeomagnetic profiles carried out on the Gorina quarry sediments northwest of the city of La Plata, Buenos Aires Province, confirm previous results from the Hernández Quarries and other Late Cenozoic sediments in Argentina. However, some adjustments have been made concerning the relationship among magnetic records, lithology and palaeoenvironments. New palaeomagnetic data indicate that the Brunhes Chron (<0.78 Ma) is found in the Holocene and Late Pleistocene La Postrera and Buenos Aires formations whereas the Matuyama Chron (>0.78 Ma) has been recorded in sediments traditionally assigned to the Ensenada Formation. Normal polarity levels recorded at the base of these sedimentary sequences can be assigned to palaeomagnetic subzones within the Matuyama Chron. Lujanian and Ensenadan fossil land-mammal ages are represented particularly in relation to humid climatic conditions alternately superimposed in the area during the Pleistocene. The Brunhes/Matuyama boundary is shown to be useful as a palaeomagnetic isochrone crossing different Quaternary lithologies allowing the determination of synchronism and diachronism of sedimentary horizons.

RESUMEN: Los perfiles paleomagnéticos realizados en la cantera de la localidad de Gorina, situada al noroeste de la ciudad de La Plata, Provincia de Buenos Aires, confirman los estudios previos realizados en las canteras de Hernández y otros sedimentos del Cenozoico tardío. No obstante, en el presente trabajo se han efectuado ajustes para una mejor relación de los registros magnéticos con las unidades litológicas y con los paleoambientes sedimentarios.

Los nuevos datos paleomagnéticos indican que los sedimentos del Holoceno y Pleistoceno tardío denominados Formación La Postrera y Formación Buenos Aires corresponden al Chron Brunhes (<0.78 Ma). Por otra parte, los sedimentos que tradicionalmente han sido estudiados como pertenecientes a la Formación Ensenada corresponden al Chron Matuyama (>0.78 Ma).

Los niveles de polaridad normal registrados en la base de las secuencias sedimentarias pueden ser referidos a subzonas de dicha polaridad dentro del Chron Matuyama.

Los fósiles de las edades mamífero Lujanian y Ensenadense están representados en las secuencias estudiadas. Los sedimentos portadores corresponden a períodos de clima relativamente húmedo que en forma alternada imperaron en la región durante el Pleistoceno. La isocrona paleomagnética Brunhes/Matuyama se registra en relación con distintas litologías. Esta particularidad resulta de utilidad a los fines de determinar el sincronismo y diacronismo de horizontes sedimentarios del Cuaternario.

1 INTRODUCTION

Palaeomagnetic data from Late Cenozoic sedimentary sequences, exposed in Hernández and the Gorina quarry, situated northwest of the city of La Plata, Buenos Aires province (Fig. 1), confirm the Brunhes and Matuyama polarity zonation established in previous works. New palaeomagnetic data reported here suggest an older age for the base of the exposed sequences (>0.78 Ma) because normal polarity levels within the Matuyama Polarity Chron can be associated with the Jaramillo and Olduvai subchrons.

Some samples from the Hernández quarry were demagnetized at Buenos Aires University's Palaeomagnetic Laboratory and others at Stockholm's University; the remanences were measured using a Digico magnetometer. All the samples from Gorina were demagnetized in Buenos Aires by using a high sensitivity cryogenic magnetometer allowing measurements of specimens with low intensity ranging down to 10^{-7} emu. Low field susceptibility measurements were made by using a Bartington susceptibility meter and a KT-9 hand-held susceptibility meter for field measurements.

Two well defined polarity zones have been determined according to records obtained from the six most complete palaeomagnetic profiles carried out near the Hernández and Gorina localities. The Holocene-Late Pleistocene sedimentation corresponding to the La Postrera and Buenos Aires formations occurred during the Brunhes Chron, whereas Early to Middle Pleistocene sedimentation represented by the Ensenada Formation occurred during the Matuyama Chron. The Brunhes/Matuyama palaeomagnetic isochrone crosses through different lithologies; it is recorded in silt-clayey palustrine environments, at a sharp palaeosol/loess contact, and also in sandy loess deposits.

The purpose of this magnetostratigraphic study is to clarify the usefulness of palaeomagnetism in Quaternary chronostratigraphic studies in

208

coastal plain areas exposed to alternating periods of deposition/erosion and intense weathering. On the other hand, intensity and susceptibility seem to be appropriate tools to be used in the study of palaeoclimatic and palaeoenvironmental changes due to their sensitivity to mineralogical changes.

2 PREVIOUS WORK IN THE REGION

The Quaternary sedimentary sequences are made up of the so-called Pampean and Post-Pampean sediments. The Pampean sediments exposed in the quarries comprise the older Ensenada Formation and the younger Buenos Aires Formation, in the sense of Frenguelli (1957). According to Cortelezzi (1978), the sandy silts with calcretes at the base of the quarries belong to the Ensenada Formation, whereas the overlying sediments correspond to the Buenos Aires Formation. The Post-Pampean (Holocene) sediments at the top of the sedimentary sequences have been named La Postrera Formation according to Tonni et al. (1988). These authors consider that the Pampean sediments comprise the Lujanian and Ensenadan land-mammal ages, according to Frenguelli (1957). Fidalgo & Martínez (1983) studied the Buenos Aires and Ensenada formations in the research area under the general denomination of Pampean sediments and they related these formations to the aforementioned Lujanian and Ensenadan land-mammal ages. According to Riggi et al. (1986), the older Ensenada Formation of the La Plata region contains vertebrate fossils which have been assigned to the Ensenadan land-mammal age established by Pascual (1965).

Mineralogical studies of Pampean loessian deposits have been carried out on samples from different sites of the Argentina loess plateau by Teruggi (1957).

The stratigraphic position of the Lujanian mammal records (obtained from Late Pleistocene silts) has been more clearly defined than the stratigraphic position of the oldest Ensenadan fossil fauna. This has traditionally been a topic of controversy in the research area. The absence of natural Quaternary geological sections has been a contributary factor and studies are limited due to the intensive exploitation of these quarries.

The present study on palaeomagnetic data in the exposed sedimentary sequences in Hernández and Gorina quarries also takes into account previous research carried out in La Plata city by Bobbio et al. (1986) and in Hernández by Bidegain (1991) and Bidegain et al. (1995,1996).

Normal polarity subzones within the Matuyama Chron have been mentioned in several palaeomagnetic profiles carried out on sedimentary sequences in different localities of the Pampean region. In the Mar del Plata coastal cliffs, Orgeira (1990) and Ruocco (1990) have mentioned records of the Jaramillo and Olduvai events in relation to the Arroyo Seco and Miramar formations, respectively. Palaeomagnetic studies carried out in the city of Buenos Aires by Nabel & Valencio (1981), and more recently by Re & Orgeira (1991), also confirm the existence of normal polarity subzones within the Matuyama Chron. In Entre Rios Province and in the previously mentioned works in the Hernández quarries, Bidegain (1991) indicates records of normal polarity levels within the Matuyama Chron. These normal polarity subzones correspond to the Jaramillo (0.99-1.07 Ma) and Olduvai (1.77-1.95 Ma) events according to Candle & Kent (1995).

3 HERNÁNDEZ QUARRIES

Three major pits in the Hernández locality (Fig.1) are situated at some 15 m a.s.l. and on the left bank of El Gato Creek, which is a tributary of the Río de La Plata. The area has traditionally been exploited for civil engi-

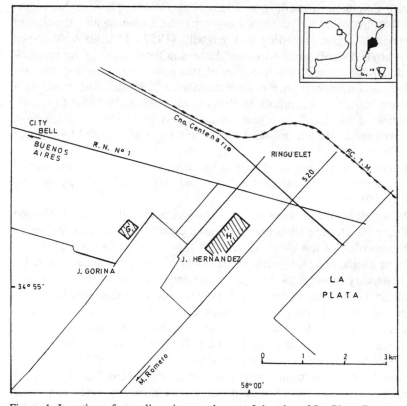

Figure 1. Location of sampling sites northwest of the city of La Plata, Buenos Aires Province, Argentina.

210

neering purposes. These sediments are sandy-silts with calcretes; their low plasticity (IP <10) makes them suitable aggregates for engineering projects, mainly roads. In order to reach these layers it is necessary to dig through younger clayey silts (Late Pleistocene-Holocene) sediments, unsuitable for the aforementioned construction purposes due to the low $CaCO_3$ and sand contents. Geotechnical grain size determination indicates that more than the 90% by weight of this material passes through the 0.177 mm sieve size and the coarser retained materials are small $CaCO_3$ and manganese-iron oxide concretions.

Older sediments corresponding to the Ensenada Formation are richer in $CaCO_3$ than the above mentioned ones. The sand-size fraction increases to more than 20% by weight, particularly in the loess horizon and palaeochannels.

Both Hernández and Gorina quarries consist of sedimentary sequences of loess and related palaeosols. Loess deposits in the area have been more strictly defined as loess-like units by Teruggi & Imbellone (1987), considering them as reworked loess deposits rather than primary ones. Beyond Teruggi & Imbellone's (1987) work on one isolated profile considering alternating palaeosols/loess sequences in Hernández, a regional study has not been done yet. According to these authors, sedimentary sequences are represented by several palaeosols affected by polygenesis; they also point out that the pedogenesis developed in one horizon has affected underlying older ones owing to the small thickness of each layer. The horizontality of some layers of palaeosols is remarkable in sectors which are closest to the watershed line but this characteristic disappears in lower topographic levels of the horizons due to lateral replacements caused by palaeochannels and palaeolagoons.

Grain-size analysis carried out on samples from the Hernández quarries are shown in Figure 2. Susceptibility values were also put in relation to chemical determination of FeO concentrations. Total iron oxide (Fe_2O_3) concentrations were not considered because they do not change significantly along the sequences, according to Bidegain et al. (1995, 1996).

Although susceptibility values are low in comparison with those obtained from rocks richer in ferromagnetic minerals, the general pattern obtained allows us to draw some interpretations about palaeoclimatic and palaeoenvironmental changes which occurred during the Quaternary. Higher susceptibility values correspond to sandy loess and loess-like with calcrete layers and lower susceptibility values to clayey palaeosols.

Sedimentary horizons labelled A, B and C (Fig. 2) show a decrease in susceptibility values downwards with the formation of calcium carbonate nodules, veins and rhizocretes developed in a crumbly reddish brown sediment (B). This horizon lies discordantly over a clayey silt bed (C) corresponding to a humid climate during the Late Pleistocene-Holocene transition.

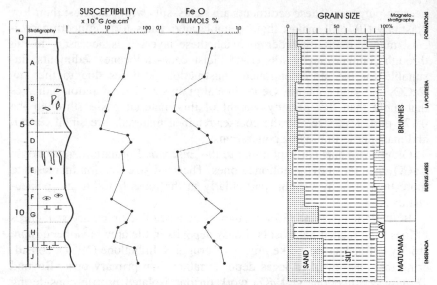

Figure 2. Correlation between volumetric susceptibility and FeO contents in Hernández quarries. Grain-size determinations corresponding to La Postrera, Buenos Aires and Ensenada formations and established magnetostratigraphy.

The sedimentary horizon labelled as D is a loess-like silt showing the higher susceptibility values than over and underlying horizons. Underlying D horizon there is a well-developed clayey palaeosol (E) with calcareous rootlets and abundant rounded manganese-iron oxide concretions (pisolites) of 2-3 mm size, particularly concentrated in the top of E.

With the exception of an interbedded palaeosol (H) and its corresponding palustrine facies, the sand fraction increases notably in the horizons indicated with G, I and J and the carbonate accumulations can acquire a platy morphology constituting convolutions in some sectors. The sandy loess (G) is associated with the Lowest Brunhes Chron, the horizon labelled H corresponds to the most conspicuous and locally developed palaeosol closely related to the Upper Matuyama Chron. The mentioned palaeosol H passes laterally to a palustrine environment with dark greenish clayey silts structured in prismatic blocks at the base and light greenish friable fine prisms (sandy) at the top. Sandy silt palaeochannels affecting these layers contain rounded clay and carbonate pebbles, which seem to indicate a sea level lowering. One of these palaeochannels shown in Figure 4 contains alternating 1-2 cm thick sandy/clayey silt layers, reflecting water-table variations. Horizon I (Fig. 2), is a calcrete floor with horizontal and subvertical $CaCO_3$ plates, volcanic ash and mud cracks filled by calcium carbonate. In this horizon, calcretes in convolutes form have also been developed at both sides of a palaeochannel bearing vertebrate fossils.

212

Although the textural and structural pattern is interrupted due to lateral replacements in the oldest sandy exposed sediments, the silt fraction increases progressively in sediments belonging to the Buenos Aires and La Postrera formations (Fig. 2). Iron oxides concentrations, chemically determined as FeO, are related to changes in susceptibility values. Iron oxide concentrations and susceptibility values were obtained from three more or less complete profiles carried out in the Hernández quarries, being the same horizon sampled more than one time in different sites. Represented values are the mean of the obtained data. The best relation between susceptibility and FeO concentrations was determined at the Brunhes/Matuyama transition by intensive sampling (more than 100 sites) in units J, I, H and G (Bidegain et al. 1995, 1996).

4 MAGNETOSTRATIGRAPHY IN HERNÁNDEZ

Alternate field demagnetization (AF) was carried out in the palaeomagnetic laboratory by using Digico and cryogenic magnetometers.

Figure 3 shows two palaeomagnetic profiles from the Hernández quarries indicating Declination, Inclination and VGP.

According to the revised calibration of the geomagnetic polarity time scale by Candle & Kent (1995), the oldest obtained records could belong to the Olduvai subchron (1.77-1.95). The Brunhes/Matuyama boundary is closely related to the loess (G) / palaeosol (H) as shown in Figure 2 corresponding to a palaeosurface on top of a clayey palaeosol.

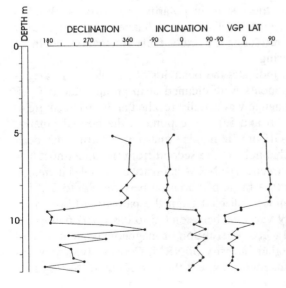

Figure 3. Declination and Inclination of stable remanent magnetization, VGP and polarity zonation in Hernández quarries.

213

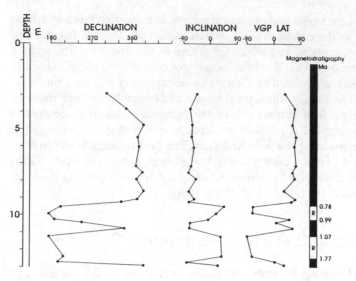

Figure 3. Continued.

Figure 4 shows a geological section from Hernández and it has been picked up in order to show palaeoenvironmental lateral replacement in the Brunhes/Matuyama transition zone. The palustrine environment – with upward increasing grain size – is indicated to the SE and the major palaeosol development to the NW. Caves, containing vertebrate fossil bones belonging to the Ensenadan land-mammal age, as indicated on the right of the scheme (krotovina in Fig. 4), can usually be found within the palaeosol. The dashed line divide the two main polarity chronozones, Brunhes (normal) and Matuyama (reverse). The palaeochannel, with alternating sandy/clayey small layers reflecting the water table fluctuation, could be referred to a sea level lowering.

The stereoplot (Fig. 5a) indicates the behavior of samples from the reverse zone. Viscous components were cleaned at in an applied 5mT AF and the characteristic remanence was usually reached at 30 mT peak field. Zijderveld diagrams (Fig. 5b and 5c), correspond to the normal polarity samples Hern-1 and Hern-3 from Hernández quarries. The first has been collected from the palaeochannel and the second from the uppermost levels of the palustrine environment. The NRM intensity of samples increases upwards in this horizon with a range of variation between 8.5 to 26,05 × 10^{-6} emu while the samples collected from the palustrine lower parts showed a range of intensity variation between 0,5 to 0,8 × 10^{-6} emu. X-ray diffractograms showed a better definition of magnetite content in relation to those levels with higher intensity of NRM. Conversely, it was impossible to determine the magnetite content through X-ray diffractograms

214

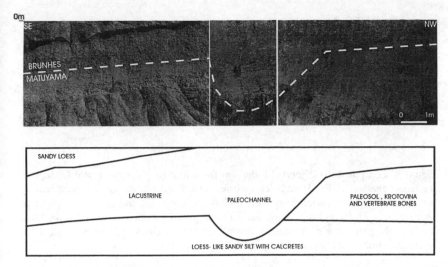

Figure 4. Photograph and scheme of the Brunhes/Matuyama transition zone recorded in the Ensenada and Buenos Aires formations in Hernández locality. The dashed line shows the contact palaeochannel/palaeosol to the right of the photo; the palustrine environment is shown to the left. The watershed line is to the NW and El Gato Creek is situated to the SE.

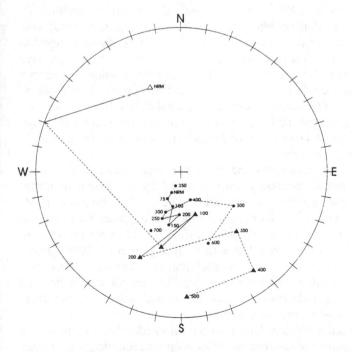

Figure 5. a) Stereoplot for samples corresponding to the Matuyama Chron.

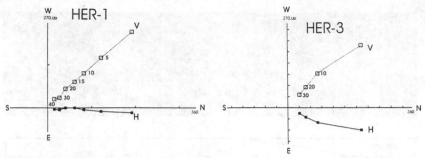

Figure 5. Continued. b) Zijderveld diagram for a normal polarity sample taken from the palaeochannel in the Hernández quarries. Open and closed squares represent the projection of magnetization vector end-points on a vertical and horizontal plane respectively, c) Zijderveld projection for a normal polarity sample taken from the upper levels of the palustrine environment shown in Figure 4. Open and closed squares represent the same as in the previous sample.

in the lower levels of the palustrine environment due to the presence of amorphous minerals (Cortelezzi et al. 1994, Bidegain et al. 1995, 1996).

Several X-ray diffractograms carried out on the clay fraction indicate magnetite as dominant ferromagnetic mineral in loess deposits in the area (Bidegain et al. 1995, 1996). Antiferromagnetic minerals as goethita (α FeO.OH) were also determined in relation to some levels of sandy silts with calcretes at the lowermost horizons of the profiles. Its formation has been indicated in relation to pedogenesis as an indicator for progressive oxidizing conditions (Duchaufour 1987). Paramagnetic silicates and iron oxides have also been determined by X-ray diffractograms carried out on sediments from the Hernández quarries. Among the first, for the iron contribution, it can be mentioned hornblenda and among the second akaganeite (β FeO.OH) and lepidocrocite (δ FeO.OH), particularly in relation to humidity and soil formation.

Some samples of palaeosols and loess deposits related to the Brunhes/Matuyama polarity boundary were analyzed by isothermal remanent magnetization (IRM) applying a peak field of 2,3 T. The saturation (IRM) has been reached below 0,5 T of applied field and the intensity of saturation (IRM) acquired by loess samples at a field of 0.5 were higher than that of palaeosols samples by a factor of 4 (Bidegain et al. 1996). However, more intensive mineralogical and rock magnetic determinations must be done in the near future in order to develop the methodology to be used in the study of soils, palaeosols, palaeoclimatic and palaeoenvironmental changes in the region during the Quaternary.

There are not older fossil records than the Ensenadan land-mammal age in the area and there are controversies between palaeontologists whether some of the fossil findings belong to the Lujanian or to the Ensenadan

216

land-mammal ages. Besides, some of the fossil records have been unin-
tentionally mixed with younger ones due to wrong determination of the
real site position. However fossil findings belonging to the Ensenadense
mammal age correspond to the oldest exposed horizons assigned to Matuyama Chron while fossil findings belonging to the Lujanian land-
mammal age correspond to the Brunhes Chron as indicated in Figure 4.
Fossil bearing horizons are mainly palaeosols, palustrine environments
and palaeochannels which generally show lower susceptibility values than
the loess deposits.

5 THE GORINA QUARRY

Geologic and palaeomagnetic studies have been also carried out in a Gorina quarry situated northwest of the city of La Plata and 2 km from the
previously mentioned Hernández quarries (Fig.1).

Figure 6 shows the sedimentology of the Holocene (La Postrera Formation) and Pleistocene (Buenos Aires and Ensenada formations) exposed
sequences of loess and interbedded palaeosols in the Gorina quarry. The
dashed line indicates the Brunhes/Matuyama boundary. Normal polarity

Figure 6. Holocene (La Postrera Formation) and Pleistocene (Buenos Aires and Ensenada Formations) sediments exposed in the Gorina quarry. The dashed line shows
the Brunhes/Matuyama polarity boundary.

217

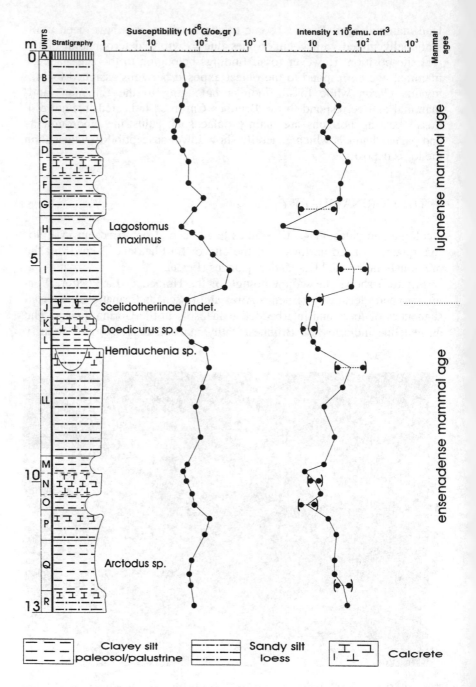

Figure 7. Specific magnetic susceptibility values and NRM intensities related to sedimentary horizons at Gorina. Fossil findings in relation to the land-mammal ages and stratigraphy are also shown.

sampling levels within the Matuyama Chron recorded in this work will be studied in detail in the near future taking advantage of new, deeper exploitation.

Figure 7 shows the relative thickness of each sampled horizon in the sequence of the Gorina quarry in relation to specific susceptibility and NRM intensity. A total of 120 samples were obtained for laboratory susceptibility measurements by using a Bartington susceptibility meter; mean values are represented in this figure. Several measurements were done directly on the geologic profile by using a K-T 9 susceptibility meter. The scattering of intensity values are indicated by error bars. Vertebrate fossil findings in relation to sedimentological layers and respective mammal ages are also indicated in this figure.

Horizon A in Figure 7 represents the recent soil developed in a silty loess (B). The horizon labelled C is a partially sandy, clayey silty horizon with $CaCO_3$ concretions and manganese-iron oxides in the form of patches as signs of humid conditions. Records of Lujanian land-mammal age fossils have been found in this horizon in other localities of the research area.

Horizons D, E and F are clayey silts (palaeosol) with calcretes at the central part (E), of about 0,30-0,40 m in thickness. The horizon labelled G is a sandy loess-like silt which lies discordantly over a clayey silt palaeosol (H). The thickness of loess deposits increases notably in Unit I. Higher susceptibility and intensity values in the latter compared to those values recorded in Unit H are directly related to the increase of wind-blown ferromagnetic particles.

The loess horizon I lies discordantly over the most intensive sampled fossil-bearing palaeosol in the Hernández-Gorina area. The palaeosol has been indicated with J, K, L, but its calcium carbonate concretions (K) arc irregularly distributed within these materials. The colour of this unit changes laterally from brownish to greenish according to the palaeoenvironmental conditions. A thick loess-like deposit (LL) consisting of partially clayey sandy-silts underlies Unit L. The abundance of calcretes constituting hard plates on the top of the LL horizon makes this layer useful as road materials. Evidence of drainage rejuvenation, predating palaeosol formation, consists of palaeochannels which partially eroded horizon LL (Fig. 7).

It seems that these structures were produced during a lower sea-level stand, as previously mentioned in the Hernández locality. Susceptibility and intensity values increase also in the LL horizon in relation to the underlying clayey palaeosol (M, N, O).

Horizon P is a loess-like deposit with high concentration of $CaCO_3$ (30%) and Horizon Q corresponds to a clayey silty palaeolagoon with patches of manganese-iron oxides and low concentration of $CaCO_3$ (<1%). The underlying R horizon is a clayey silty layer with calcium carbonate on the upper levels in the form of nodules and plates.

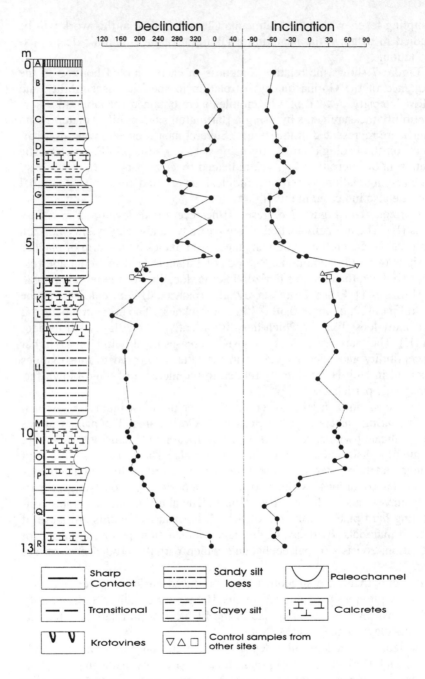

Figure 8. Declination and inclination of stable remanent magnetization of samples from the Gorina Quarry. Normal polarity levels at the base of the sequence could be assigned to the Jaramillo or Olduvai subzones.

220

The measured susceptibility and intensity values shown in Figure 7 seem to be closely related to changes in the ferromagnetic mineral content. Such changes seem also to indicate changes in the grade of weathering, throughout the sequences. Fine-grained sediments in palaeosols show generally lower susceptibility and intensity (NRM) values than coarse-grained (sandy) sediments. Chemical determination of Fe_2O_3 and FeO are not yet available. However -according to chemical analysis that have been carried out so far- palaeosols and palustrine environments show lower concentration of Fe^{2+} than loess deposits as indicated in Hernández.

6 MAGNETOSTRATIGRAPHY IN GORINA

All samples oriented in the field were taken by using square section core sampler made of non-magnetic materials. The sediment obtained was put in cubic plastic boxes of a volume of 8 cm^3. Alternating Field Demagnetization (AF) was carried out in the palaeomagnetic laboratory by using a cryogenic magnetometer. The Gorina records are synthesized in Figure 8 which gives also the lithostratigraphic subdivision, sampling levels declination and inclination after demagnetization and obtained magnetostratigraphy. Scattering in declination values in the normal polarity zone (Brunhes) may be due to low intensity, reworking and bioturbation of materials.

All samples and double-samples were demagnetized stepwise up to values well above the SRM (Stable Remanent Magnetization) levels. The SRM seems usually to be reached between 20 and 30 mT peak field.

Figure 9 shows stereoplots and demagnetization curves of some samples from the Gorina profile. Normal polarity samples in the stereogram to the left indicate the behaviour of declination and inclination during demagnetization of (G1) and (G117). The first sample belongs to the normal subzone within the Matuyama Chron and was collected from sandy silts at the base of the sequence. The second sample (G117) belongs to the Brunhes Chron and it was collected from the loess I. Alternating field demagnetization (AF) up to only 10 mT removes about 70 percent of the initial NRM intensity suggesting a soft magnetization. On the right, the figure shows the stereoplot and curves for samples G9 and G11. The first corresponds to the clayey silt palaeosol (J) and the second to the lowermost levels of the loess labelled I. The middle demagnetization field (MDF) is reached applying AF higher than 20 mT indicating a more stable remanence than in the previous samples G1 and G117. Applying alternating AF up to 60 mT on samples G9 and G11 about 20% of initial intensity (Jo) remains.

Soft and hard demagnetization has been observed as well in loess as in palaeosols. Figure 10 shows the horizontal projection (H) by solid black

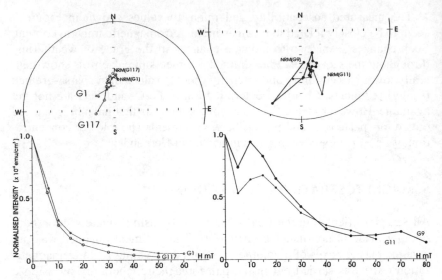

Figure 9. Stereoplot and demagnetization curves for normal and reversed polarity samples from Gorina.

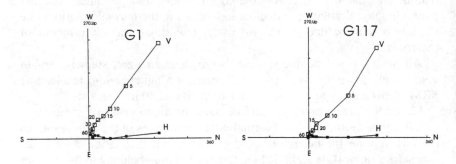

Figure 10. Zijderveld diagrams for normal polarity samples (G1) and (G117). The first was collected from the oldest exposed sediments at the base of the sedimentary sequence in Gorina, the second was collected from the loess sediments assigned to the Brunhes Chron.

squares and the vertical projection (V) by open squares for samples G1 and G117. The origin seems to be reached at 60 mT peak field. Figure 11 shows Zijderveld diagrams for samples G9 and G11 with higher coercivity of remanence than in the previous ones. The highest applied peak field was 80 mT on G9 and 60 mT on G11; viscous components of magnetization were generally 'cleaned' at 5 mT of applied field.

The Brunhes/Matuyama boundary (Fig. 8) recorded in Gorina is related to the onset of loess deposits and not to the palaeosurface on the horizon J. Samples indicated by squares and triangles have been collected from

222

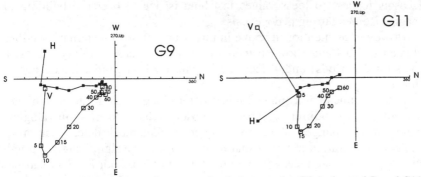

Figure 11. Zijderveld diagrams for reversed polarity samples G9 (palaeosol J) and G11 (loess I).

other sites of I horizon in the Gorina quarry and they have been considered as control samples.

The major clayey palaeosol sampled in Gorina (J, K, L in Figs 7 and 8) shows reverse polarity as in the Hernández quarries (H in Figs 2 and 3). The underlying loess and palaeosols, assigned to the Matuyama Chron, show a thickness of about 6 meters.

Normal polarity levels at the base of the studied sequence must be studied intensively in the future, because there are still uncertainties whether they belong to the Jaramillo or Olduvai subzones. However, according to the International Standard Geomagnetic Polarity Time Scale (GPTS) the palaeomagnetic obtained data indicate that Gorina and Hernández sedimentary sequences cover at least the last 0.99 Ma including the Upper Jaramillo subchron.

7 DISCUSSION

The studied loess sections contain clayey palaeosols which represent warmer and wetter climatic conditions corresponding to interglacial periods, in contrast to the cold, arid environments in which loess accumulated during glacial intervals.

Heller & Liu (1982, 1985) and Heller et al. (1991), while studying loess and palaeosols in China, found that susceptibility values increase in palaeosols and decrease in loess deposits. In addition, these authors state that susceptibility values depend strongly on paramagnetic grains in the loess, whereas in the palaeosols, susceptibility values are produced by the growth of ultrafine magnetite during pedogenesis. In agreement with this, Maher & Thomson (1991) consider that the higher susceptibility of pa-

223

laeosols respect to loess values in China is due to bacterial building of magnetosomes during pedogenesis.

However, as mentioned above in this paper, all the Quaternary sections studied in La Plata show an increase of specific susceptibility values in loess (arid periods) and a decrease in palaeosols (humid climatic conditions).

Both chemical and mineralogical determinations indicate that higher and lower susceptibility records are closely related to the mineralogical composition changes along the sequence. Chemical determination of Fe_2O_3 concentration indicates that it does not change significantly through the Quaternary sequences in the area (around 6%), both in loess and palaeosols. Although in lower concentrations, the FeO rate of variation is greater than the total iron (Fe_2O_3) rate of variation. Concentration of FeO decreases down to 0.02% in intensively weathered horizons, while it may increase up to 0.40% by weight in loessic units (Bidegain 1995, 1996).

The disagreement between susceptibility values of Chinese and Argentine loess/palaeosols sequences must be related to the nature and origin of the materials. Consequently, the palaeosols can be referred to warmer and wetter climatic conditions and the loess layers to dry and cold climate in different sites throughout the world, but magnetic susceptibility values cannot be referred to global climatic changes, without taking into account the parent rock composition in each region.

According to Teruggi (1957), the mineralogical composition of loess in Argentina differs from the European and North America, loess and probably from the Chinese as well. Teruggi (1957), analyzing sediments of different grain-size from several sites of the Argentina loess plains found that the major mineralogical difference among the mentioned deposits lay in the contribution of volcanic minerals. The major contribution of wind-blown pyroclastic materials in Argentine loess has also been produced during dry and arid periods. Teruggi (1957) states also that the primary fraction of heavy minerals consists of opaque iron minerals and that magnetite is by far the predominant species.

Besides paramagnetic contribution, according X-rays determinations, it seems that magnetite is the main contributor to the obtained susceptibility. Susceptibility values increase in relation to loess deposits where magnetite is clearly defined by X-rays determination. Conversely, in a more weathered horizon with lower FeO concentration it is impossible to determine magnetite by X-rays. Moreover, amorphous minerals seem to mask any other iron minerals which is most probably in a very low concentration and poorly crystallized.

Figure 6 shows that susceptibility values are higher in the loess deposits (by $100-300 \times 10^{-6}$ c.g.s.; 1,26 to $3,77 \times 10^{-3}$ SI) than in palaeosols where they decrease (around 50×10^{-6} c.g.s.; $0,6 \times 10^{-3}$ SI). According to Tarling & Hrouda (1993), it is possible that the low susceptibility values in sedi-

224

ments can be controlled by paramagnetic minerals. However, this does not seem to be the case in the studied sequences from the La Plata area. Magnetosome due to bacterial activity have not been determined yet and specific studies concerning quantitative contribution of super-paramagnetic minerals for the susceptibility values obtained must be done in the near future.

Teruggi (1957), studying the fine sand and silt fraction of Argentine loess, indicated that magnetite is the main heavy mineral in these fractions. X-ray diffractograms carried out in Hernández quarries show also a better definition of magnetite in loess rather than in palaeosols considering the clay fraction (Bidegain et al. 1995, 1996). Moreover, laboratory measurements of IRM in loess and palaeosols also show that intensity of saturation level increases in loess (120×10^{-6} emu) and decreases in palaeosols (30×10^{-6}). The saturation (IRM) is reached in both cases in fields less than 0.5 T and the enhancement of intensity seems to be due to the higher concentration of magnetite in loess. The increasing amounts of ferromagnetic minerals can also be observed visually in the loess and palaeosols by using a hand magnet.

Antiferromagnetic and paramagnetic iron oxyhydroxides (FeO.OH) have also been determined by X-ray diffractograms in the research area. The former was goethite, determined in association with sandy silts with calcretes at the base of the sequences (Ensenada Formation). The latter consists of akaganeite and lepidocrocite associated with palaeosols that are suggestive of more humid conditions during the Late Pleistocene/Holocene sediments.

From the magnetostratigraphic studies carried out in Entre Ríos, Bidegain (1991) obtained records of Kaena-age events related to a Pliocene high sea-level stand interpreted as related to global climatic changes at about 3 Ma ago. Progressive continental deposition was dominated at the beginning of the Matuyama epoch by wind-blown mineral particles and particularly volcanic ash, giving sharp polarity changes in the silts of Entre Ríos Province. Cold climatic conditions related to a glacial period of that age in Patagonia have been mentioned by Sylwan (1990). This climatic change seems to have given rise to the first loess deposition on the Pampean plains as has been indicated in the Upper Gauss to Lower Matuyama palaeomagnetic records obtained from Villa Urquiza and Aldea Brasilera in Entre Ríos. According to the aforementioned data and to those obtained by Orgeira (1991) and Ruocco (1991) in Mar del Plata, the time for the onset of dry climatic conditions can be placed in the Late Gauss or Early Matuyama times.

The Hernández and Gorina loess/palaeosols sequences are much shorter ones; the older sediments belong to the Early to Middle Pleistocene and the younger ones to the Holocene. Palaeomagnetic records are assigned to the Brunhes and Upper Matuyama Chrons.

8 CONCLUSIONS

Palaeomagnetic profiles carried out in the area indicate the existence of a polarity zonation. Sediments belonging to the La Postrera and Buenos Aires formations were deposited during the Brunhes Chron and the sediments assigned to the Ensenada Formation have accumulated during the Matuyama Chron. Normal polarity levels determined within the reverse chronozone are not yet clearly defined. However, according to the international polarity time scale (GPTS), the base of the studied sequences could have been deposited during the Jaramillo (0.99-1.07 Ma) or at the end of the Olduvai (1.77 Ma) subchrons.

According to the last palaeomagnetic profile carried out in Gorina, the Brunhes/Matuyama boundary is closely related to the beginning of the loess deposition (I in Fig. 7), where susceptibility and intensity values increase due to the highest amount of ferromagnetic minerals. The present study has revealed that the Brunhes/Matuyama boundary has also been recorded in different environments. This particularity makes the palaeomagnetic isochrone a useful tool for the interpretation of synchronism and diachronism of Quaternary horizons in the studied area (Fig. 12).

The changes of magnetic susceptibility recorded in the present study are evidence of the alternation between wet and dry periods occurred

SYNCHRONISM AND DIACHRONISM

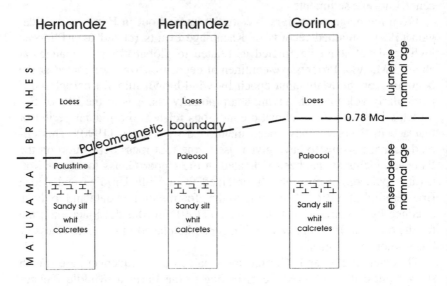

Figure 12. Palaeomagnetic isochrone determined in relation to different palaeoenvironments, northwest of the city of La Plata.

during the Pleistocene and the Holocene in the region. Wind-blown iron, bearing minerals were carried to the area during drier climatic conditions (glacial periods), whereas humid climatic conditions alternatively super-imposed on them, allowed the building of clayey palaeosols (interglacial periods). Susceptibility values closely related to climatic sensitive Quaternary lithologies reflect these changes through high respective low records. According to paleontological determinations that have been carried out in several quarries, Lujanian and Ensenadan fossil mammal have been found in palaeosols and palustrine environments.

ACKNOWLEDGEMENTS

I would like to acknowledge the Comisión de Investigaciones Científicas de la Provincia de Buenos Aires (CIC) and the University of La Plata for the support to the present study. I am also very grateful to the 'Daniel Valencio' Palaeomagnetic Laboratory in Buenos Aires, and to the Palaeomagnetic Laboratory in Stockholm University; Dr. Maizza for X-ray diffraction analysis and Dra. A.M. Sinito for the magnetic susceptibility measurements. Many thanks also to Francisco Prevosti for the determination of fossil mammals. The comments and suggestions of two anonymous reviewers greatly improved the final version of the manuscript.

REFERENCES

Bidegain, J.C. 1991. Sedimentary development, magnetostratigraphy and sequence of events of the Late Cenozoic in Entre Ríos and surrounding areas in Argentina. Ph.D. Thesis, Stockholm University. Akademitryck AB, Edsbruk. Sweden, p.128.

Bidegain, J.C., R.R. Iasi, R.H. Pérez & R. Pavlicevic 1995. Correlación de parámetros magnéticos con la concentración de óxido ferroso en sedimentos cuaternarios de la localidad de Hernández, La Plata, Provincia de Buenos Aires. *IV Jornadas Geológicas y Geofísicas Bonaerenses*, Actas I:177-185. Junín.

Bidegain, J.C., R. Pavlicevic, R.R. Iasi & R.H. Pérez 1996. Susceptibilidad magnética y concentraciones de FeO en loess y paleosuelos cuaternarios como indicadores de cambios paleoambientales y paleoclimáticos. *XIII Congreso Geológico Argentino y III Congreso de Explotación de Hidrocarburos*, Actas II: 521-535. Buenos Aires.

Bobbio, M.L., S.M. Devincenzi, M.J. Orgeira & D.A.Valencio 1986. La magnetoestratigrafía del Ensenadense y Bonaerense de la ciudad de La Plata; su significado geológico. *Revista de la Asociación Geológica* 51(1-2): 7-22. Buenos Aires.

Candle, S.C. & D.V. Kent 1995. Revised calibration of the geomagnetic polarity time scale for the late Cretaceous and Cenozoic. *Journal of Geophysical Research*, 100: 6093-6095.

Cortelezzi, C.R. 1978. Estratificación convoluta en sedimentos de la Formación Ensenada, Pleistoceno de los alrededores de La Plata, Provincia de Buenos Aires, República Argentina. *VII Congreso Geológico Argentino*, Actas, II: 683-693. Neuquén.

Cortelezzi, C.R., J.C. Bidegain & A.V. Parodi 1994. La presencia de alofano en sedimentos de los alrededores de la Ciudad de La Plata, Provincia de Buenos Aires. *Segunda Reunión Argentina de Mineralogía y Metalogenia, Instituto de Recursos Minerales*. UNLP 3: 443-448. La Plata.

Duchaufour, Ph. 1987. *Manual de Edafología*. Barcelona: Masson. SA. 21-24. Spain.

Fidalgo, F. & O. Martínez 1983. Algunas características geomorfológicas dentro del partido de La Plata, Provincia de Buenos Aires. *Revista de la Asociación Geológica Argentina* 38(2): 263-273. Buenos Aires.

Frenguelli, J. 1957. Geografía de la República Argentina. Neozoico. T. II, Tercera Parte. *Sociedad Argentina de Estudios Geográficos*. Buenos Aires.

Heller, F. & T. Liu 1982. Magnetostratigraphy dating of loess deposits in China. *Nature* 300(5891): 431-433.

Heller, F. & T. Liu 1984. Magnetism of Chinese loess deposits. *Geophysical Journal of Astronomy Society*, 77: 125-141.

Heller, F., L. Xiuming, T. Liu & X. Tongchun 1991. Magnetic susceptibility in loess in China. *Earth and Planetary Science Letters*, 103: 301-310. Amsterdam: Elsevier.

Maher, A.B. & R. Thomson 1991. Mineral magnetic records of the Chinese loess and palaeosols. *Geology* 19: 3-6. USA.

Nabel, P. & D.A. Valencio 1981. La magnetoestratigrafía del Ensenadense de la ciudad de Buenos Aires, su significado geológico. *Revista de la Asociación Geológica Argentina* 36(1): 7-18. Buenos Aires.

Orgeira, M.J. 1990. Palaeomagnetism of late Cenozoic fossiliferous sediments from Barranca de los Lobos (Buenos Aires Province, Argentina). The magnetic age of the South America land-mammal ages. *Physics of the Earth and Planetary Interiors*, 64: 121-132. Amsterdam: Elsevier Science Publishers B.V.

Pascual, R., E.J. Ortega Hinojosa, D. Gondar & E. Tonni 1965. Las edades del Cenozoico mamalífero de la Argentina, con especial atención a aquellas del territorio bonaerense. *Anales de la Comisión de Investigaciones Científicas de la Provincia de Buenos Aires* 1(VI): 165-193. La Plata.

Re, G. & M.J. Orgeira 1991. Estudio paleomagnético de una secuencia de sedimentos del 'Ensenadense-Bonaerense' del subsuelo de la ciudad de Buenos Aires. *Revista de la Asociación Geológica Argentina* 46(3-4): 159-166. Buenos Aires.

Riggi, J.C., F. Fidalgo, O. Martínez & N. Porro 1986. Geología de los 'Sedimentos Pampeanos' en el partido de La Plata. *Revista de la Asociación Geológica Argentina* 41(3-4): 316-333. Buenos Aires.

Ruocco, M.I. 1990. Palaeomagnetic analyses of continental deposits of the last 3 Ma from Argentina: Magnetostratigraphy and fine structures of reversals. Ph.D. Thesis University of Stockholm, Akademitryck, AB, Edsbruk. Sweden.p.100.

Sylwan, C. 1990. Palaeomagnetism, Paleoclimate and Chronology of Late Cenozoic Deposits in Southern Argentina. Ph.D. Thesis. University of Stockholm, Akademitryck, A.B. Edsbruk, Sweden. p. 110. Unpublished.

Tarling, D.H. & F. Hrouda 1993. *The magnetic anisotropy of rocks*. Chapman & Hall, p. 217.

Teruggi, M. E. 1957. The nature and origin of Argentine Loess. *Journal of Sedimentary Petrology* 27(3): 322-332.

Teruggi, M.E. & P.A. Imbellone 1987. Paleosuelos loéssicos superpuestos en el Pleistoceno Superior-Holoceno de la región de La Plata, Provincia de Buenos Aires, Argentina. *Ciencia del Suelo* 5(2): 175-188. Buenos Aires.

Tonni, E.P, W.D. Berman, F. Fidalgo, O. Gentile & H. Correa 1988. La fauna local Hernández (Pleistoceno tardío), Partido de La Plata, Provincia de Buenos Aires; y sus sedimentos portadores. *Segundas Jornadas Geológicas Bonaerenses*, Actas: 67-78. Bahía Blanca.

VI Congreso Latinoamericano de Ciencias del Mar
Mar del Plata, October 1995: Selected Papers

Edited by

FEDERICO I. ISLA
Centro de Geologica de Costas y del Cuaternario,
Universidad Nacional de Mar del Plata, Mar del Plata, Argentina

Foreword

'No Geology, without Marine Geology'. These words of P.H. Kuenen (1950) envisaged the Geological or Wegenerian Revolution in knowledge. Today the history of the Third Planet could not be handed without considering the processes that occurred in the oceans.

It is assumed that man's activity is altering the biogeochemical cycles of carbon (global warming) and water (sea level rise). When we are supposedly going into the end of the Present Interglacial, we are extending the warmth into unexpected conditions.

Sea-level projections are forecasting a threat to our coastal cities while planners and decision-makers seemed deaf, and coastal defences are failing one after another. In this scenario, coastal geologists should not only point out regional problems but propose effective solutions. In doing so, the Holocene trend or perspective should not be disregarded. We ought to discriminate the natural component from man's impact on coastal systems.

In our South American continent, the Holocene coastal evolution surges up as necessary to understand problems not referred to in our short history. We should learn from 'developed' experiences that failed to solve coastal problems until regional or long-term processes were comprehended. During the VI Latin American Congress on Marine Sciences (Mar del Plata, October 23-27, 1995), this Symposium on Coastal Holocene Evolution was held in order to review the knowledge about sea-level trends, erosion rates and environmental changes related to the coast. These papers wish to be only a guide to encourage new studies, and to stress the advantages of coordinating them into programs, such as the present IGCP 367 (International Union of Geological Sciences), LOICZ (International Geosphere-Biosphere Program), OSNLR (Intergovernmental Oceanographic Commission) and the INQUA's Shoreline Commission.

Federico I. Isla
Mar del Plata, May 23, 1996

Pleistocene wave-built terraces of Northern Rio de Janeiro State, Brazil

13

LOUIS MARTIN
ORSTOM – Centre de Bondy, Bondy, Cedex, France
JEAN-MARIE FLEXOR
Observatório Nacional/CNPq, Rio de Janeiro, Brazil
KENITIRO SUGUIO
Instituto de Geociências/USP, São Paulo, Brazil

ABSTRACT: Previous studies of a great portion of eastern and southeastern coasts of Brazil have frequently shown the presence of two distinct generations of sandy terraces of marine origin. The most recent is Holocene in age, being related to the last sea-level rise, whose maximum elevation was attained about 5100 years BP. The most ancient is Pleistocene in age, being related to the penultimate sea-level rise, whose culmination stage occurred about 123,000 years BP. In Northern Rio de Janeiro State, these two generations of marine terraces are present. However, it has not been possible yet to obtain absolute ages of these sandy deposits.

RESUMO: Os estudos prévios de grande trecho das costas leste e sudeste do Brasil mostraram freqüentemente a presença de duas gerações distintas de terraços arenosos de origem marinha. O mais recente é de idade holocênica, estando relacionado à última subida de nível do mar, cujo máximo ocorreu há aproximadamente 5100 anos AP. O mais antigo é de idade pleistocênica, estando ligado à penúltima subida de nível do mar, cujo máximo foi verificado há cerca de 123,000 anos AP. Na porção norte do Estado do Rio de Janeiro, essas duas gerações de terraços marinhos estão presentes, embora não tenha sido possível obter as idades absolutas desses depósitos arenosos.

1 INTRODUCTION

Studies performed along the Brazilian coast, during the last decades, indicated the existence of records related to several high sea-levels during the Quaternary (Villwock et al. 1986, Martin et al. 1988). The most recent is quite well known thanks to numerous radiocarbon dates, which have allowed to delineate relative sea-level fluctuation curves during the last 7000 years, for several sectors of the Brazilian coast. In spite of their differences in amplitude, these curves show an essential fact that, during the last 7000 years, the relative sea-level was higher than the present in most part of the Brazilian coastline. This coast was under submergence conditions until about 5100 years BP, but it reversed into emergence after 5100 years BP, with the intercalation of brief submergence episodes (Suguio et al. 1985, Martin et al. 1987).

Evidence of the penultimate high sea-level position is given by huge sandy terraces extending, at least, from the Paraíba State to the Rio Grande do Sul State. The marine origin of these sandy deposits is confirmed by the presence of fossilized *Callichirus* burrows (Suguio & Martin 1976, Suguio et al. 1984), as well as by the occurrence of typical syngenetic sedimentary structures (Suguio & Tessler 1987, Tessler & Suguio 1987). At the surface of these terraces there are ancient beach-ridge alignments, with some differences when compared with the Holocene terraces (Martin et al. 1981). These sands are in general whitish in surface but in subsurface they may show a dark brownish colour due to humic and/or fulvic acids impregnation. In opposition to the Holocene terraces, where mollusk shells are frequently found, they are absent in the Pleistocene terraces, probably because they were dissolved by organic acids (Dehira & Suguio 1994). However, preserved shells were sampled from a clayey layer at the base of a Pleistocene terrace, which gave a radiocarbon age older than 30,000 years BP. In Brazil, only in one location in southern Bahia State, more-or-less preserved coral samples have been randomly collected from the basal portion of a Pleistocene terrace. They belong to the genus *Siderastrea* and only five of them, composed of pure aragonite, have been dated by the uranium/ionium method (Martin et al. 1982).

Samples CP-1, CP-2, CP-6 and CP-8 indicated ages close to their overall mean: 123,500 + 5700 years BP. Only sample CP-7 showed an age significantly older, probably due to slight contamination. Therefore, it is possible to assume that the maximum of the penultimate high sea-level occurred about 123,000 years BP, in agreement with available data from the majority of localities around the world.

In most of Brazilian eastern and southeastern coastlines the relative sea-level corresponding to this transgressive episode is situated 8 ± 2 m above the present level. Nevertheless, in some sectors like in the Todos os

Table 1. U/Io dates of coral samples from the State of Bahia (Martin et al. 1982).

Sample	$^{234}U/^{238}U$	U(ppm)	^{234}U(dpm/g)	Io(^{230}Th)(dpm/g)	Age(yr BP)
CP-1	1.07	2.59	2.08	1.43	122,000 ± 6100
CP-2	1.08	2.70	2.17	2.42	116,000 ± 6900
CP-6	1.11	3.09	2.57	1.85	132,000 ± 9000
CP-7	1.11	2.58	2.14	1.61	142,000 ± 9700
CP-8	1.08	2.56	2.30	1.60	124,000 ± 8700

Santos Bay (State of Bahia), these records could be lower as a consequence of modern tectonic movements (Martin et al. 1984a).

2 RECORDS OF THE 123,000 YEARS BP HIGH SEA LEVEL AT NORTHERN RIO DE JANEIRO STATE

A preliminary geologic mapping of the Quaternary deposits was enough to make evident the existence of at least two distinct generations of sandy marine terraces in several sectors of the northern Rio de Janeiro State (Fig. 1). Despite the absence of absolute datings, some arguments based on previous experience allow to suggest that the higher terraces are probably Pleistocene in age. Considering the inadequacy of the radiocarbon dating method, as well as the absence of adequate samples for U/Io dating, the following criteria have been used:

a) ages beyond the reliability of the radiocarbon method (more than 30,000 years BP);

b) morphological features of the deposits, particularly the alignment characteristics of the ancient beach-ridges;

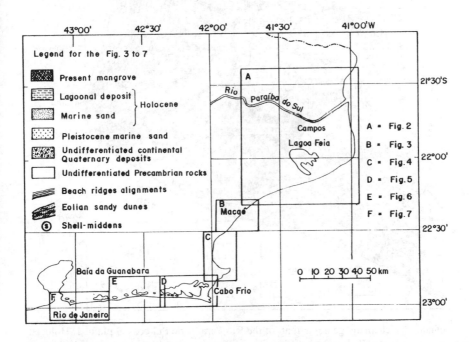

Figure 1. Location map of the studied area, geographical position corresponding to Figures 2-7 and legend (Figs 3-7) of the maps.

235

c) the absence of mollusk shells, and sands commonly impregnated by humic and/or fulvic acids with a slight cohesion;

d) their positions in the coastal plain, as well as their heights; and,

e) overconsolidation of clay beds beneath the sands (Massad 1985).

Previous works revealed that in the coastal plain of the Rio Paraíba do Sul estuary (Fig. 2) there are two distinct generations of sandy wave-built terraces (Martin et al. 1984b, 1987). In its southern portion, between Ma-

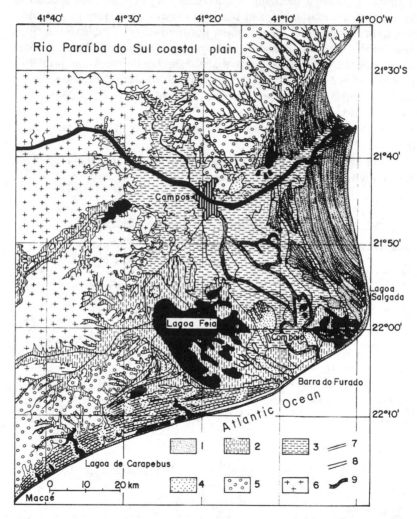

Figure 2. Schematic geologic map of the Rio Paraíba do Sul coastal plain. 1. Holocene marine terrace, 2. lagoonal sediments, 3. fluvial sediments (intralagoonal delta), 4. Pleistocene marine terrace, 5. Barreiras Formation, 6. Precambrian rocks, 7. alignments of Pleistocene beach-ridges, 9. palaeochannels of the Rio Paraíba do Sul (from Martin et al. 1987).

236

caé and Barra do Furado, there is a narrow sand barrier landward from the sea, followed by an almost desiccated lagoonal lowland and an extensive sand flat overlying Barreiras Formation or Precambrian crystalline rocks.

The morphological characteristics of the regressive beach-ridges covering this terrace is quite different from the beach-ridges found around the Rio Paraíba do Sul outlet and, on the other hand, perfectly comparable to that occurring at southern Bahia State. Moreover, it is possible to recognize that the surface of this terrace was intensively dissected by a drainage net during the last Pleistocene relative low sea-level (i.e., Oxygen Isotope Stages 2-4). For example, the Lagoa de Carapebus (Fig. 2) is situated in a valley excavated within this terrace, and was subsequently drowned during the last relative sea level-rise. Obviously, this valley was formed before 7000 years BP, when the present sea-level was overpassed during the last transgression.

Radiocarbon ages of 6590 ± 250, 6620 ± 240, 6000 ± 230 and 5930 ± 240 years BP, obtained in mollusk shells sampled from lagoonal deposits situated between sandy beach-ridges, suggest that the inland deposits covered by regressive beach-ridges are older than 7000 years BP (Martin et al. 1984b). In the remaining portion of the coastal plain, always situated landward of the Holocene deposits, there are more-or-less eroded records of these ancient deposits. Thus, in the Comboio region, at the center of the Rio Paraíba do Sul outlet coastal plain (Fig. 2), the elongated lowlands within sandy deposits correspond to areas situated between ancient beach-ridges, which were partially excavated during the last sea-level drop. These sandy deposits cannot be called chêniers, as it was done by Dias (1981), because they are not covering clayey sediments. A radiocarbon age of 6000 ± 200 years BP, obtained from mollusk shells sampled from lagoonal deposits, is also indicating that the fluvial excavation was before 7000 years BP, and the drowning of the lowlands occurred during sea-level rise. Considering that in this area only two Quaternary high sea-levels have been recognized until today and that the most recent is Holocene in age, it is possible to draw the conclusion that the older ancient deposits must correspond to the high sea-level of 123,000 years BP.

The Pleistocene marine deposits forming an essential part of the southern coastal plain of the Rio Paraíba do Sul extend until north of the site of Macaé. However, from the Lagoa de Carapebus to Macaé (Fig. 2), the intermediate lowland disappears and the Holocene deposits are reduced to present beaches, which have a tendency to advance over the Pleistocene sands; as they have different colours, this phenomenon is clearly visible. The lower elevation of the Pleistocene terrace seems to suggest that this area was under permanent subsidence during the Quaternary. This hypothesis is apparently confirmed by a gravimetric anomaly indicated in the Bouguer map of the area (Martin et al. 1984b).

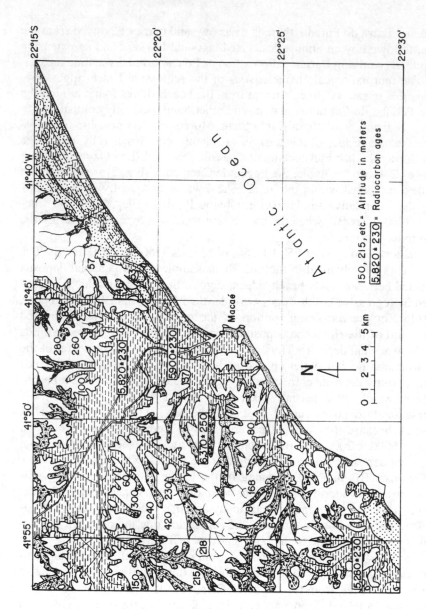

Figure 3. Schematic geologic map of the Macaé region.

238

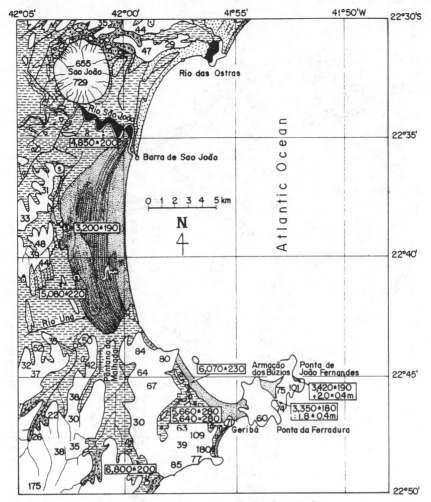

Figure 4. Schematic geologic map of the Barra de São João region.

In the Rio das Ostras area (Figs 3 and 4), most of the sandy terraces are assigned to the Pleistocene. To the north, these deposits are relatively important, and the Lagoa Salgada shows morphological characteristics typical of a lagoon installed within a zone excavated during a sea-level lower than the present, and invaded during the transgression. Between Rio das Ostras and Barra de São João, the Pleistocene deposits are reduced to a narrow band in the oceanic margin besides some continental remains, frequently impregnated by humic and/or fulvic acids. In this area, the Holocene deposits are represented only by the present beaches.

In the internal portion of the large valleys excavated by the Rio São

239

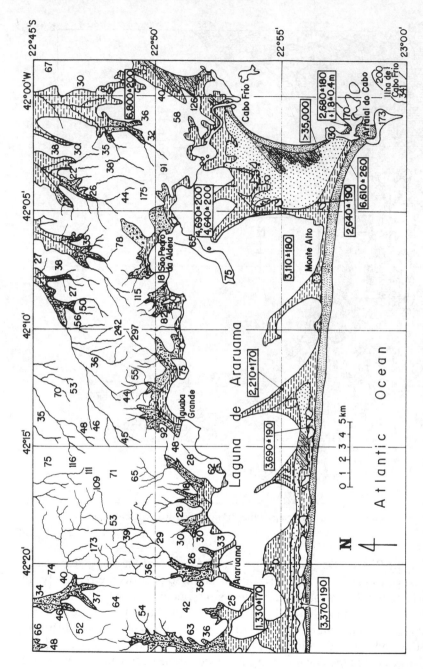

Figure 5. Schematic geologic map of the Araruama lagoon region.

240

João and the Rio Una (Fig. 4), similarly as it takes place in the Rio Itabapoana valley in the limit of the states of Rio de Janeiro and Espírito Santo, there are two generations of sandy marine terraces. The external deposits are more developed and covered by typical Holocene beach-ridges. More inlandwards, there are eroded relicts of Pleistocene terraces.

Between Cabo Frio and Arraial do Cabo (Fig. 5), most of the sandy deposits are of Pleistocene age. This statement is based on a radiocarbon date of more than 30,000 years BP, obtained from a wood fragment sampled in a sand with abundant impregnation by organic matter. These Pleistocene deposits are partially covered by Holocene aeolian dunes and they were preserved from erosion during the last transgression behind crystalline rock hills of the Arraial do Cabo.

From Arraial do Cabo to Niterói (Figs 5, 6 and 7) there are, frequently, two sandy barriers separated by a depression, sometimes occupied by small lagoons, like the Lagoa Vermelha (Fig. 6). Muehe (1982) and Coe Neto et al. (1986), based on relative sea-level fluctuation curves proposed by Suguio & Martin (1981) and Martin et al. (1983), assigned a Holocene age for both units. The inner barrier would have been formed during the high sea-level of 5100 years BP, and the second could be related to the high sea-levels of 3600 and 2500 years BP (Coe Neto et al. 1986). However, a sample of mollusk shells collected from the bottom of Lagoa Vermelha indicated a radiocarbon age of 4830 ± 280 years BP. This age shows that the lagoon and, consequently, the external barrier is older than 5000 years BP. Moreover, drilling made in the eastern extreme lowland of the Brejo do Espinho attained an organic matter saturated sand, probably Pleistocene in age, at a depth of 3 m. Finally, in the Praia Seca area, there are outcrops of organic matter saturated sand from the internal barrier.

On the other hand, in the Itaipu-Açu area, Ireland (1987), in his work on the sedimentary history of the lagoons in this region, stated: 'It is clear that the present barrier (external barrier) is older than 2770 years BP. It is suggested that the fossil barrier (internal barrier) and the sedimentary sequence behind it is Pleistocene in age. This formation is considered to be related to the Cananéia Formation which has been described in São Paulo State...'

3 CONCLUSIONS

In the absence of absolute ages, it seems reasonable to attribute a Holocene age to both sandy barriers occurring between Arraial do Cabo and Itaipu-Açu (Muehe 1982, Coe Neto et al. 1986), like the barriers found in the Jacarepaguá coastal plain (Maia et al. 1984). The relative sea-level oscillations during the last 7000 years, which are clearly exposed in deposits

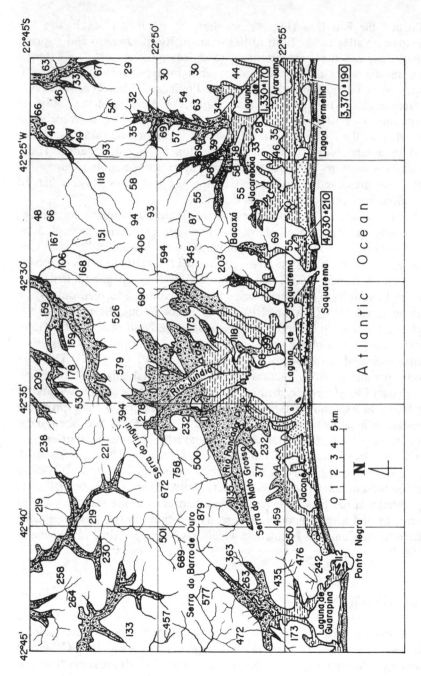

Figure 6. Schematic geologic map of the Saquarema lagoon region.

242

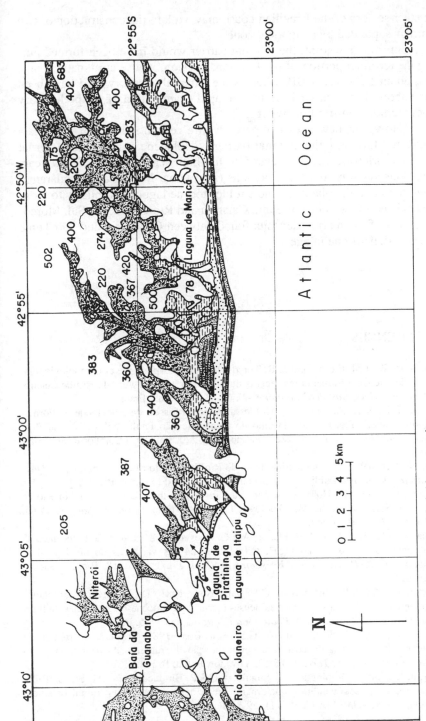

Figure 7. Schematic geologic map of the Maricá lagoon region.

in other sectors of the Brazilian coast, may explain the construction of two barriers separated by a lagoonal zone.

From this viewpoint, the internal barrier would have been formed during the terminal portion of the last transgression, whose culmination stage was about 5100 years BP, whereas the external barrier would have been built about 3600 years BP. In fact, a rapid examination of the preliminary ages seem to confirm this hypothesis.

However, new ages compared with a certain number of other data suggested that only the external barrier is Holocene, whereas the internal barrier is Pleistocene, and related to the high sea-level of 123,000 years BP. Consequently, the large lagoons (Araruama, Maricá and Saquarema) were settled at the places of ancient Pleistocene lagoons, similarly to those that existed in the states of Santa Catarina and Rio Grande do Sul. Moreover, most of the marine deposits found between Cabo Frio and São Tomé are also Pleistocene in age.

REFERENCES

Coe Neto, R., J.M. Froidefond & B. Turcq 1986. Géomorphologie et chronologie relative des dépôts sédimentaires récents du littoral brésilien a l'est de Rio de Janeiro. *Bull. Inst. Géol. Bassin d'Aquitaine*, 40: 67-83, Bordeaux, France.

Dehira, L.K. & K. Suguio 1994. Moldes de conchas fósseis de pelecípode na Formação Cananéia, Pleistoceno Marinho do Estado de São Paulo. *38° Congresso Brasileiro de Geologia*, Resumos expandidos: 219-220. Balneário Camboriú (SC), Brazil.

Dias, G.T.M. 1981. O complexo deltaico do Rio Paraíba do Sul. *IV Simpósio do Quaternário no Brasil*, Publicação Especial 2: 38-88, Rio de Janeiro. Brazil.

Ireland, S. 1987. The Holocene sedimentary history of the coastal lagoons of Rio de Janeiro State, Brazil. In: M.J. Tooley & I. Shennan (Eds), *Sea-level changes: 25-65,* London: *Basil Blackwell.*

Maia, M.A.C., L. Martin, J.M. Flexor, A.E.G. Azevedo 1984. Evolução holocênica da planície costeira de Jacarepaguá, R.J.: Influência das variações do nível do mar. *Anais do XXXIII Congresso Brasileiro de Geologia*, 1: 105-118. Rio de Janeiro. Brazil.

Martin, L., A.C.S.P. Bittencourt, G.S. Vilas-Boas 1981. Différentiation sur photographies aériennes des terrasses sableuses marines pléistocènes et holocènes du littoral de l'état de Bahia, Brésil. *Photointerprétation* 3, fasc. 4/5. Paris.

Martin, L., A.C.S.P. Bittencourt & G.S. Vilas-Boas 1982. Primeira ocorrência de corais pleistocênicos da costa brasileira: datação do máximo da penúltima transgressão. *Ciências da Terra* 1: 16-17, Salvador, Bahía. Brazil.

Martin, L., J.M.L. Dominguez, K. Suguio, A.C.S.P. Bittencourt & J.M. Flexor 1983. Schéma de la sédimentation quaternaire sur la partie centrale du littoral brésilien. *Cahiers ORSTOM, Sér. Géol.* 8(1): *59-81.* Paris.

Martin, L., A.C.S.P. Bittencourt, J.M. Flexor, G.S. Vilas-Boas 1984a. Evidência de um tectonismo quaternário nas costas do Estado da Bahia. *Anais do XXXIII Congresso Brasileiro de Geologia* 1: 19-35. Rio de Janeiro.

Martin, L., K. Suguio, J.M.L. Dominguez, J.M. Flexor, A.E.G. Azevedo 1984b. Evolução da Planície Costeira do Rio Paraíba do Sul durante o Quaternário: Influência das variações do nível do mar. *Anais do XXXIII Congresso Brasileiro de Geologia* 1: 84-97. Rio de Janeiro.

Martin, L., K. Suguio, J.M. Flexor, J.M.L. Dominguez & A.C.S.P. Bittencourt 1987. Quaternary evolution of the central part of the Brazilian coast. The role of relative sea level variation and of shoreline drift. *UNESCO Reports in Marine Science* 43: 97-145.

Martin, L., K. Suguio & J.M. Flexor 1988. Hauts niveaux marins pléistocènes du littoral brésilien. *Palaeogeography, Palaeoclimatology, Palaeoecology*, 68(2/4): 231-239.

Massad, F. 1985. *Progressos recentes dos estudos sobre as argilas quaternárias da Baixa- da Santista*. Associação Brasileira de Mecânica dos Solos (ABMS). São Paulo. 21 pages.

Muehe, D. 1982. The coastline between Niterói and Ponta Negra. *International Geographical Union. Commission on Coastal Environment Field Trip*: 23-27. Rio de Janeiro.

Suguio, K. & L. Martin 1976. Presença de tubos fósseis de *Callianassa* nas formações quaternárias do litoral paulista e sua utilização na reconstrução paleoambiental. *Boletim IG, Instituto de Geociências*, Vol. 7: 17-26, São Paulo.

Suguio, K. & L. Martin 1981. Progress in research on Quaternary sea-level changes and coastal evolution in Brazil. *Proceedings of the Symposium: Holocene Sea Level Fluctuations, Magnitude and Causes*. Department of Geology, University of South Carolina, 166-181, Columbia, USA.

Suguio, K. & M.G. Tessler 1987. Characteristics of a Pleistocene nearshore deposit: an example from southern State of São Paulo coastal plain. *Quaternary of South America and Antarctic Peninsula* 5: 257-268. Rotterdam: Balkema Publishers.

Suguio, K., S.A. Rodrigues, M.G. Tessler & E.E. Lambooy 1984. Tubos de 'ophiomorphas' e outras feições de bioturbação na Formação Cananéia, Pleistoceno da planície costeira Cananéia-Iguape. In: L.D. Lacerda et al. (orgs.) *Restingas: origem, estruturas, processos:* 111-122. Niterói, Rio de Janeiro, Brazil.

Suguio, K., L. Martin, A.C.S.P. Bittencourt, J.M.L. Dominguez, J.M. Flexor & A.E.G. Azevedo 1985. Flutuações do nível relativo do mar durante o Quaternário Superior ao longo do litoral brasileiro e suas implicações na sedimentação costeira. *Revista Brasileira de Geociências* 15(4): 273-286.

Tessler, M.G. & K. Suguio 1987. Características sedimentológicas da Formação Cananéia (Pleistoceno Superior) na área de Paranaguá-Antonina (Estado do Paraná, Brasil). *Proceedings of the IGCP-Project 201 (Quaternary of South America) meeting*: 43-54. Mérida, Venezuela.

Villwock, J.A., L.J. Tomazelli, E.L. Loss, E.A. Dehnhardt, N. Horn Filho, F.A. Bachi & B.A. Dehnhardt 1986. Geology of the Rio Grande do Sul Coastal Province. *Quaternary of South America and Antarctic Peninsula* 4: 79-97. Rotterdam: Balkema Publishers.

Holocene coastal evolution of Tierra del Fuego, Argentina

14

GUSTAVO G. BUJALESKY
CONICET, Centro Austral de Investigaciones Científicas, Ushuaia,
Tierra del Fuego, Argentina

ABSTRACT: The analysis of the evolutive trend of the Fuegian coast during the Holocene must consider the following facts: a) the Beagle Channel is located at the active seismotectonic setting of the Fuegian Andes (Scotia Plate Domain). The northeastern Atlantic coast is situated at the extraandean lowlands, a more stable tectonic environment conformed over a Mesozoic undeformed platform (South-American Plate Domain). b) The last Pleistocene glaciation covered entirely the Beagle Channel area; the Atlantic coast was ice-free during this episode. c) The Beagle Channel (5 km wide) is under microtidal range and small short-period wind waves conditions. It presents an indented rocky shore where pocket gravel beaches develop. The macrotidal Atlantic coast shows cliffs carved on Quaternary glacial deposits and Tertiary sediments, with extensive gravel beaches. This coast is affected by long-period swells or significant storm waves. The northern coast of the Beagle Channel is characterized by a system of terraces where at least three levels have been established at 8-10 m, 4-6 m, 1.5-3 m (over MSL). These Holocene raised beaches are reaching maximum elevations of nearly 10 m above MSL with ages of approximately 6000 ^{14}C yr BP. The estimated average tectonic uplift is of approximately 1.5 to 2.0 mm/yr for this period. At the northeastern Atlantic coast, the marine deposits of Bahía San Sebastián of about 5000 ^{14}C yr BP are at an altitude of 1.8 m above MSL. This supratidal seaward gradient would be partly due to diminishing wave height within the bay as a consequence of El Páramo spit growth and a gradual protection from Atlantic waves.

The tectonic uplift during Holocene times reached a maximum at the western Beagle Channel area, diminishing northwards and eastwards. It seems to be negligible at the Bahía San Sebastián and Bahía Thetis areas, where also littoral forms have developed under relatively stable eustatic conditions.

247

RESUMEN: El análisis de las tendencias evolutivas de la costa de Tierra del Fuego durante el Holoceno debe tener en cuenta las siguientes premisas: a) la costa del Canal Beagle se ubica en el área sismotectonicamente activa de los Andes Fueguinos (dominio de la Placa de Scotia) y la costa atlántica nororiental se sitúa en un ambiente más estable, extrandino, desarrollado sobre una plataforma de rocas mesozoicas no deformadas (dominio de la Placa Sudamericana); b) La última glaciación cubrió el Canal Beagle; mientras que la costa atlántica estuvo libre de hielo durante ese evento; c) El Canal Beagle, de 5 Km de ancho, presenta un régimen micromareal y olas de viento de período corto y poca altura, y presenta una costa rocosa dentada con desarrollo de playas de grava de bolsillo en bahías. La costa atlántica (macromareal) muestra acantilados labrados en depósitos glacigénicos y sedimentos del Terciario y extensas playas de grava. Está sometida a olas de mar de leva de período largo o fuertes olas de tormenta.

La costa norte del Canal Beagle está caracterizada por sistemas de terrazas, en los que se han reconocido al menos tres niveles 8-10 m, 4-6 m, 1,5-3 m. Las playas elevadas del Holoceno alcanzan una cota máxima de 10 m s.n.m. con una edad de aproximadamente 6000 ^{14}C años AP. La tasa promedio de ascenso tectónico estimada es de 1,5 to 2,0 mm/año para los últimos 6000 años. En la costa atlántica nororiental, los depósitos litorales supramareales de la Bahía San Sebastián tienen una edad de 5000 ^{14}C años AP y una altitud de 1,8 m s.n.m. Este gradiente se debería parcialmente a la disminución gradual de la energía de olas que alcanzaba la cabecera de la bahía como consecuencia del crecimiento de la espiga El Páramo.

El levantamiento tectónico durante el Holoceno alcanzó un máximo en la sección oriental del Canal Beagle, disminuyendo gradualmente en dirección este y norte y podría considerarse despreciable en el área de Bahía San Sebastián, donde las formas litorales habrían evolucionado bajo condiciones eustáticas relativamente estables.

1 INTRODUCTION

The Argentine sector of the Isla Grande de la Tierra del Fuego is located between lat. 52°40' S-55°7' S and long. 65°05'-68°40'W (Fig. 1). The eastern Atlantic coast extends for 330 km along, affected by a NW-SE trend. It presents a macrotidal range, and it is exposed to high energy Atlantic waves and strong and intense westerly winds. Extensive and wide beaches and littoral forms are composed of gravel and coarse sand. Pleistocene glacigenic deposits conform high and cliffs at its northern section; these and other submerged glacial deposits have supplied the sediments for beach generation.

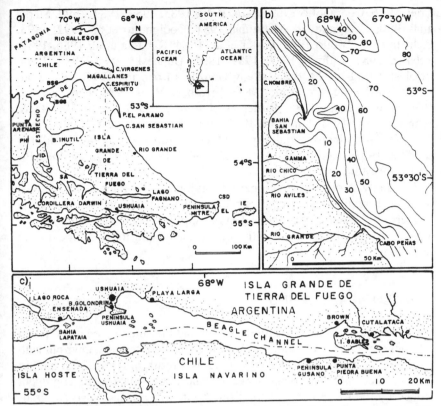

Figure 1. Location maps. a) BSG Bahía San Gregorio, BGG Bahía Gente Grande, PH Puerto Hambre, ID Isla Dawson, SA Seno del Almirantazgo, EL Estrecho de Lemaire, IE Isla de los Estados, CSD Cabo San Diego; b) Northern Atlantic coast of Tierra del Fuego, depths in meters referred to spring low tide level (SLTL); c) Beagle Channel.

The southernmost coasts of Tierra del Fuego (northern Beagle Channel coast and southern Atlantic coast) extend for 220 km in a W-E trend. It presents an indented rocky shoreline, where pocket gravel beaches develop in the embayments. Beagle Channel occupies a drowned glacial valley and connects the Atlantic and Pacific oceans at this latitude. This channel is 5 km wide, its average depth ranges between 100 and 450 m and it presents a microtidal range. Holocene raised beaches can be recognized in many places along the southern coast of Tierra del Fuego and their elevations vary considerably.

Several authors have studied the Holocene and Present coastal deposits of Tierra del Fuego (Nordenskjöld 1898, Popper 1887, 1891, Andersson 1906, Halle 1910, Bonarelli 1917, Methol & Sister 1947, Petersen & Methol 1948, Petersen 1949, Feruglio 1950, Auer 1956, 1959, 1970, 1974,

249

Urien 1966, Codignotto 1969, 1976, 1979, 1983, 1984, 1987, 1990, Markgraf 1980, Codignotto & Malumián 1981, Codignotto et al. 1992, 1993, Etchichury & Tófalo 1981, Porter et al. 1984, Rabassa et al. 1986, 1989, Rabassa 1987, Mörner 1987, 1989, 1991, Bujalesky et al. 1987, 1995, Bujalesky 1988, 1990, Bujalesky & González Bonorino 1991, Vilas et al. 1987a, 1987b, Ferrero et al. 1987, Ferrero & Vilas 1988, Ferrero et al. 1989, Kokot et al. 1988, Rutter et al. 1989, Isla 1989, 1994, Isla & Schnack 1989, 1991, Isla et al. 1991, 1994, Isla & Bujalesky 1993, 1995, González Bonorino & Bujalesky 1990, Gagliardo 1990, 1994, Gordillo et al. 1992, 1993, Gordillo 1989, 1990a,b, 1991, 1992, 1993; Gordillo & Piñero 1989).

The analysis of the evolutionary behaviour of the Fuegian coast must consider dissimilar characteristics between the northern and southern parts of Tierra del Fuego, concerning tectonic uplift, glacio-isostatic rebound, wave and tide climate, availability of sediments and local hydrodynamic conditions.

2 GEOLOGICAL SETTING

The northern coast of the Beagle Channel and the southern coast of Tierra del Fuego are located within the Andean Cordillera tectonic environment. The Fuegian Andes present a W-E trend as a result of a transform motion between the South American, Antarctic and Scotia plates. The alignment formed by the western end of the Estrecho de Magallanes, Seno Almirantazgo and Lago Fagnano marks the South American-Scotia plates boundary and the northern limit of the left lateral transpressional transform motion. Recent fault scarps, sag ponds and landslides along this alignment indicate an important tectonic activity (Winslow 1982, Dalziel 1989). The basement is composed of pre-Jurassic highly deformed metamorphic rocks (Lapataia Formation, Borrello 1969), covered by Late Jurassic deformed acidic volcanic and pyroclastic rocks (Lemaire Formation, Borrello 1969). These formations have been interpreted as the Early Cretaceous marginal basin floor. The stratigraphic scheme follows with Tithonian to Early Cretaceous marine sedimentary rocks, with low-grade metamorphism (Yaghan Formation, Kranck 1932). The Mesozoic rocks are overlain (angular unconformity) by Palaeocene marine beds (Río Claro Formation, Caminos et al. 1981) and continental deposits (Sloggett Formation; Caminos et al. 1981). Several deformation phases related to the Gondwanic (Late Palaeozoic-Early Mesozoic), Patagonidic (Cretaceous) and Andic (Eocene-Miocene) movements affected the sedimentary sequence (Caminos et al. 1981).

The northern Atlantic coast of Tierra del Fuego lies on a more stable setting, away from the Andean foldbelt. The Atlantic lowlands developed on a stable platform, composed of non-deformed rocks of the Springhill Formation (Late Jurassic-Early Cretaceous; Thomas 1949). The oldest exposed sediments in this area are Tertiary rocks of continental or marine origin (Codignotto & Malumián 1981). Plio-Pleistocene glacial deposits overlie the Tertiary rocks and form coastal cliffs up to 90 m high at Cabo Espíritu Santo. This region is supposed to be free of Holocene tectonic uplift.

The northern Atlantic coast of Península Mitre is located at the foot-hill belt (southern boundary of Magallanes or Austral Basin), composed of early to middle Cretaceous marine wackes, shales and sandstones of the Beauvoir Formation (Camacho 1948, Petersen 1949, Furque 1966), Late Cretaceous marine sandstones of Policarpo and Leticia formations (Camacho 1967; Furque & Camacho 1949) to early Tertiary sedimentary rocks (marine limestones and sandstones of Río Bueno Formation, Furque & Camacho 1949; Petersen 1949; Camacho 1967). These formations were deformed during the middle Tertiary (Caminos 1980; Winslow 1982).

The seismicity of the Scotia Arc is well reported by Pelayo & Wiens (1989). Taking into account the probably influence of earthquakes in the evolution of the littoral forms, it is worth to mention between others the following episodes: a) the earthquake of December 17, 1949 (6:53 hs UT), when sinking movements occurred on certain shores of Lago Fagnano, (7.75 degrees in Richter's scale), location of the epicenter according to Castano (1977) lat. 54°06'S, long. 70°30'W; relocated at lat. 53°24'S, long. 69°13'12"W, with error elipse major semi-axis of 112 km, azimuth N78°W, and minor semi-axis of 57 km (Pelayo & Wiens 1989); b) the earthquake of December 17, 1949 (15:07 hs. UT), relocated epicenter at lat. 53°59'24"S, long. 68°46'12"W, error elipse major semi-axis of 62 km, azimuth N75°W, and minor semi-axis of 31 km (Pelayo & Wiens 1989); c) the earthquake of June 15, 1970, with a magnitude of 7.0, epicenter located northwards of Isla de los Estados (lat. 54°18'S, long. 63°36'W) at a focal depth of 6 km (Unesco 1972, Pelayo & Wiens 1989); d) the earthquake of December 29, 1975, magnitude of 6.5, epicenter located in the Drake Passage (lat. 56°48'S, long. 68°30'W) at focal depths of 11 km (Unesco 1979, Pelayo & Wiens 1989).

3 GLACIATIONS

The Pliocene-Quaternary glaciations in Tierra del Fuego have been studied by many authors (Nordenskjöld 1898, Bonarelli 1917, Caldenius 1932, Feruglio 1950, Auer 1956, Codignotto & Malumian 1981, Rabassa et al. 1988, 1989, 1990, 1992, Rabassa & Clapperton 1990, Porter 1989, Meglioli et al. 1990a,b, Meglioli 1992, Coronato 1990, 1993, 1995a,b, Clapperton 1993). Nordenskjöld (1898), Bonarelli (1917) and Codignotto

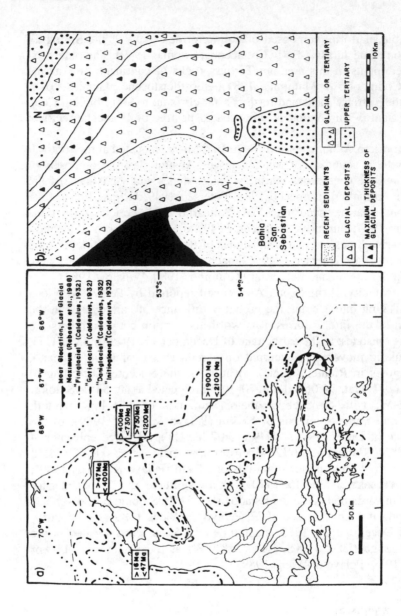

Figure 2. a) Glaciation boundaries in Tierra del Fuego (Caldenius 1932). Dating of glaciations in northern Tierra del Fuego after Meglioli (1992). The outer limit of the Last Glacial Maximum (Moat Glaciation) in the Beagle Channel after Rabassa et al. (1988). b) Distribution of offshore glacial deposits in northern Tierra del Fuego (modified from Total Austral & Geomatter 1980).

252

& Malumián (1981) thought that none of the oldest glaciations totally covered the island. Caldenius (1932) and Auer (1956) considered that the pre-Wisconsinan glaciations extended over the whole of Tierra del Fuego and the glaciers flew northwards and eastwards from the Cordillera Darwin along deep valleys or corridors (Estrecho de Magallanes, Bahía Inútil Bahía San Sebastián, Lago Fagnano and Beagle Channel) reaching the Atlantic Shelf (Fig. 2). Meglioli et al. (1990a,b) and Meglioli (1992) recognized several Plio-Pleistocene glaciations in northern Tierra del Fuego and the oldest one (older than 1.9 Ma BP, Late Pliocene-Nebraskan?) covered the entire island with the exception of a small area (32 km^2 or less) between Río Grande and Bahía San Sebastián. This hypothesis is supported by the evidence of highly weathered till deposits and erratic boulders 9 km westwards of the city of Río Grande. The absence of other till deposits in the drainage basin of the Río Grande is explained considering its geographic position where the melting channels of the major ices lobes flew. Meanwhile, the area northwards of Bahía San Sebastián was located in a higher interlobated position and it was almost undissected by rivers (Meglioli 1992, 1994). The last glaciation (older than 16 and younger than 47 ^{14}C kyr BP, Late Wisconsin) in northern Tierra del Fuego was restricted to the western Estrecho de Magallanes and Bahía Inútil, in the Chilean territory (Porter 1989, Meglioli et al. 1990a,b, Meglioli 1992). The last two Quaternary glaciations (Wisconsinan and Illinoian in age, or older) were recognized along the Beagle Channel, reaching approximately 1400-1500 m thick during Late Wisconsinan times (Rabassa et al. 1990). The Last Glacial Maximum was attained around 18-20 ^{14}C kyr BP and the ice recession started before 14.7 ^{14}C kyr BP (Rabassa et al. 1990).

4 CLIMATE

The climate of Tierra del Fuego Island lacks marked continental influences and is determined by its upper mid-latitude location within the belt of prevailing westerlies (lat. 40° to 60°S), in the path of eastward moving cyclones and not far from the Antarctic ice (Tuhkanen 1992). The Andean Cordillera causes a steep climatic gradient from west to east and from south to north. The mean annual temperature in Ushuaia and Río Grande is around 5 to 6°C, the rainfall is 499 mm/yr in Ushuaia and 340 mm/yr in Río Grande (Servicio Meteorológico Nacional 1986). The most frequent winds have a west to northwest direction in the Magellanic region, a west direction in Río Grande (annual frequency: 39.3%, average velocity: 33 km/h), but in Ushuaia the relief causes deviations, where southwest is the prevailing direction (23.6%, 31 km/h).

The winds are more persistent and stronger in spring and summer than in winter. Köppen (1936) identified three climatic types for Argentine

253

Tierra del Fuego: steppe climate (BSk, north of Bahía San Sebastián), humid temperate climate (Cfc, Bahía San Sebastián and Río Chico area, Ushuaia and eastern margin of Lago Fagnano area), and tundra climate (ETC, Río Grande drainage basin and Península Mitre). According to the system of 'seasonal climates' of Troll & Paffen (1964) the southern part of the island presents an oceanic climate (III.2.) and the northern part displays a dry steppe climate with mild winters (III.10.a.). Walter (1976) considered that almost the whole of Tierra del Fuego belongs to the Antarctic zone (oceanic variant).

5 HYDRODYNAMIC CONDITIONS

The water masses that surround Tierra del Fuego are the result of intense mixture processes because of vertical convective currents between the Pacific and Atlantic oceans waters, that develop at the Drake Passage, northwards of the Antarctic convergence and in region that extends between the Islas Malvinas and the eastern Tierra del Fuego coast (Capurro 1981, Boltovskoy 1981). Northwards of lat. 44°00'S, the Malvinas Current mixes with tropical waters (Brazilian Current). This oceanographic environment between the Antarctic and the Subtropical convergences is named Subantartic.

The Atlantic coast of Tierra del Fuego presents a macrotidal range with a semi-diurnal regime. The mean tidal range at the Atlantic coast increases from south to north (Bahía Thetis: 2.5 m; Cabo San Pablo: 4.6 m; Río Grande: 5.2 m; Bahía San Sebastián: 6.6 m). The flood and ebb currents reach a velocity of 2 knots in NW-SE directions, respectively. The Beagle Channel has a microtidal range and a semi-diurnal regime with diurnal inequalities. Mean tidal range is of 1.1 m at Ushuaia and the tidal wave moves from west to east (Servicio de Hidrografía Naval 1981, 1995).

The wave climate is relatively bening at the Atlantic coast due to the dominance of strong winds from the west. IMCOS Marine Limited (1978) reports data from ship's observations obtained from the British Meteorological Office, covering the sea area from the coast to long. 65°W and between lat. 50°S and 55°S (1949-1968). This data reports that: a) the frequency of wave heights higher than 3.5 m was very low; b) around 20% of waves were less than 1 m in height on average throughout the year; c) the long period waves have low frequency, and wave periods higher than 10 seconds come from the east to northeast; d) gales of 41-47 knots from any direction between N and ESE (with a return period of 50 years) were estimated to generate an extreme wave of height of 12 m and a period 11.5 seconds in a depth of 50 m (referred to spring tide level); e) this estimated extreme wave would break in a water depth of 15 m (chart depth + tidal height above chart datum + storm surge), and would be near breaking point in 10 m depth even at spring high water.

A one year record at the Cullen area (lat. 52°49'19,1"S-long. 68°13'52,3"W, 110 km northwards of the city of Río Grande) gave the following results: a) a maximum wave height of 5.86 m (Compagnie de Recherches et d'Etudes Oceanographiques & Geomatter 1985); b) a maximum significant wave height of 3.43 m; c) an average significant wave height of 1.02 m; d) a maximum period of 17.5 seconds; e) a maximum significant period of 12.9 seconds; f) an average significant period of 5.5 seconds; g) the waves higher than 3 m corresponded to periods of 7 to 9 seconds; h) the longer periods were associated to wave heights of 1.25 m; i) the stronger swell were associate to north-norteastern winds; j) the estimated extreme wave height would be of 5.8 m for NE to E winds and of 7 m for winds from the north considering a return period of 50 years.

The Beagle Channel offers a short fetch to the main southwestern winds and the waves are choppy with periods of 1 to 3 seconds. High wind velocities yield small plunging breakers with heights of up to 0.5 m. The southern Atlantic coast of Península Mitre receive strong open ocean swell from the south but unfortunately there are no available records.

6 COASTAL GEOMORPHOLOGY

6.1 Northern cliffs

Cliffs extend along 40 km from Cabo Espíritu Santo (90 m high) to Cabo Nombre (10 m high) at the northern part of Tierra del Fuego. They are composed of glacial deposits (Drift Tapera Sur sensu Codignotto 1979, Codignotto & Malumián 1981) older than Illinoian (older than 400 kyr BP; Pampa de Beta Drift, Río Cullen Drift and Serranías de San Sebastián Drift, sensu Meglioli et al. 1990, Meglioli 1992). There are also Tertiary continental silty-sandy deposits (Cullen Formation, Petersen & Methol 1948, Codignotto & Malumián 1981). These cliffs are affected by rapid erosion. At Cabo Nombre, cliffs had not retreated between February 1987 and February 1988, but after that, strong swells at spring high tide level (on February 20, 1988, breaker height was greater than 3 m with periods of 9 to 12 seconds) caused a retreat of 3.7 m (Bujalesky 1990). The effective cliff retreat would be produced by these episodic events although there is also a continuous percolation, freezing and melting action on these weak deposits. These cliffs supplied the gravels that by longshore growth and landward recession have formed the El Páramo spit. Submerged moraines related to these deposits have been found at 40 m depth. Another marine system (related to Banco Sarmiento and Cabo Vírgenes Moraines,

Caldenius 1932) has been geophysically recognized at 70 m depth (Fig. 2b; Isla & Schnack 1989, 1991).

6.2 *Bahía San Sebastián*

This semicircular embayment (55 km × 40 km) occupies a wide, low-relief valley formed by glaciers during the Pleistocene and reshaped by the sea during the Holocene transgression. It shows different environments:

1. Fossil marsh. Inactive marsh located landwards of the National Route 3 (Fig. 3; Ferrero et al. 1987, Vilas et al. 1987a, Ferrero & Vilas 1989, Isla et al. 1991). A flat surface, formerly drowned seasonally by the

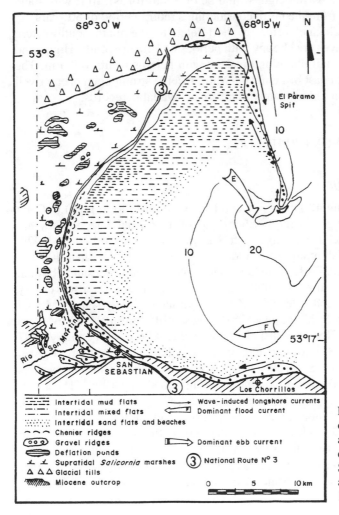

Figure 3. Morphology, sedimentology and dominant processes in Bahía San Sebastián (modified after Isla et al. 1991).

sea, was subject to strong deflation after the road construction. Very shallow ponds developed and the roots of small shrubs have been deflated by the westerlies causing their death. Erosion continues and an erosive scarp recedes. When the water table emerges, erosion ceases, but the pond continues enlarging towards the east. The cycle closes when the pond or lagoon is naturally silted up by wind-blown material. Blown silt and fine sand are minimum during the winter season, when snow and ice cover the entire area. Blow-outs from this fossil marsh are the principal source of sediment for the marsh and mudflat.

2. Upper marsh. It is controlled by deflation processes. Circular clumps 1-2 m across, dominated by *Salicornia*, are scattered over the mudflat and evolving in the wind direction: *Salicornia* is progressively buried on the windward sides, *Lepidophyllum* shrubs colonize the top, whereas grass grows downwind (eastern sides).

3. Gravel ridges. Waves entering the bay are strong enough to transport gravels along its southern coast. They also approach obliquely, causing a gravel beach drift towards the northwest. The Río San Martín inlet forms the northern limit of these gravel and sand deposits. At the southern coast of the bay, occasional large waves move pebbles up to 20 cm in diameter across the tidal berm. Fossil ridges were constructed during the regressive phase of the sea-level fluctuation. Attached to a palaeocliff sculptured into the Carmen Silva Formation (Miocene siliciclastic deltaic deposits; Codignotto & Malumián 1981), older ridges span in at least 3 stages, separated by northerly oriented depressions.

4. Cheniers. North of the Río San Martín, littoral processes have reworked shells and sands and constructed cheniers over the mudflat. Pelecypods, gastropods and equinoids from the subtidal zone are transported to the coast during easterly storms. There are three chenier lineations. The oldest is continuous, and extends farther north than the others. The most recent consist of 0.8-1 m high ridges separated every 400 m by washover channels.

Both, gravel ridges and cheniers, are the result of storms in the bay, and differ only by virtue of the fact that the Río San Martín limits the longshore transport of gravels.

5. Tidal flats. Tides move clockwise within the bay, and grain size progressively decreases in the same direction. This transition from sand to mudflats towards the north reverses in the tidal flat slope: sand increases to the lower portions with persistent wave action. The sand flat extends in front of the gravel ridges and constitutes a monotonous rippled or flat surface without channels. The mixed flat, extending from the village San Sebastián to the Río San Martín inlet, is characterized by flaser, wavy and lenticular bedding. The mudflat occupies the widest area (10 km) of the tidal flat. It comprises an upper zone, very flat and uniform, and a lower zone, where meandering tidal channels are more frequent.

6. Tidal channels. The mudflat has a drainage of meandering tidal channels reaching sizes over 3 m depth and 50 m width. Within the channels, there is a rapid evolution of point bars, and slumping. Sediment transport in these channels increases significantly during spring tides, and during winter, when cold water and ice are able to carry higher sediment loads and larger particles (fine sand).

6.3 *Península El Páramo*

Península El Páramo is a 20 km long gravel spit barrier that closes partially the San Sebastián Bay by the east (Fig. 4). It ends at a 36 m deep channel. El Páramo Spit not only has grown longitudinally to the south but receded to the west over the tidal flat; in this way it behaves as a transgressive spit. The northern part of Península El Páramo shows a beach ridge plain, where the oldest ridges are parallel to a palaeocliff with a west-east orientation and the younger ones developed a concave shoreline in plan view. Another beach plain of over 200 ridges (1200 m wide, 8 km long) develops to the south showing a north south alignment and a landward convex shoreline.

The beach ridges of a northeastern trend are the oldest ones and are cut by marine erosion. They are composed of coarse gravel. At the bay shore, the ridges are growing to the north and represent an episodic onshore progradation of gravel over the tidal flat. These two beach-ridge complexes represent different stages of the spit growth under progressive wave energy diminishing to the bay side as consequence of the tidal flat progradation.

The central sector (7 km long) presents parallel shorelines and is 50 m wide at high tide and 200 m wide at low tide. This sector has washover channels sloping towards the bay. They are active during sea storms or during spring tides with strong Atlantic swells. Overwash contributes to the lateral spit migration over the tidal flat.

The southern sector has a triangular shape. Its maximum width is 900 m and its length is 2 km. The western side comprises 60 beach ridges of coarse gravel. The youngest ones are asymptotic to the present shoreline while the oldest ones have been cut by erosion. On the eastern side, there is an equivalent number of beach ridges. They are cut by erosion in its northern section. The beach ridges on both sides are imbricated along a central line that is almost parallel to the present western shoreline. The shape of the southern sector is controlled by the relief of the abrassion platform. The channel that bounds the spit probably has a glaciofluvial origin. The gravel thickness in this sector is estimated in 20 m.

The Atlantic beach is formed by boulders, gravel and sand. The difference in elevation between the beach crest and the toe of the beach is 9.5 m to the north increasing gently to the south because of the topography of the abrassion platform. Five zones can be differenciated in this beach (Bujalesky

258

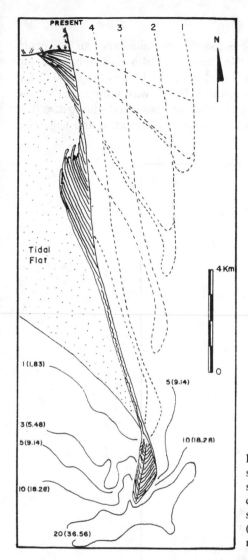

Figure 4. Morphologic features and
sedimentary evolution of El Páramo
spit (Bujalesky 1990). 1, 2, 3, 4: successive stages of growing of El Páramo
spit as from an initial cuspate foreland
(1). 1(1.83): depths in fathoms and
meters, respectively, referred to SLTL.

et al. 1987, Bujalesky 1988, 1990): storm berm (width: 15 m, gradient:
3.5°), storm swash terrace (width: 25 m, gradient: 2°), tidal berm (width:
10 m), high intertidal (width: 50 m, gradient: 7°, reflective beach), low
intertidal (width: 70 m, subhorizontal, dissipative beach).

The bay beach is 100 m wide to the south with 10 m of difference of
elevation and 50 m wide at 10 km to the north of the spit point with a difference of elevation of 7 m. It is formed mainly by gravel. Four zones are
differenciated: storm ridge, tidal ridge, intertidal (gradient: 4-6°), gravel
pavement.

6.4 *Río Chico-Río Grande area*

Different Pleistocene and Holocene coastal environments can be recognized south of Bahía San Sebastián, between Cabo San Sebastián and Cabo Peñas (Fig. 5). A large elongated deposit (14 km long and 2 km wide) of Pleistocene gravelly beachface facies was recognized at the northern part of this region (La Sara Formation sensu Codignotto 1969). It reaches an altitude of 23 m a.m.s.l. Its surface is smoothed and vegetated

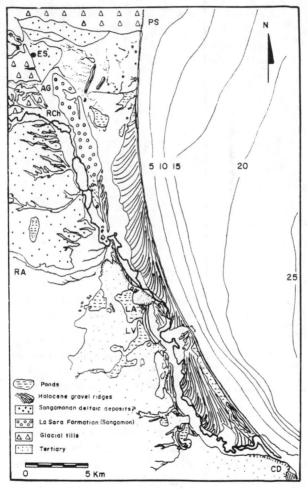

Figure 5. Morphology and Pleistocene and Holocene sedimentary environments in the Río Chico area (modified from Codignotto 1990, after aerial photographs of the Servicio de Hidrografía Naval 1970-1971). AG Arroyo Gamma, ES Estancia Sara, PS Punta Sinaí, RCh Río Chico, RA Río Avilés, LA Laguna Arcillosa, LV Laguna de las Vueltas, CD Cabo Domingo. Depths in meters referred to SLTL.

by grass, but it is possible to recognize NW-SE beach alignments from aerial photographs. Its eastern margin conforms a former erosive scarp.

Low grade seaward sloping wet lowlands and seasonal ponds develop in between Tertiary highlands, westward of the Río Chico. Other saltuary reduced Pleistocene raised beaches appears following the eastern Tertiary highland boundary, westward of Río Chico. Holocene small gravelly beaches, barriers and spits developed attached to the base of this line and at the base of the eastern margin of the main exposure of La Sara Formation. The shape of some of these former Holocene features suggests a northward growth, meanwhile others, located in more restricted embayments, do not show a prevailing longshore transport direction, appearing as swash-aligned beaches. An extensive beach-ridge plain develops eastward the Río Chico enclosing Holocene estuarine facies.

The beach ridges represent successive stages of growth to the south and have caused the Río Chico inlet migration in that direction (Codignotto & Malumián 1981, Codignotto 1990). The oldest beach ridges were totally eroded, the northern ones are cut by erosion and the southern and younger ones tend to be asymptotic to the present shoreline.

The younger and distal beach ridges were eroded and the growing of a later nonattached spit caused the formation of a lagoon. This process would indicate discontinuos pulses of sediment supply and a relatively scarcity of sediment supply to maintain beach stability under a strong longshore drift. The development of the beach ridge plain has accompanied the palaeorelief of the erosion platform of relatively uniform depth all along the N-S direction of progradation.

A narrow Holocene gravelly beach ridge plain (250 m wide) develops between Cabo Domingo and the Río Grande inlet, attached to the base of a Pleistocene marine terrace (18 m a.m.s.l.). This terrace represents a former delta of the Río Grande (Halle 1910) and can be correlated to the La Sara Formation. The Holocene beach ridges have grown to the southeast drifting long 5 km of the La Misión rivulet mouth and growing towards inside the Río Grande estuary (Fig. 6). A wide abrasion platform (up to 3 km wide) carved on Tertiary sandstones develops seawards of the shoreline. A narrowing beach ridge plain develops southward of the Río Grande inlet, showing a northward grow direction and erosion of its oldest and southern components. An interlacing beach ridge complex is attached to the northern base of Cabo Peñas cliff.

The Río Grande inlet is controlled by macrotides and high-energy waves, where processes and littoral forms related to both domains can be recognize depending upon the state that the system is representing. A gravel spit barrier (Punta Popper) is located at the downdrift side of the inlet. It has grown in a direction opposite to the regional drift. This behaviour is generated by the interaction between strong ebb tidal currents and the refraction of northeastern waves on a broad abrasion platform.

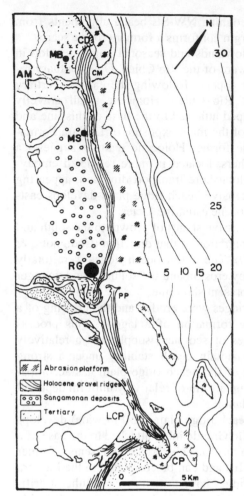

Figure 6. Morphology of Holocene beach ridges between Cabo Domingo CD) and Cabo Peñas (CP). RG Río Grande city, MS Misión Salesiana, MB La Misión bog (Auer's drilled site), CM Caleta La Misión, AM Arroyo La Misión, PP Punta Popper spit, LCP Laguna Cabo Peñas. Depths in meters referred to SLTL.

Under these conditions, a clockwise sediment circulation cell operates seaward from the inlet. The historical evolution of the spit involves 10 to 13 years cycles of longshore growth, landward retreat and breakdown. Longshore growth is mainly due to sediment transport and erosion of the backbarrier beach by ebb tidal currents. Wind waves develop in the estuary waters at high tide and favour the initiation of sediment motion. Littoral drift on the seaward side of the spit has a secondary role. Sluicing overwash provides sediment to the backbarrier beach and causes a landward migration. When the spit reaches its maximum possible length, overwashing sediment supply is insufficient to compensate the eroded volume at the backbarrier. Then the landward migration rate of the shoreface could be faster than the backbarrier. The throat of the inlet is

262

anchored in bedrock; it does not migrate significantly, becoming unstable with its minimum flow area reduction. These facts cause the breakdown of the spit barrier. A transverse bar fixed to the shoreface of the spit forms as consequence of the reduction of the minimum flow area, where stronger ebb tidal currents meet the waves. After the spit barrier breakdown, minimum flow area increases, ebb tidal currents velocities diminish and waves recycle the transverse bar sediments.

6.5 *Northern coast of the Península Mitre*

The northern coast of the Península Mitre presents low cliffs mainly carved on Late Cretaceous to Early Tertiary sandstones and small shallow and sediment-filled embayments. Some of them have been choked by Holocene sandy spits and barriers. The Península Mitre lowlands present peat bogs of ombrotrophic *Sphagnum magellanicum* (Rabassa et al. 1996). Their slopes dip gently seawards generating a shallow water table. Unfortunately, only a descriptive study was carried out on the Holocene coastal features of the eastern part of Península Mitre (Isla 1994).

The following features deserve to be mentioned as good examples of littoral forms: a) beach ridges between Cabo Leticia and Punta Noguera that drifted the Río Leticia inlet to the east (Fig. 7); b) beach ridges at the Río Bueno inlet where the growth of oldest ridges caused an eastward inlet migration; the youngest ones do it so in an opposite direction; c) the lagoon located at Bahía Policarpo, northward of Lago Luz was blocked out by a narrow barrier; d) Caleta Policarpo estuary, where a small ebb-tidal delta develops and an extensive beach ridge plain bounds it by its western

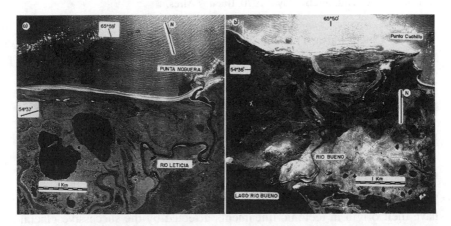

Figure 7. Distinctive coastal sedimentary environments of northern Península Mitre. From west to east: a) Río Leticia beach ridges, b) Río Bueno beach ridges.

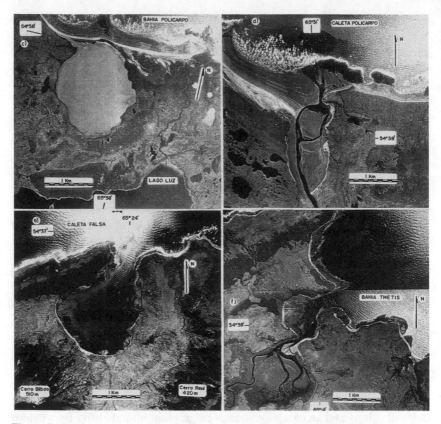

Figure 7. Continued. c) Bahía Policarpo lagoon, d) Caleta Policarpo estuary and beach ridges, e) Caleta Falsa, f) Bahía Thetis. Aerial photographs from the Armada Argentina, Servicio de Hidrografía Naval 1970, Buenos Aires.

flank; e) Caleta Falsa; f) Bahía Thetis. Bahía Thetis comprises two structural controlled embayments. The inner embayment presents a silty tidal flat carved by tidal channels; the outer one is dominated by sand deposits. The early Cretaceous low-grade metamorphic rocks constitute an abrasion platform and the inner sediment deposition is conditioned by a natural rocky 'jetty'. A flood tidal delta develops between the inner and the outer embayments indicating the boundary to the Atlantic wave influence (Isla 1994).

6.6 *The Beagle Channel northern coast*

The embayments of the indented rocky shoreline of the Beagle Channel have their origin in tectonic lineaments affected by the successive glacial modelling action. They present a restricted hydrodynamic environment where small pocket gravel beaches develop. The distinctive features that can be recognized along the Beagle Channel coast are Holocene terraces reaching up to 10 m a.m.s.l. and attached to metamorphic rocks or glacial deposits. These Holocene raised beaches are often capped by anthropogenic shell midden deposits. The Playa Larga ('Long Beach') represents a good example of well-developed terraces, located a few hundred meters east of the Río Olivia inlet, near the eastern boundary of Ushuaia city (Gordillo et al. 1992). This site presents a sequence of five superimposed raised beaches developed at 1.6 m (405 ± 55 ^{14}C yr BP), 3.8 m (3095 ± 60 ^{14}C yr BP), 5.2 m (4335 ± 60 ^{14}C yr BP), 7.5 m (5615 ± 60 ^{14}C yr BP) and 10 m a.m.s.l. (still undated).

The Bahía Lapataia-Lago Roca valley (20 km west of Ushuaia) is a palaeofjord that was occupied by a lateral and tributary valley-glacier system during the Last Glacial Maximum (18-20 ^{14}C kyr BP; Gordillo et al. 1993). Well-rounded glacially formed rocky hills, 'roches moutonnées', lateral moraines and kame landforms are present in this area and *Sphagnum* peat bogs develop at the lowlands. Holocene marine deposits are scattered along Bahía Lapataia, Archipiélago Cormoranes, Río Ovando, Río Lapataia and the eastern shoreline of Lago Roca, overlying glacial landforms and reaching a maximum altitude of at least 8.4 m a.s.l. (Fig. 8; Gordillo et al. 1993).

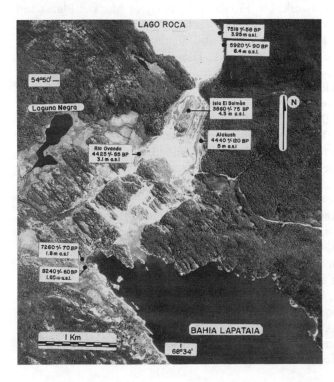

Figure 8. Lago Roca-Bahía Lapataia palaeofjord. Radiocarbon dating and altitudes of raised beaches after Gordillo et al. (1993). Aerial photographs from the Armada Argentina, Servicio de Hidrografía Naval 1970, Buenos Aires.

Codignotto & Malumián (1981) and Codignotto (1983, 1984) reported radiocarbon dates from shells of La Sara Formation as older than 43,000 yr BP. Radiocarbon dating on shells from the 18 m height marine terrace located at Río Grande indicated an age of 29,650 ± 1450 ^{14}C yr BP (Isla & Selivanov 1993). Aminoacid racemization techniques applied to shells of La Sara Formation reported D/L ratio of aspartic acid of 0.36 (Rutter et al. 1989). It was considered that this deposit is older than the last interglacial stage (Sangamon, Oxygen Isotope Substage 5e; Rutter et al. 1989). Rutter and Meglioli (in Meglioli, 1992) carried out new aminoacid racemization analysis on other valves collected from this formation and concluded that this raised beach is probably Oxygen Isotope Substage 5e.

The last interglacial marine deposits of La Sara Formation and those recognized at the northern margin of the Río Grande estuary would represent regressive beach-face facies composed of gravels that were highly exposed to the Atlantic waves.

Codignotto (1984) suggested that the northern drift bounded the embayment and supplied the sediments that integrate La Sara Formation. It is more probably that the main sediment supply came from the wave reworking of the deltaic glacifluvial deposits of the Chico and Avilés rivers.

During recent fieldwork, Dr. F. Isla (University of Mar del Plata) and the present author recognized estuarine facies of the Río Chico palaeoembayment reaching Laguna O'Connor and Laguna de la Suerte (28 km landwards from the present shoreline). These areas would represent the inner landward boundary of the maximum Sangamonan transgression (Fig. 9). The estuarine deposits of the Río Chico palaeoembayment show a gently slope seawards and a transition to the Holocene deposits.

At the present alluvial plain of Arroyo La Misión (10 km north of the city of Río Grande), a former Pleistocene glaciofluvial valley carved on Tertiary rocks was filled with sediments of varied origins. The uppermost portion of the sequence recorded the Holocene transgression (Auer 1959, 1974; Deevey et al. 1959, Markgraf 1980, Porter et al. 1984). A Holocene lake (at a present level of 5.7 m below high tide level) was flooded by the marine transgression about 9000 ^{14}C yr BP and a tidal flat developed at an altitude of 0.9 m a.h.t.l. at 4000 to 2000 ^{14}C yr BP (Mörner 1991).

At Bahía San Sebastián, the Holocene sedimentation took place after the sea-level rise and later stillstand. The initiation of this sequence is not precisely dated but a radiocarbon date indicates a minimum age of 5270 ± 190 yr BP (Fig. 10; Vilas et al. 1987b, Ferrero et al. 1989, Isla et al. 1991). The 8 km wide sequence of cheniers developed between 5270 and 1080 ^{14}C yr BP, suggests a sea-level fall of 1.8 m (0.363 m/1000 yr). Each of these chenier alignments originated every 300-400 yr (Ferrero et al. 1989, Isla 1994). The postglacial transgression sculptured a cliff at the present

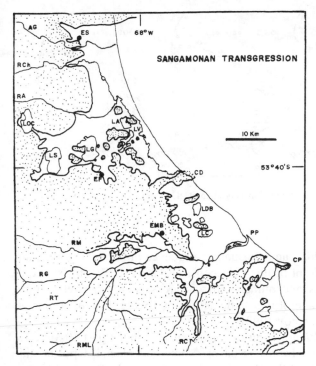

SANGAMONAN TRANSGRESSION

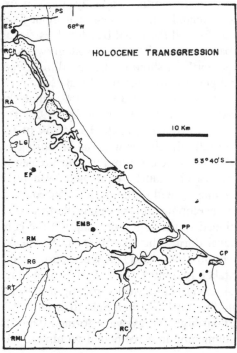

HOLOCENE TRANSGRESSION

Figure 9. Sangamonan and Holo-
cene transgression boundaries at
Río Chico (RCh) and Río Grande
(RG) drainage basins. AG Arroyo
Gamma, ES Estancia Sara, RA
Río Avilés, LOC Laguna
O' Connor, LS Laguna de la Suer-
te, LG Laguna Grande, LA
Laguna Arcillosa, LV Laguna de
las Vueltas, CD Cabo Domingo,
LDB Laguna Don Bosco, LC
Laguna de los Cisnes, PP Punta
Popper spit, EF Estancia Los
Flamencos, RM Río Moneta,
EMB Estancia María Behety, RT
Río de la Turba or Menéndez, RC
Río Candelaria, RML Río Mac
Lennan, CP Cabo Peñas.

267

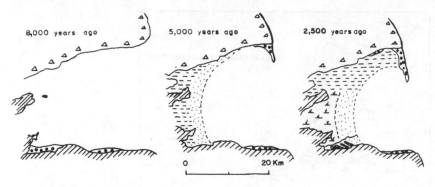

Figure 10. Evolution of San Sebastián Bay (Isla et al. 1991). Ages expressed in radio-carbon years.

southern coast of Bahía San Sebastián. A sequence of barrier/lagoonal deposits developed in front of this palaeocliff (Vilas et al. 1987b, Isla et al. 1991). Shell middens and human artifacts are scattered between dunes (Borrero 1989). The age of this settlement is of 1620 ± 140 to 1190 ± 90 ^{14}C yr BP (Isla & Selivanov 1993).

The growth of El Páramo Spit took place at least in the past 5000 years. Sediment supply came from coastal or submerged glacial deposits. Initially, a cuspate foreland developed in the north. Gradually, the spit extended from the tip of the cuspate foreland due to a sediment input from the Atlantic shoreline. This detritus formed the fossil beach ridges facing toward the bay in the northern part of the spit. As the spit grew along shore, provision of sediment to the northern shore of the bay diminished. The Atlantic shore was eroded and gravels from the fossil ridges were recycled (Figs 4 and 11; Bujalesky 1990; González Bonorino & Bujalesky 1990; Bujalesky & González Bonorino 1991; Isla & Bujalesky 1995). The levelling across the older beach ridges originated at the bay flank of the spit suggests that relative sea level fell slightly, or was stable, during its growth. The approximately 1 m-altitude decrease from the oldest (eastern) beach ridge to the youngest (western) one can be explained by a fetch shortening and a decrease in wave energy within the bay as consequence of the tidal flat progradation. The development of wave-cut platforms off Bahía San Sebastián and the continued spit transgression support the conclusion of a relative stable sea level during the last 5000 yr (Bujalesky 1990, Bujalesky & González Bonorino 1990, 1991). The accretion of the southern extreme of the spit have taken roughly 1000 yr, involving a long-term southward progradation (2.4 m/yr) and a westward migration (1.2 m/yr; Bujalesky & González Bonorino 1991). At San Sebastián tidal flat, the seaward gradient of the supratidal high-energy wave formed deposits would be partially the result of diminishing in wave height within the bay

268

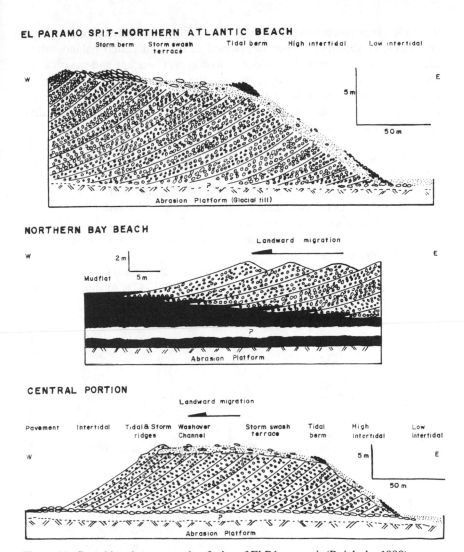

Figure 11. Gravel beach transgressive facies of El Páramo spit (Bujalesky 1990).

in response to spit growth and progressive protection from Atlantic swells (Bujalesky 1990; Bujalesky & González Bonorino 1990). On the other hand, Isla (1994) considered that the key to establish the highest level of the Holocene postglacial transgression could be found in the beach ridges attached to the base of the southwestern palaeocliff of Bahía San Sebastián (close to the road between the Argentine and Chilean customs).

The geomorphology and evolutive trend differences between Bahía San Sebastián Bay and the Río Chico palaeoembayment essentially arise from

269

the underlying palaeorelief carved during Pleistocene glaciations on Tertiary rocks. A shallower relief developed at the Río Chico palaeoembayment, allowed the formation of regressive gravel beaches and marsh environments in wave protected areas. At the Atlantic flank, an extended beach ridge plain grew from the northern margin of the embayment diminishing the wave attack. Its backbarrier side was stabilized by a gradual estuarine infilling. Growth of this beach ridge plain took place under limited sediment input conditions as in Península El Páramo. The progressive elongation was maintained by erosion and sediment recycling (cannibalisation), resulting in a significant landward retreat. The downdrift extreme of the beach ridge plain evidences a pulse of sediment scarcity where a lagoon developed between a beach ridge plain and the outer beach ridge (Fig. 12). This fact shows a senile stage of the system where recycled sediments result scarce to support downdrift growth under a significant longshore currents.

The development of gravel barriers mainly depends on the accumulation rate and its elongation capacity (Carter et al. 1987). At mature stages,

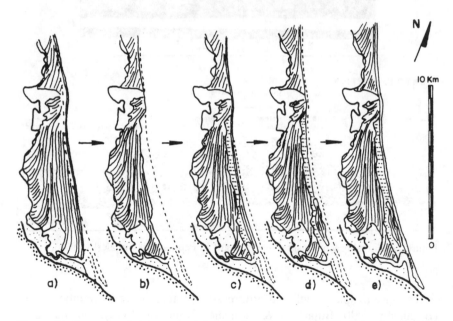

Figure 12. Recent evolutive trend of the distal extreme of the Río Chico beach ridge plain. a) Development of beach ridge plain under appropriate sediment supply. b) Erosive stage due to insufficient sediment supply under strong southward longshore transport. c) Increase in sediment supply (pulse) as a consequence of cannibalisation of the northern beach ridges. Development of a flying spit and a backbarrier marsh. d) Erosive stage due to scarcity of sediment supply. e) New pulse of sediment supply and present spit formation.

270

the growth of the gravel barriers is conditioned by the sediment volume that reaches their distal points. The elongation is maintained when the extreme part of the barrier is within the wave transport capacity and the sediment availability is appropriate. At longer distances to the distal point of the barrier, the possibility that a clast reaches it decreases. Then, the sediment along the barrier is recycled causing a thinning of its proximal part (cannibalisation, Carter et al. 1987). The gravel barriers that develop in areas with limited sediment supply tend to migrate landwards, even under stable sea level conditions, and they turn into transgressive features. This results in a landward gravel transport, due to the beach reflective behaviour, high infiltration rates and washover processes (Carter 1983, Orford & Carter 1982, 1984, Carter & Orford 1984, 1991, Carter et al. 1987, 1989, Forbes 1984, Taylor et al. 1986, Forbes & Taylor 1987). The landward migration of gravel barriers is due to rollover (Orford et al. 1990). The beach-face sediments are transported over the beach crest to the back-barrier beach by storm waves. There, the gravel stays passive until it is buried by the following overwash deposit. While the gravel barrier migrates landwards, the sediments of the inner flank emerge on the beach-face and incorporate again to the cycle. The balance between the accretion of the barrier crest due to overtopping and the breakdown by overwash determines the rollover and migration rates (Orford et al. 1990).

At the northern and eastern coast of Península Mitre (Bahía Thetis, Caleta San Mauricio, Bahía Buen Suceso), Isla (1994) recognized fossil beaches at a maximum altitude of 2 m above high tide level. Unfortunately, the absence of datable shells did not permit to confirm their ages. Present storm waves deposits attached to hill slopes or cliffs reach an elevation up to 2 m a.h.t.l. This fact does not allow to consider glacioisostatic or tectonic uplift that would have affected the glacioeustatic or dynamic components recognized in other areas (Isla 1994).

Porter et al. (1984) considered that the Holocene relative sea level along the Estrecho de Magallanes and the Beagle Channel reached a maximum altitude of at least 3.5 m above the present sea level about 5000-6000 ^{14}C yr BP and then fell progressively to its present level. These authors proposed a glacioisostatic and hydroisostatic warping as the cause of this sea level behaviour. It is worth mentioning that Porter et al. (1984) records of Puerto Hambre are located at the mainland opposite to Dawson Island and more than 100 km behind the outer limit of the last glaciation. This site presents two raised beaches: a) at 2.05 ± 0.5 m height about 7980 ± 50 ^{14}C yr BP, and b) at 3.5 ± 0.5 m height about 3970 ± 70 ^{14}C yr BP. Porter et al. (1984) considered these data as 'apparently anomalous', and interpreted that this area may reflect isostatic response to deglaciation. These authors did not take into account the tectonic setting of the area and that Puerto Hambre is located next to the Magellan Fault alignment. To the North and in an area not affected by tectonic uplift other data were reported: Bahía Gente Grande (3.36 ± 0.15 m, 5860 ± 40 ^{14}C yr BP) and Bahía San Gregorio (3.0 ± 0.2 m, 3860 ± 40 ^{14}C yr BP; Porter et al. 1984).

The present Beagle Channel was occupied by a glacial lake at about 9400 ^{14}C yr BP, with a level up to 30 m above the present sea level. The Beagle Channel opened before 8200 ^{14}C yr BP and the lake water was replaced by seawater. The marine environment was fully established along the channel at least by 7900 ^{14}C yr BP (Rabassa et al. 1986). Holocene raised beaches were recognized along the northern Beagle Channel coast reaching maximum elevations of nearly 10 m a.s.l with ages of approximately 6000 ^{14}C yr BP (Lago Roca, Bahía Golondrina, Playa Larga; Rabassa et al. 1986, Rabassa 1987, Gordillo et al. 1992, 1993). The estimated uplift rate is of approximately 1.5 to 2.0 mm/yr for the last 6000 yr (Rabassa et al. 1986, Rabassa 1987) and it increased up to 2.9 mm/yr for the last 1000 yr (Gordillo et al. 1993). The youngest terraces at Playa Larga (405 ± 55 ^{14}C yr BP at 1.7 m above the present counterpart) and Bahía Brown (985 ± 135 ^{14}C yr BP at 1.8 m) suggest that the last coseismic uplift movements could have been quite recent, in relation to its average return period and it would have been probably followed by a long quiescent time to allow the stress accumulation according to the prevailing longterm tectonic trend (Gordillo et al. 1992). It is considered that the oldest Holocene coastal deposits may partially be the result of isostatic recovery and the younger levels have been due to recent tectonic uplift (Rabassa et al. 1990, Gordillo et al. 1992, 1993).

Mörner (1987, 1991) pointed out that the Estrecho de Magallanes (Magellan Straits) and Beagle Channel coasts were under different uplifting behaviours and the area has not undergone any significant glacioisostatic warping during the Holocene. Mörner (1991) considered, for the regional Holocene eustatic changes, a sea level rise from 9000 to 4000 ^{14}C yr BP to a level only slightly above the present (ranging from 0.0 m or 0.5-1.0 m up to 1-2 m), but he argued that these higher levels seems to be the effects of storm waves rather than tidal flat levels. One of the assumptions on which Mörner (1987, 1991) sustained his conclusion is the incorrect affirmation of 'the absence of elevated Holocene terraces' and the general horizontality along the Beagle Channel. Its coast is characterized by a terrace system and at least three levels have been established at 8-10 m, 4-6 m and 1.5-3 m (Gordillo et al. 1992).

Taking into account the hydrodynamic setting and the morphology at the Beagle Channel, it would be difficult that the highest wind waves or storm surges entering form the Atlantic side would have been able to build beach ridges 2 m higher than present wave conditions. Earthquakegenerated standing waves would have left beach deposits of the same age but different elevations, depending on the shoreline configuration. This fact has not been observed (Gordillo et al. 1992).

The comparison of the Holocene raised beaches between the northern Atlantic coast of Tierra del Fuego (La Misión: Auer 1974 and Mörner 1991, Bahía San Sebastián: Ferrero et al. 1989) and the northwestern coast of Beagle Channel (Punta Pingüinos: Auer 1974, Bahía Golondrina: Gordillo 1990b; Playa Larga: Gordillo et al. 1992) indicate differential tectonic uplifting rates of 1.2 ± 0.2 mm/yr for the last 7760 ^{14}C yr BP, 1.3 ± 0.3 mm/yr for the last 5400 ^{14}C yr BP and 1.2 ± 1 mm/yr for the period comprised between 7760 and 5400 ^{14}C yr BP, respectively.

The cold and shallow-water mollusk assemblages associated to the raised beaches have not shown significant climatic changes during the Holocene, although minor temperature fluctuations could not be ruled out (Gordillo & Piñero 1989, Gordillo 1992, 1993, Gordillo et al. 1992, 1993).

8 CONCLUSIONS

1. The tectonic uplift during the last 8000 yr was maximum at the western Beagle Channel (approx. 1.2 ± 0.2 mm/yr), diminishing northwards and eastwards. It seems to be negligible at Bahía San Sebastián. The glacioisostatic rebound at Beagle Channel seems to have operated during deglaciation or in a 1-2 milennia after the final ice recession.

2. The extensive and gentle supratidal gradient of San Sebastián Bay (1.8 m for the last 5270 ^{14}C yr; Ferrero et al. 1989) would be partly due to progressive diminishing wave set-up, as a consequence of spit growth operating like a natural jetty. Littoral forms have developed under relatively stable eustatic conditions since 5000 yr BP.

3. The distinctive characteristic of the Holocene coastal deposits at the northern Tierra del Fuego (Bahía San Sebastián and Río Chico area) is the presence of regressive-like sequences at protected areas, meanwhile transgressive-like beach facies have developed at exposed areas.

4. The dissimilarities in geomorphological and evolutionary trends of the littoral deposits of the northeastern Atlantic coast mainly arise from the underlying palaeorelief, dipping northwards and carved during the Pleistocene glaciations.

5. The growth of spits and beach ridge plains took place under limited sediment supply. The progressive elongation was sustained by erosion and sediment recycling at the seaward flank, resulting in a significant landward transport.

6. In the northern coast, several evidences indicate a senile stage of evolution: shallow palaeoembayments infilled (Río Chico), progressive thinning of spits (El Páramo), cannibalization of spits (El Páramo, Río Chico).

ACKNOWLEDGEMENTS

The author wish to thank Dr. Federico Isla and an anonymous reviewer for comments on the original manuscript. Thanks also to the great number of colleagues and friends that shared field and laboratory work with the author.

REFERENCES

Andersson, G.J. 1906. Geological fragments from Tierra del Fuego. *Bulletin of the Geological Institute of the University of Uppsala*, 8: 169-183.

Auer, V. 1956. The Pleistocene of Fuego-Patagonia. Part I: The Ice and Interglacial Ages. *Annales Academiae Scientarium Fennicae, A III Geologica-Geographica* 48: 1-226.

Auer, V. 1959. The Pleistocene of Fuego-Patagonia. Part III: Shorelines displacements. *Annales Academiae Scientarium Fennicae, A III Geologica-Geographica* 60: 1-247.

Auer, V. 1970. The Pleistocene of Fuego-Patagonia. Part V: Quaternary problems of Southern South America. *Annales Academiae Scientarium Fennicae, A III Geologica-Geographica* 100: 1-194.

Auer, V. 1974. The isorhythmicity subsequent to the Fuego-Patagonia and Fennoscandian ocean level transgressions and regressions of the latest glaciation. *Annales Academiae Scientarium Fennicae, A III Geologica-Geographica* 115: 1-88.

Boltovskoy, E. 1981. Masas de agua del Atlántico Sudoccidental. In D. Boltovskoy (ed.), *Atlas del zooplancton del Atlántico Sudoccidental y métodos de trabajo con el zooplancton marino:* 227-237. Instituto Nacional de Desarrollo Pesquero, Special Publication, Mar del Plata, Argentina.

Bonarelli, G. 1917. Tierra del Fuego y sus turberas. *Anales del Ministerio de Agricultura de la Nación*, Sección Geología, Mineralogía y Minería, XII(3): 1-119. Buenos Aires.

Borrello, A.V. 1969. Los Geosinclinales de la Argentina. *Anales de la Dirección de Geología y Minería* 14: 1-188. Buenos Aires, Argentina.

Borrero, L.A. 1989. Relevancia de los estudios tafonómicos para interpretar los sistemas adaptativos humanos del Holoceno en Patagonia y Tierra del Fuego. *Simposio Internacional sobre el Holoceno en América del Sur*, Resúmenes: 37. Paraná.

Bujalesky, G.G. 1988. Estudio de la zonación de la playa atlántica de la espiga Península El Páramo mediante el análisis estadístico multivariado de gravas. *II Reunión Argentina de Sedimentología:* 41-45. Buenos Aires.

Bujalesky, G.G. 1990. Morfología y Dinámica de la Sedimentación Costera en la Península El Páramo, Bahía San Sebastián, Isla Grande de la Tierra del Fuego. Ph.D. Dissertation, Facultad de Ciencias Naturales y Museo, Universidad Nacional de La Pata, 188 pp, 2 app. Unpublished.

Bujalesky, G. G., G. González Bonorino, A. Arche, F. Isla & F. Vilas 1987. La espiga Península Páramo, Isla Grande de la Tierra del Fuego, Argentina. *X Congreso Geológico Argentino*, Actas I: 115-117. San Miguel de Tucumán.

Bujalesky, G. & G. González Bonorino 1990. Evidence for stable sea level in the Late Holocene in San Sebastián Bay, Tierra del Fuego, Southernmost Argentina. *International Symposium of Quaternary Shorelines: Evolution, Processes and Future Changes (IGCP-274)*, 9. La Plata, Argentina.

Bujalesky, G. & G.González Bonorino 1991. Gravel spit stabilized by unusual (?) high-energy wave climate in bay side, Tierra del Fuego, southernmost Argentina. In N.C. Kraus, K.J. Gingerich & D.L. Kriebel (eds), *Coastal Sediments '91. Proceedings of a Special Conference on Quantitative Approches to Coastal Processes* 1: 960-974. Seattle, Washington, June 25-27. American Society of Civil Engineers, New York.

Bujalesky, G., M. Salemme, F. Isla & M. Ferrero 1995. Physical environment at Los Chorrillos Archaeological Site, Bahía San Sebastián, Tierra del Fuego, Argentina. *Terra Nostra, XIV INQUA Congress*, Abstracts: 39. August 3-10, 1995, Berlin.

Caldenius, C.C. 1932. Las glaciaciones cuaternarias en la Patagonia y Tierra del Fuego. *Geografiska Annaler* 14: 1-164. Stockholm.

Camacho, H.H. 1948. Geología de la cuenca del Lago Fagnano o Cami, Gobernación Marítima de Tierra del Fuego. Ph.D. Dissertation, Facultad de Ciencias Exactas y Naturales, Universidad de Buenos Aires. Unpublished.

Camacho, H.H. 1967. Las transgresiones del Cretácico Superior y Terciario de la Argentina. *Revista de la Asociación Geológica Argentina* 22(4): 253-260. Buenos Aires.

Caminos, R. 1980. Cordillera Fueguina. *Segundo Simposio de Geología Regional Argentina*, Academia Nacional de Ciencias de Córdoba 2: 1461-1501. Córdoba.

Caminos, R., M. Haller, O. Lapido, A. Lizuain, R. Page & V. Ramos 1981. Reconocimiento geológico de los Andes Fueguinos, Territorio Nacional de Tierra del Fuego. *VIII Congreso Geológico Argentino*, Actas 3: 759-786. San Luis. Argentina.

Capurro, L.R. 1981. Características físicas del Atlántico Sudoccidental. In D. Boltovskoy (ed.), *Atlas del zooplancton del Atlántico Sudoccidental y métodos de trabajo con el zooplancton marino:* 219-225. Special Publication, Instituto Nacional de Desarrollo Pesquero, Mar del Plata, Argentina.

Carter, R.W. 1983. Coastal landforms as products of modern process variations, and their relevance in eustatic sea-level studies: examples from eastern Ireland. *Boreas* 12: 167-182.

Carter, R.W. & J.D. Orford 1984. Coarse clastic barrier beaches: a discussion of the distinctive dynamic and morphosedimentary characteristics. In Greenwood, B. & R.A. Davis (eds), Hydrodynamics and sedimentation in wave dominated coastal environments. *Marine Geology* 60: 377-389.

Carter, R.W. & J.D. Orford 1991. The sedimentary organisation and behaviour of drift-aligned gravel barriers. In Kraus, N.C., K.J. Gingerich & D.L. Kriebel (eds), *Coastal Sediments '91: proceedings of a Special Conference on Quantitative Approches to Coastal Processes* 1: 934-948. Seattle, Washington, June 25-27. American Society of Civil Engineers, New York.

Carter, R.W., J.D. Orford, D.L. Forbes & R.B. Taylor 1987. Gravel barriers, headlands, and lagoons: an evolutionary model. *Proccedings Coastal Sediments '87:* 1776-1792. American Society of Civil Engineers, New York.

Carter, R.W., D.L. Forbes, S.C. Jennings, J.D. Orford, R.B. Taylor, & J. Shaw 1989. Barrier lagoon coast evolution under differing relative sea-level regimes: examples from Ireland and Nova Scotia. *Marine Geology*, 88: 221-242.

Castano, J.C. 1977. Zonificación sísmica de la República Argentina. *Instituto Nacional de Prevención Sísmica, San Juan*, Publicación Técnica 5, 40 pp. San Juan, Argentina.

Clapperton, C.H. 1993. *Quaternary Geology and Geomorphology of South America*. Elsevier Science Publishers, 779 pp.

Codignotto, J.O. 1969. Nota acerca de algunos aspectos geológicos de la costa patagónica comprendida entre Punta Loyola y el Cabo Vírgenes. *Servicio de Hidrografía Naval, Boletín* 6(3): 257-263. Armada Argentina, Buenos Aires.

Codignotto, J.O. 1976. Geología y rasgos geomorfológicos de la Patagonia extraandina, entre el río Chico de Gallegos (Santa Cruz) y la bahía de San Sebastián (Tierra del Fuego). Ph.D. Dissertation, Facultad de Ciencias Exactas y Naturales, Universidad de Buenos Aires. Unpublished.

Codignotto, J.O. 1979. Hojas Geológicas 63a Cullen, 64a Bahía San Sebastián y 65b Río Grande. *Servicio Geológico Nacional.* Unpublished.

Codignotto, J.O. 1983. Depósitos elevados y/o de acreción del Pleistoceno-Holoceno en la costa fueguino patagónica. *Simposio Oscilaciones del nivel del mar durante el Último hemiciclo deglacial en la Argentina*: 12-26. Mar del Plata.

Codignotto, J.O. 1984. Estratigrafía y geomorfología del Pleistoceno-Holoceno costanero entre los paralelos 53°30' Sur y 42°00' Sur, Argentina. *IX Congreso Geológico Argentino*, Actas III: 513-519. San Carlos de Bariloche.

Codignotto, J.O. 1987. Cuaternario marino entre Tierra del Fuego y Buenos Aires. *Revista de la Asociación Geológica Argentina* 42(1-2): 208-212. Buenos Aires.

Codignotto, J.O. 1990. Evolución en el cuaternario alto del sector de costa y plataforma submarina entre Río Coig, Santa Cruz y Punta María, Tierra del Fuego. *Revista de la Asociación Geológica Argentina* 45(1-2): 9-16. Buenos Aires.

Codignotto, J.O. & N. Malumián 1981. Geología de la región al norte del paralelo 54° Sur de la Isla Grande de la Tierra del Fuego. *Revista de la Asociación Geológica Argentina* 36(1): 44-88. Buenos Aires.

Codignotto, J., R. Kokot & S. Marcomini 1992. Neotectonism and sea-level changes in the coastal zone of Argentina. *Journal of Coastal Research* 8(1): 125-133.

Codignotto, J., R. Kokot & S. Marcomini 1993. Desplazamientos verticales y horizontales de la costa argentina en el Holoceno. *Revista de la Asociación Geológica Argentina* 48(2): 125-132. Buenos Aires.

Compagnie de Recherches et d'Etudes Oceanographiques & Geomatter 1985. Campagne meteo-oceanographique, site Río Cullen/Hydra, periode de Fevrier 1984 a Fevrier 1985. Technical Report CREO/1249, TOTAL AUSTRAL, Buenos Aires, Argentina, 399 pp. Unpublished.

Coronato, A.M. 1990. Definición y alcance de la última glaciación pleistocena (Glaciación Moat) en el Valle de Andorra, Tierra del Fuego. *XI Congreso Geológico Argentino*, Actas 1: 286-289. San Juan.

Coronato, A.M. 1993. La Glaciación Moat (Pleistoceno superior) en los valles Pipo y Cañadón del Toro, Andes Fueguinos. *XII Congreso Geológico Argentino*, Actas 6: 40-47. Mendoza.

Coronato, A.M. 1995a. Geomorfología glacial de valles de los Andes Fueguinos y condicionantes físicos para la ocupación humana. Ph.D. Dissertation, Facultad de Filosofía y Letras, Universidad de Buenos Aires, 318 pp. Unpublished.

Coronato, A.M. 1995b. The last Pleistocene Glaciation in tributary valleys of the Beagle Channel, Southernmost South America. *Quaternary of South America & Antarctic Peninsula*, 9(1993): 153-172. Rotterdam: A.A. Balkema Publishers.

Dalziel, I.W.D. 1989. Tectonics of the Scotia Arc, Antarctica. *Field Trip Guidebook T 180*, 28th International Geological Congress, 206 pp.

Deevey, E.S., L.J. Gralenski & V. Hoffrén 1959. Yale natural radiocarbon measurements IV. *American Journal of Sciences Radiocarbon Supplement*, 1: 144-172.

Etchichury, M.C. & R.M. Tófalo 1981. Sedimentología de muestras litorales de Tierra del Fuego entre Cabo Espíritu Santo y Mina María. *Revista de la Asociación Geológica Argentina* 36(4): 333- 357. Buenos Aires.

Ferrero, M.A., G. González Bonorino, A. Arche, F. Isla & F. Vilas 1987. La llanura intermareal de la Bahía San Sebastián, Isla Grande de la Tierra del Fuego, Argentina. *X Congreso Geológico Argentino*, Actas I: 111-113. San Miguel de Tucumán.

Ferrero, M.A. & F. Vilas 1988. Secuencia vertical ideal generada por la progradación de una llanura intermareal fangosa, Bahía San Sebastián, Tierra del Fuego. *II Reunión Argentina de Sedimentología:* 95-99. Buenos Aires.

Ferrero, M.A. & F. Vilas 1989. Hidrodinámica, procesos deposicionales y erosivos en canales intermareales de la Bahía San Sebastián, Tierra del Fuego, Argentina. *Thalassas* 6: 89-94.

Ferrero, M.A., F. Vilas & A. Arche 1989. Resultados preliminares sobre la variación relativa del nivel del mar en la Bahía San Sebastián, Tierra del Fuego, Argentina. *II Reunión del Cuaternario Ibérico*, 2. Madrid.

Feruglio, E. 1950. Descripción Geológica de la Patagonia. *Dirección General de Yacimientos Petrolíferos Fiscales* 3: 1-431. Ministerio de Industria y Comercio de La Nación. Buenos Aires.

Forbes, D.L. 1984. Coastal geomorphology and sediments of New Foundland. *Current Research part B*, Geological Survey of Canada, paper 84-1B, 11-24.

Forbes, D.L. & R.B. Taylor 1987. Coarse-grained beach sedimentation under paraglacial conditions, Canadian Atlantic Coast. In: D. Fitzgerald & P. Rosen (eds), *Glaciated Coasts:* 51-86. New York: Academic Press.

Furque, G. 1966. Algunos aspectos de la Geología de Bahía Aguirre, Tierra del Fuego. *Revista de lu Asociación Geológica Argentina* 21: 61-66. Buenos Aires.

Furque, G. & H.H. Camacho 1949. El Cretácico Superior de la costa atlántica de Tierra del Fuego. *Revista de la Asociación Geológica Argentina* 4: 263-297. Buenos Aires.

Gagliardo, M. 1990. Estudio de minerales pesados (sedimentología, mineralogía, estadística y distribución) de la playa de la espiga San Sebastián, Tierra del Fuego. Liccnciatura (Master Sc.-like degree) Thesis, Facultad de Ciencias Exactas y Naturales, Universidad de Buenos Aires. Unpublished.

Gagliardo, M. 1994. Minerales pesados de la playa de la espiga San Sebastián, Tierra del Fuego. *II Reunión de Mineralogía y Metalogenia*, Instituto de Recursos Minerales, Universidad Nacional de La Plata, Publicación 3: 101-108. La Plata.

González Bonorino, G. & G.G. Bujalesky 1990. Spit growth under High-Energy wave climate on bay and ocean flanks, Tierra del Fuego, Southernmost Argentina. *International Symposium on 'Quaternary Shorelines: Evolution, Processes and Future Changes' (IGCP 274):* 35. La Plata, Argentina.

Gordillo, S. 1989. Moluscos marinos del Holoceno Medio en el extremo SO del Parque Nacional Tierra del Fuego (Argentina). *Jornadas Nacionales de Ciencias del Mar*, Resumen 509.

Gordillo, S. 1990a. Braquiópodos del Holoceno Medio del Canal Beagle, Tierra del Fuego, Argentina. *XI Congreso Geológico Argentino*, 2: 215-218. San Juan.

Gordillo, S. 1990b. Malacofauna de los niveles marinos holocenos de la Península Ushuaia y alrededores (Canal Beagle, Argentina). *III Reunión de Campo de Geología del Cuaternario*: 24-25. Bahía Blanca, Argentina.

Gordillo, S. 1991. Paleoecología de moluscos marinos del Holoceno Medio en Isla Gable, Canal Beagle (Tierra del Fuego, Argentina). *Ameghiniana* 28(1-2): 127-133. Buenos Aires.

Gordillo, S. 1992. Tafonomía y paleoecología de moluscos bivalvos del Holoceno del Canal Beagle, Tierra del Fuego. Ph.D. Dissertation, Universidad Nacional de Córdoba. Unpublished.

Gordillo, S. 1993. Las terrazas marinas holocenas de la región del Beagle (Tierra del Fuego) y su fauna asociada. *XII Congreso Geológico Argentino*, Actas 6: 34-39. Mendoza.

Gordillo, S. & L. Piñero 1989. Macro y micropaleontología de depósitos marinos holocenos en el Archipiélago Cormoranes, Parque Nacional Tierra del Fuego. *Simposio Internacional sobre el Holoceno en América del Sur:* 106-111. Paraná, Argentina.

Gordillo, S., G. Bujalesky, P. Pirazzoli, J. Rabassa & J. Saliège 1992. Holocene raised beaches along the northern coast of the Beagle Channel, Tierra del Fuego, Argentina. *Palaeogeography, Palaeoclimatology, Palaeoecology* 99: 41-54.

Gordillo, S., A. Coronato & J. Rabassa 1993. Late Quaternary evolution of a subantarctic paleofjord, Tierra del Fuego. *Quaternary Science Reviews*, 12: 889-897.

Halle, T. 1910. On Quaternary deposits and changes of level in Patagonia and Tierra del Fuego. *Bulletin of the Geological Institution of the University of Uppsala* 9(17-18): 93-117. Upsala.

IMCOS Marine Limited 1978. Meteorological and Oceanographic Study: Offshore Tierra del Fuego. Technical Report 78/111, TOTAL AUSTRAL: 17pp., 4 app. Buenos Aires, Argentina. Unpublished.

Isla, F.I. 1989. Holocene sea-level fluctuation in the southern hemisphere. *Quaternary Science Reviews* 8: 359-368. Pergamon Press, London.

Isla, F.I. 1994. Evolución comparada de bahías de la Península Mitre, Tierra del Fuego. *Revista de la Asociación Geológica Argentina*, 49(3-4): 197-205. Buenos Aires.

Isla, F., F. Vilas, G. Bujalesky, M. Ferrero, G. González Bonorino & A. Arche 1991. Gravel drift and wind effects over the macrotidal San Sebastian Bay, Tierra del Fuego. *Marine Geology*, 97: 211-224.

Isla, F. & E. Schnack 1989. Dos morenas sumergidas en el norte de Tierra del Fuego. *Jornadas Nacionales de Ciencias del Mar*, 98. Puerto Madryn.

Isla, F. & E. Schnack 1991. Submerged moraines offshore northern Tierra del Fuego, Argentina. *Quaternary of South America & Antarctic Peninsula* 9: 205-222. Rotterdam: A.A.Balkema Publishers.

Isla, F. & G. Bujalesky 1993. Saltation on Gravel Beaches: Tierra del Fuego, Argentina. *Marine Geology*, 115: 263-270.

Isla, F.I. & G. Bujalesky 1995. Tendencias evolutivas y disponibilidad de sedimento en la interpretación de formas costeras: casos de estudio de la costa argentina. *Revista de la Asociación Argentina de Sedimentología* 2(1-2): 75-89. Buenos Aires.

Isla, F. & A. Selivanov 1993. Radiocarbon contributions to the Quaternary eustatism of Buenos Aires, Chubut and Tierra del Fuego, Argentina. *Taller Internacional 'El Cuaternario de Chile'*, Abstracts: 47. Universidad de Chile, Santiago.

Isla, F., G. Bujalesky, M. Ferrero, G. González Bonorino, F. Vilas & A. Arche 1994. Facies relationships between the transgressive environments of San Sebastián Bay, Tierra del Fuego. *Symposium and field meeting 'The Termination of the Pleistocene in South America', IGCP Project 253*, Abstracts, 32. Tierra del Fuego, Argentina.

Kokot, R.R., S.C. Marcomini & J.O. Codignotto 1988. Evolución holocena en espigas de barrera; Caleta Valdés-Bahía San Sebastián. *Simposio Internacional sobre el Holoceno en América del Sur:* 57-60. Paraná, Argentina.

Köppen, W. 1936. Das geographische System der Klimate. In W. Köppen & R. Geiger (eds), *Handbuch der Klimatologie* 1C: 1-44. Verlag von Gebruder Borntraeger, Berlin.

Kranck, E.H. 1932. Geological investigations in the Cordillera of Tierra del Fuego. *Acta Geographica* 4(2): 1-231. Helsinki.

Markgraf, V. 1980. New data on the late and postglacial vegetational history of 'La Misión', Tierra del Fuego, Argentina. *Proceedings of the IV International Palynological Congress* 3: 68-74. Lucknow, India (1976-1977).

Meglioli, A. 1992. Glacial geology and chronology of Southernmost Patagonia and Tierra del Fuego, Argentina and Chile. Ph.D. Dissertation, Lehigh University, Bethlehem, U.S.A., 216 pp. Unpublished.

Meglioli, A. 1994. Glacial Stratigraphy of central and northern Tierra del Fuego, Argentina. In Rabassa, J., M. Salemme, A. Coronato, C. Roig, A. Meglioli, G. Bujalesky, M. Zarate & S. Gordillo (eds), *Field Trip Guidebook, Symposium and Field Meeting 'The Termination of the Pleistocene in South America'*: 9-21. IGCP Project 253, Ushuaia, March 15-25, 1994.

Meglioli, A., E.B. Evenson, P. Zeitler & J. Rabassa 1990a. Cronología relativa absoluta de los depósitos glaciarios de Tierra del Fuego, Argentina y Chile. *XI Congreso Geológico Argentino,* Actas 2: 457-460. San Juan.

Meglioli, A., E.B. Evenson & J. Rabassa 1990b. Multiple relative and absolute dating techniques applied to the glacial history of Tierra del Fuego. *Geological Society of America, Northeastern Section Meeting*, Syracuse, Abstracts.

Methol, E. & R. Sister 1947. Informe preliminar al estudio de los aluviones auríferos de la Gobernación Marítima de Tierra del Fuego entre río Gamma y cabo Espíritu Santo. Ministerio de Industria y Comercio de la Nación, Dirección General de Industria y Minería, Buenos Aires. Unpublished report.

Mörner, N.A. 1987. Sea level changes and tectonics in Tierra del Fuego. *Bulletin of the International Union for Quaternary Research Neotectonics Commission*, 10: 31.

Mörner, N.A. 1989. Holocene sea level changes in the Tierra del Fuego region. *Bulletin of the International Union for Quaternary Research Neotectonics Commission*, 12: 85-87.

Mörner, N.A. 1991. Holocene sea level changes in the Tierra del Fuego region. *Boletin IG-USP*, Special Publication 8: 133-151. Sao Paulo.

Nordenskjöld, O. 1898. Notes on Tierra del Fuego. An account of the Swedish Expedition of 1895-1897. *Scottish Geographical Magazine*, Edinburgh, 12: 393-399.

Orford, J.D. & R.W. Carter 1982. Crestal overtop and washover sedimentation on a fringing sandy gravel barrier coast, Carnsore Point, southeast Ireland. *Journal of Sedimentary Petrology* 52: 265-278.

Orford, J.D. & R.W. Carter 1984. Mechanisms to account for the longshore spacing of overwash throats on a coarse clastic barrier in southeast Ireland. *Marine Geology* 56: 207-226.

Orford, J.D., R.W. Carter, & D.L. Forbes 1990. Gravel migration and sea level rise: some observations from Story Head, Nova Scotia, Canada. *Journal of Coastal Research* 7(2): 477-488.

Pelayo, A.M. & D. Wiens 1989. Seismotectonics and relative plate motions in the Scotia Sea region. *Journal of Geophysical Research* 94(86): 7293-7320.

Petersen, C. 1949. Informe sobre los trabajos de relevamiento geológico efectuados en Tierra del Fuego entre 1945-1948. Ministerio de Industria y Comercio de la Nación, Dirección General de Industria y Minería, Buenos Aires. Unpublished report.

Petersen, C. & E. Methol 1948. Nota preliminar sobre rasgos geológicos generales de la porción septentrional de Tierra del Fuego. *Revista de la Asociación Geológica Argentina* 3(4): 279-291. Buenos Aires.

Popper, J. 1887. Exploración de la Tierra del Fuego. Conferencia dada en el Instituto Geográfico Argentino, el 5 de marzo de 1887. *Boletín del Instituto Geográfico Argentino*, 8. Buenos Aires.

Popper, J. 1891. Apuntes geográficos, etnológicos, estadísticos e industriales sobre la Tierra del Fuego. Conferencia dada en el Instituto Geográfico Argentino, el 27 de julio de 1891. *Boletín del Instituto Geográfico Argentino*, 12. Buenos Aires.

Porter, S.C. 1989. Character and ages of Pleistocene drifts in a transect across the Strait of Magellan. *Quaternary of South America & Antarctic Peninsula*, 7: 35-49. Rotterdam: A.A. Balkema Publishers.

Porter, S., M. Stuiver & C.J. Heusser 1984. Holocene sea- level changes along the Strait of Magellan and Beagle Channel, Southernmost South America. *Quaternary Research* 22: 59-67.

Rabassa, J. 1987. Lago Roca, Tierra del Fuego: the highest Holocene marine beach in Argentina? *IGCP Project 200, Late Quaternary Sea-Level Correlations and Applications*, Abstracts 21, Dalhousie University-NATO Advanced Study Institutes Programme, Halifax.

Rabassa, J. & C.H. Clapperton 1990. Quaternary Glaciations of the Southern Andes. *Quaternary Science Reviews*, 9: 153-174.

Rabassa, J., C.J. Heusser & R. Stuckenrath 1986. New data on Holocene sea transgression in the Beagle Channel: Tierra del Fuego, Argentina. *Quaternary of South America & Antarctic Peninsula*, 4: 291-309. Rotterdam: A.A.Balkema Publishers.

Rabassa, J., D. Serrat, C. Marti & A. Coronato 1988. Estructura interna de drumlins, Isla Gable, Canal Beagle, Tierra del Fuego. *II Reunión Argentina de Sedimentología*, 222-226. Buenos Aires.

Rabassa, J., C.J. Heusser & N. Rutter 1989. Late Glacial and Holocene of Argentina, Tierra del Fuego. *Quaternary of South America & Antarctic Peninsula* 7: 327-351. Rotterdam: A.A.Balkema Publishers.

Rabassa, J., D. Serrat, C. Marti & A. Coronato 1990. El Tardiglacial en el Canal Beagle, Tierra del Fuego, Argentina. *XI Congreso Geológico Argentino*, Actas 1: 290-293. San Juan.

Rabassa, J., G. Bujalesky, A. Meglioli, A. Coronato, S. Gordillo, C. Roig & M. Salemme 1992. The Quaternary of Tierra del Fuego, Argentina: the status of our knowledge. *Sveriges Geologiska Undersökning*, Ser. Ca 81: 241-256. Sweden.

Rabassa, J., A. Coronato & C. Roig 1996. The peat bogs of Tierra del Fuego, Argentina. In Lappalainen, E. (ed.), *Global Peat Resources, International Peat Society*, Geological Survey of Finland, 261-266.

Rutter, N., E. Schnack, J. Del Río, J. Fasano, F. Isla. & U. Radtke 1989. Correlation and dating of Quaternary littoral zones along the Patagonian Coast, Argentina. *Quaternary Science Reviews* 8: 213-234. London: Pergamon Press.

Servicio de Hidrografía Naval. 1981. Derrotero Argentino. Parte III: Archipiélago Fueguino e Islas Malvinas. *Publicación H.203*, 4a edición. Armada Argentina. 304 pp. Buenos Aires.

Servicio de Hidrografía Naval. 1995. Tablas de marea para el año 1995. Puertos de la República Argentina y puertos principales de Brasil, Uruguay y Chile. *Publicación H 610*. Armada Argentina, 494 pp. Buenos Aires.

Servicio Meteorológico Nacional. 1986. Estadísticas meteorológicas (1971-1980). *Estadística 36*. Fuerza Aérea Argentina, 338 pp. Buenos Aires.

Taylor, R.B., R.W. Carter, D.L. Forbes & J.D. Orford 1986. Beach Sedimentation in Ireland: contrasts and similarities with Atlantic Canada. *Current Research, Part A*, Geological Survey of Canada, Paper 86-1A: 55-64.

Thomas, E.R. 1949. Manantiales Field. Magallanes Province, Chile. *Bulletin of the American Association of Petroleum Geologists* 33(9): 1579-1589.

Total Austral & Geomatter 1980. Geophysical and geotechnical soil survey. Offshore Tierra del Fuego, Argentina. Unpublished Technical Report.

Troll, C. & K. Paffen 1964. *Die Jahreszeitenklimate der Erde*. Erdkunde 18: 5-28, 1 map.

Tuhkanen, S. 1992. The climate of Tierra del Fuego from a vegetation geographical point of view and its ecoclimatic counterparts elsewhere. *Acta Botanica Fennica*, 145: 1-64.

Unesco 1972. Résumé annuel d'informations sur les catastrophes naturelles, 1970. 66 pp. Paris.

Unesco 1979. Annual summary of information on natural disasters, 1975. 104 pp. Paris.

Urien, C.M. 1966. Edad de algunas playas elevadas en la Península de Ushuaia y su relación con el ascenso costero postglaciario. *III Jornadas Geológicas Argentinas*, 2: 35-41.

Vilas, F., A. Arche, G. González Bonorino, F. Isla & M. Ferrero 1987a. Sedimentación mareal en la Bahía San Sebastián, Tierra del Fuego, Argentina. *Acta Geológica Hispánica*, 21-22: 253-260. Madrid.

Vilas, F., A. Arche, M. Ferrero, G. Bujalesky, F. Isla & G. González Bonorino 1987b. Esquema evolutivo de la sedimentación reciente en la Bahía San Sebastián, Tierra del Fuego, Argentina. *Thalassas*, 5(1): 33-36.

Walter, H. 1976. *Die Ökologischen System der Kontinente. Prinzipien ihrer Gliederung mit Beispielen*. Gustav Fischer Verlag, Stuttgart, 132 pp.

Winslow, M.A. 1982. The structural evolution of the Magallanes basin and neotectonics in the southernmost Andes. In C. Craddock (ed.), *Antarctic Geoscience*, University of Wisconsin Press, Madison, 143-154.

Holocene coastal evolution in Rio Grande do Sul, Brazil

15

JORGE ALBERTO VILLWOCK & LUIZ JOSÉ TOMAZELLI
Centro de Estudos de Geologia Costeira e Oceânica (CECO), Instituto de Geociências,Universidade Federal do Rio Grande do Sul, Porto Alegre, Brazil.

ABSTRACT: The main morphological features and the sedimentary sequences accumulated in alluvial fan and lagoon-barrier depositional systems preserved on the broad Rio Grande do Sul coastal plain, southern Brazil, were developed under the control of Late Quaternary climatic changes with its associated glacio-eustatic sea-level fluctuations. The records of three significant moments of this coastal plain Holocene history (the last postglacial transgression, the Holocene regression and the present transgression) allowed to outline the geological evolution of this portion of the South American Atlantic coast.

The last postglacial transgression added a new lagoon-barrier system to the Pleistocene multiple barrier that keeps apart the Patos-Mirim lagoonal system from the sea. The ancient coastline related to its highstand, reached about 5 ka, is preserved as a palaeocliff at several places along the ocean side of the multiple barrier and along the margins of the Patos-Mirim lagoonal system where it is sometimes replaced by beach ridges and raised sandy spits. Pleistocene incised valleys were drowned during the Holocene transgression creating a submergence coastline displaying rias and bays partially filled with sediments bearing estuarine and marine fossil assemblages.

The Holocene regression, forced by a fall in relative sea-level, induced progradation of the last barrier by addition of accretionary beach ridges. Stepped raised lagoonal terraces, truncation of spits and beach ridge sets indicate that the Holocene drop was episodic rather than uniform, punctuated by stillstands and minor sea-level fluctuations of higher frequency.

There is evidence strongly suggesting that the Rio Grande do Sul coast is being submitted to a transgression that has reversed its former regressive tendency: (a) a widespread erosion of present coastline that in several places allows the coastal exposition of peat (showing radiocarbon ages between 1000 to 2000 years BP) and lagoonal sediments; (b) the same widespread erosion along the Patos-Mirim lagoonal shoreline and, (c) the presence of a drowned lagoonal terrace bounded by a scarp at a depth of −1 to −2 m, situated up to 2 km of the present shoreline. Probably this ongoing transgresive event is similar to the preceding high frequency fluctuations that have modulated the general regressive trend of this coast during the late Holocene.

RESUMO: As principais feições morfológicas e as seqüências sedimentares acumuladas em um sistema deposicional de leques aluviais e quatro sistemas deposicionais do tipo laguna- barreira, preservados na ampla planície costeira do Rio Grande do Sul, Brasil, desenvolveram-se sob o controle das mudanças climáticas e das decorrentes flutuações glacioeustáticas do nível do mar ocorridas no Quaternário Superior. Os registros de três momentos significativos da história holocênica dessa planície costeira (a última transgressão pós-glacial, a regressão holocênica e a transgressão atual), permitiram delinear a evolução geológica deste segmento da costa atlântica da América do Sul.

A última transgressão pós-glacial adicionou o último sistema barreira-laguna à barreira múltipla pleistocênica que separava o sistema lagunar Patos -Mirim do mar. A antiga linha de costa relacionada a este nível de mar alto, atingido a aproximadamente 5 ka, aparece preservada como uma paleofalésia ao longo de vários pontos tanto do lado oceânico da barreira múltipla como das margens do sistema lagunar Patos-Mirim onde, muitas vêzes, ela é substituida por cristas de praia e pontais arenosos suspensos. Vales incisos pleistocênicos foram afogados durante esta transgressão resultando em uma costa de submergência com rias e baías parcialmente preenchidas por sedimentos contendo assembléias fósseis de ambiente estuarino e marinho.

A regressão holocênica, forçada por uma queda no nível relativo do mar induziu à progradação da última barreira mediante a adição de conjuntos acrescionais de cristas de praias. Terraços lagunares elevados escalonados, truncamento de pontais e de conjuntos de cristas de praia indicam que o rebaixamento do nível do mar no Holoceno foi mais episódico do que uniforme, marcado por estabilizações e pequenas oscilações de freqüência mais elevada.

A erosão generalizada da linha de costa atual que permite a exposição em diversos locais, de turfas (mostrando idades de radiocarbono entre 1000 e 2000 anos AP) e sedimentos lagunares no pós-praia ou na base das dunas frontais, a mesma erosão visível ao longo da costa do sistema lagunar Patos-Mirim e a presença de um terraço lagunar submerso limitado por uma escarpa em profundidade de −1 a −2 m, situada a mais de 2 km da linha de costa atual, são evidências fortemente sugestivas de que a costa do Rio Grande do Sul esta sendo submetida a uma transgressão que reverteu a tendência regressiva anterior. Provavelmente este evento transgressivo atual é semelhante às anteriores oscilações do nível do mar, de alta freqüência, que modularam o curso regressivo da costa durante o Holoceno tardio.

1 INTRODUCTION

The Holocene is the latest interval of geological time, covering approximately the last 10,000 years of the Earth's history. It is also the latest division of the Quaternary Period, the last 2,000,000 years of geologic time which is characterized by cycles of alternate cold and warm periods producing glacial and interglacial stages. The cyclic nature of the fluctuating climate is clearly show by the temperature fluctuations interpreted from the changing oxygen isotope content of pelagic foraminifera from deep-sea cores (Emiliani 1955, Shackleton & Opdyke 1973, 1977, Imbrie et al. 1984).

The glacio-eustatic sea-level changes related to these climatic fluctuations left remarkable geomorphological and sedimentological records preserved across the Rio Grande do Sul coastal plain (Southern Brazil) a broad sandy coastal province developed on a marginal sedimentary basin (Pelotas Basin) set in an Atlantic-type continental margin.

Geological mapping of the morphological features and the sedimentary facies of the main depositional systems of this coastal plain allowed the research team of the Oceanic and Coastal Geology Center of the Geosciences Institute, Federal University of Rio Grande do Sul (UFRGS) to outline its geological evolution (Villwock et al. 1986, Villwock & Tomazelli 1995).

The purpose of the present paper is to update the final chapter of this long history: the Holocene coastal evolution of the Rio Grande do Sul coastal plain.

2 PHYSICAL AND GEOLOGICAL SETTING

The Rio Grande do Sul coastal plain is located approximately between lat. 29°S and 34°S. Its straight shoreline has a NE-SW general orientation and is about 600 km long, extending from Torres, at its northern extreme, to Barra do Chuí, at the southern end (Fig.1). As the coastal plain reaches, in some places, more than 100 km wide, it covers a total area of about 33,000 km² embracing a great number of coastal water bodies, some of them of giant dimensions, like the Lagoa dos Patos (area of 10,000 km²) and Lagoa Mirim (area of 3770 km²) lagoons.

The climate of the region is temperate, humid, with an even distribution of rain throughout the year. Average annual rainfall is around 1300 mm. The coastal plain is dominated by a bimodal high-energy wind regime

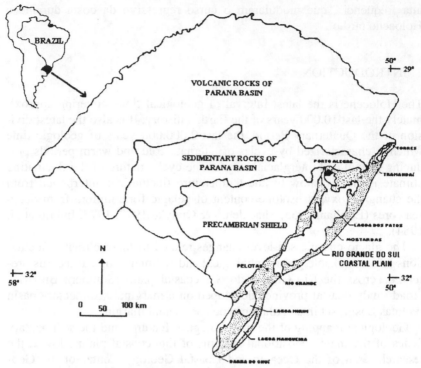

Figure 1. Geographical setting of Rio Grande do Sul Coastal Plain.

(Tomazelli 1993). The dominant wind comes from NE and although it blows throughout the year, it is more active during spring-summer months. The secondary W-SW wind becomes more important in the autumn-winter months.

The Rio Grande do Sul coast is a wave-dominated, microtidal coast with semidiurnal tides having a mean range of about 0.5 m. The region is affected by swell waves approaching from SE and producing a net northerly longshore transport of sediments. Besides the swell action, sea waves from E and NE and episodic storm waves from E and SE control the erosional and depositional processes along the sea shore (Tomazelli & Villwock 1992).

The adjacent continental shelf shows a quite regular morphology. It has an average width of 150 km and the shelf-break line is situated at a depth around 170 m. The shelf-bottom sediments are predominantly terrigenous clastic sediments with some biodetritical concentrations (Martins et al. 1967).

At the northern part of the region, the coastal plain is adjacent to highlands, which are formed by Palaeozoic and Mesozoic sedimentary, and

286

volcanic rocks of the Paraná Basin that, in some places reach nearly 1000 m a.s.l. At the southern section, the lower highlands are formed by igneous and metamorphic rocks of the Precambrian shield. These highlands were the main source of over than 10,000 m thick terrigenous sediments accumulated in the Pelotas Basin since its origins in the Early Cretaceous, guided by the geotectonic events related to the opening of the South Atlantic Ocean (Villwock 1984, Fontana 1990). At present, all sandy sediments eroded from these highlands and transported by rivers to the coast are trapped in the coastal lagoons and other backbarrier environments, and none of them reach the oceanic shoreline.

Under the major control of Quaternary climatic variations and sea-level fluctuations, the younger section of this sedimentary record exposed on the Rio Grande do Sul Coastal Plain, have been accumulated in two kinds of depositional systems: (1) alluvial fans, and (2) lagoon-barriers (Villwock et al. 1986).

1. *Alluvial fan systems*. The sediments of the alluvial fan system began to be deposited probably during the Tertiary and the processes responsible for their genesis continued active during the whole of the Quaternary with an intensity controlled by cyclic climatic changes from humid to arid. The system embraces all sedimentary facies formed next to the slopes of the highlands as a response to gravitational and alluvial transport processes. Nowadays the facies of alluvial fans occur all over the western strip of the coastal plain adjacent to the highlands.

2. *Lagoon-barrier systems*. The most distal portions of the alluvial fan deposits were greatly reworked by marine and lagoonal processes during several transgressive events that occurred during the Quaternary. These glacio-eustatic fluctuations produced important lateral displacements of the shoreline on the very-low gradient of Rio Grande do Sul coastal plain and continental shelf. As a result of these shoreline shifting, a series of lagoon-barrier systems were created. Landward of the present coastline, four lagoon-barrier systems have been identified. From the oldest to the youngest, they were named as Lagoon-Barrier System I, II, III and IV, respectively. Probably each barrier defines the former limits of each transgressive event occurred during Late Pleistocene and Holocene times.

The present paper is concerned to the most recent lagoon-barrier system (Barrier IV), which has been developed during the Holocene. Its morphological features and sedimentary deposits identified along the area and along the margins of the Patos-Mirim lagoonal system, created during the Pleistocene with the development of the Barrier II and Barrier III, are records of three significant moments of the Holocene history (the last postglacial transgression, the Holocene regression and the present transgression).

3 THE LAST POSTGLACIAL TRANSGRESSION

The latest interglacial high sea-level event, responsible for construction of the most recent lagoon-barrier system of Rio Grande do Sul coastal plain, has been developed during the Holocene, corresponding to the ongoing oxygen isotopic stage 1 highstand. Beginning about 18 ka (kiloanno = 10^3 years), when sea-level at minimum lowstand was near the continental shelf edge, the postglacial transgression shifted the shoreline across the subaerially exposed continental shelf, until its maximum level was attained, around 5.1 ka. At this peak, sea-level was at 4-5 m above the present level.

Correa et al. (1991) indicated that the rates of horizontal coastline shifting over this continental shelf have varied from 8.6 to 16.7 m/year. The presence of several submerged terraces, interpreted as ancient shorelines, has been discussed by Kowsmann et al. (1977) and by Correa (1986). Some of them are marked by beach rocks and shell beds that show radiocarbon ages older than 6470 years BP, revealing that during the last transgression the sea rose fast and intermittently. In some places, winnowing of older sediments left a transgressive lag deposit composed of reworked shells, gravelly sized beach rocks rounded fragments and bone fragments of Pleistocene fossil mammals (Figueiredo Jr. 1975, Correa & Ponzi 1978, Buchmann 1994). Otherwise, Martins et al. (1967), when discussing the distribution of continental shelf bottom sedimentary facies, show that, with the exception of the sandy mud deposits accumulating today in front of the Rio Grande inlet and the muds proceeding from the Río de la Plata, all the other facies are relictual, and showing a variety of grainsizes incompatible with the on-going hydrodynamic conditions at their present location.

Reaching its highstand, this transgressive event adds a new lagoon-barrier system (Barrier IV) to the Pleistocene multiple barrier that limits the Patos-Mirim lagoonal system from the sea. Some remains of this Holocene transgressive lagoon-barrier system are preserved at the northern part of the coastal plain as a string of vegetated palaeodunes, over 20 m high, lying at the eastern side of the lagoonal system, near the city of Tramandaí (Fig. 2).

In some places, the coastal erosion of the Pleistocene barrier left a palaeocliff, 4 to 6 m high, which now separates Pleistocene from Holocene terranes. This ancient coastline is rectilinear and marked by small transgressive aeolian dunes as it may be seen near the city of Rio Grande (Fig. 3) and at the backbarrier of Lagoa do Peixe lagoon, near the town of Mostardas (Fig. 4). In other sites, it changes its configuration showing an indented profile, reentrances configuring ancient estuaries and bays partially closed by the Holocene barrier island system, as it may be seen near the city of Tramandaí (Fig. 5).

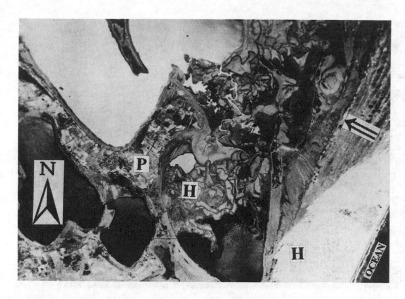

Figure 2. The arrow points to some remains of the Holocene transgressive barrier (5.1 ka) which was prograded, during the Holocene regression, through sets of accretionary beach ridges. P = Pleistocene. H = Holocene. (Northern part of Rio Grande do Sul coastal plain. Scale: 1 cm = 1.1 km).

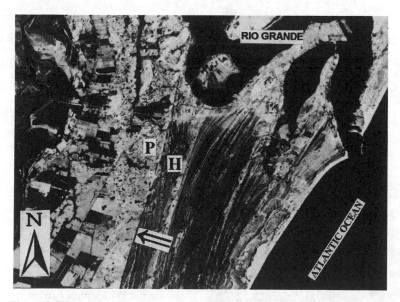

Figure 3. The arrow points to the ancient 5.1 ka coastline, a palaeocliff at the boundary of the Pleistocene (P) terrace and the Holocene (H) beach ridge plain (Coastal plain near the city of Rio Grande. Scale: 1 cm = 4 km).

289

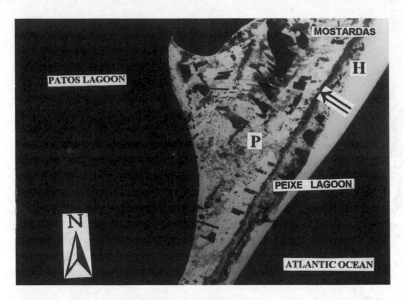

Figure 4. The arrow points to the ancient 5.1 ka coastline, a palaeocliff at the boundary of the Pleistocene (P) Barrier III terrace and the Holocene (H) lagoon-barrier system, Barrier IV. (Multiple barrier near the town of Mostardas; Holocene Lagoa do Peixe lagoon. Scale: 1 cm = 5.8 km).

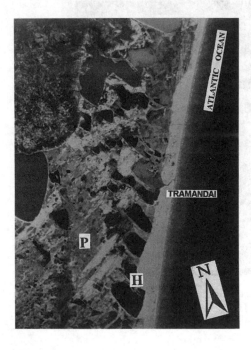

Figure 5. The ancient 5.1 ka coastline, a palaeocliff at the boundary of the Pleistocene (P) Barrier III, showing an indented profile configuring ancient estuaries and bays partially closed by the Holocene (H) Barrier IV (Multiple barrier close to the city of Tramandí; Holocene lagoon-barrier system. Scale: 1 cm = 4.5 km.).

290

At this time, there were two large inlets, Rio Grande and Taim, connecting the Patos and Mirim lagoons to the sea and exposing the lagoonal margins to the same transgressive effects. At present, an ancient cliff, 4 to 6 m high, marks the boundary between the Pleistocene and Holocene terraces all over the coast of these lagoons (Fig. 6). In several places the scarp is replaced by beach ridges, partially reworked by the wind, sometimes constituting raised sandy spits, which are additional evidence of the ancient lagoonal coastline (Fig. 7).

On the west side margin of the giant lagoonal system the lower parts of several river valleys were drowned by the rising waters forming a succession of ancient rias and bays typical of a submerging coast (Fig. 8). Oyster banks and other estuarine sediments described by Bianchi (1969) about 20 km landward in the Piratini river valley are related to this highstand. Other incised valleys, as the Fragata and Pelotas creeks, and the São Gonçalo channel alluvial plain, show sediments with estuarine and marine fossil assemblages where mollusk, echinoderma, foraminifera, ostracoda, bryozoa and coral fragments can be found (Godolphim et al. 1989, Villwock et al. 1990).

Figure 6. The arrow points to an ancient cliff – 5.1 ka – boundary between Pleistocene (P) terranes and the Holocene (H) lagoonal terrace. (Eastern coast of Lagoa dos Patos lagoon. Scale: 1 cm = 1 km).

Figure 7. The arrow points to an ancient coastline and a raised sandy spit (5.1 ka), boundary between Pleistocene (P) terranes and the Holocene (H) lagoonal terrace. (Northern coast of Lagoa dos Patos lagoon. Scale: 1 cm = 1 km).

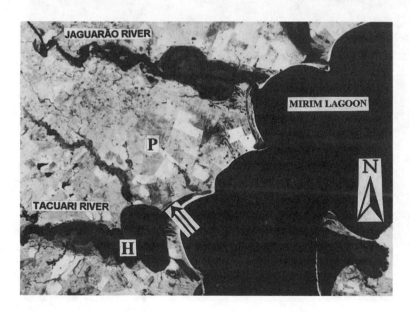

Figure 8. The arrow points to an ancient submergence coastline (5.1 ka) displaying incised valleys in Pleistocene (P) terraces filled by Holocene (H) estuarine and lagoonal sediments. (Western coast of Lagoa Mirim. Scale: 1 cm = 3.1 km).

292

4 THE HOLOCENE REGRESSION

After the maximum of the last post-glacial transgression was reached ca. 5 ka, the shoreline began to move in a seaward direction, forced by a fall in relative sea-level. The former narrow transgressive barrier became a regressive one showing, at its surface, sets of accretionary beach ridges (Fig. 2).

Geomorphic features of Holocene age such as raised lagoonal terraces (Fig. 6), and truncation of spits and beach ridges sets (Fig. 3), strongly indicate that the Holocene sea-level drop was episodic rather than uniform, punctuated by stillstands and minor sea-level fluctuations of higher frequency, similarly to those documented for other regions of the Brazilian coast (e.g. Martin et al. 1980, Suguio et al. 1985). The occurrence of very well defined stepped terraces of Holocene age along the Patos lagoon margin is clearly related to these events. Each scarp that bounds a terrace marks a stillstand event or a minor reversal during the general falling of the lagoonal waters. Likewise, the main truncations of beach ridges as those occurring south of Patos lagoon inlet probably document these temporary pauses or reversals in the general fall in sea level during the Late Holocene (Fig. 3).

During this progradation of Barrier IV the sediments supplied by rivers promoted a general aggradation in the associated lagoons and other depositional backbarrier environments. As the barrier beach did not received coarse-grained sediments from rivers, it is highly probable that sand supply for progradation was furnished mainly by the erosion and reworking of the lower shoreface and inner continental shelf, as it has been suggested by Dominguez et al. (1987) for regressive Holocene sediments of the eastern-southeastern Brazilian coast. According to this mechanism, a corollary of the 'Bruun Rule', abundant sand for seaward shoreline migration could be available by erosion of the shoreface and inner continental shelf during a forced regression caused by a drop in relative sea-level. The humid climate and the presence of appropriate vegetal species have played an important role in trapping the available sand near the shoreline as multiple parallel beach-dune ridges. As a result, a rapid seaward shift of the shoreline has occurred during this regressive phase, and the broadening of Barrier IV has reached, in same places, more than 20 km wide.

5 THE PRESENT TRANSGRESSION

Probably as a consequence of a reversal in sea-level trend, the seaward progradation of Barrier IV appears to have ceased in a time tentatively assigned between 1000 and 2000 years BP, when sea-level was apparently around 1 to 2 m below present position. Since then, the shoreline has been migrating landwards.

The sedimentological, geomorphological, and geochronological supporting evidence for this ongoing transgression, controlled largely by a new episode of relative sea-level rise, was first documented by Tomazelli & Villwock (1989). Evidence in support of this hypothesis includes: 1. A widespread erosion of the coastline that in several places allows the exposition of peat and lagoonal sediments directly at the foreshore or at the base of the foredunes (Fig. 9). These backbarrier sediments show an extended lateral continuity (in some places more than 20 km along the shoreline). Radiocarbon dating of peats have revealed ages between 2000 to 1000 years BP; 2. The occurrence of similar widespread erosion along most of the Lagoa dos Patos shoreline; 3. Bathymetric surveys and aerial photographs studies made at the Lagoa dos Patos lagoon have revealed very clearly the presence of a submerged terrace bounded by a scarp at a depth of –1 to –2 m, located up to 2 km of the present lagoonal shoreline. This drowned terrace is a strong evidence of a recent rise in lagoonal waters as a result of a concomitant sea-level rise.

In general terms, the available evidence strongly suggests that, unlike what has been proposed by Suguio et al. (1985) for other segments of the Brazilian coast, at present the Rio Grande do Sul coast is submitted to

Figure 9. Exposition, at the foreshore, of a peat and lagoonal sediments due to coastal erosion, near Barra do Chui.

transgressive conditions that have reversed a former regressive trend. Probably, this ongoing transgressive event is similar to the preceding high frequency fluctuations that have modulated the general regressive trend of the coast during the Late Holocene.

REFERENCES

Bianchi, L.A. 1969. Bancos de ostreídeos pleistocênicos da Planície Costeira do Rio Grande do Sul. *Iheringia*, Geologia, 2: 3-40. Porto Alegre, Brazil.

Buchmann, F.S.C. 1994. Distribuição dos Fósseis Pleistocênicos na Zona Costeira e Plataforma Continental Interna no Rio Grande do Sul. *Acta Geologica Leopoldensia* 39/1(27): 355-364.

Corrêa, I.C.S. 1986. Evidences of Sea-Level Fluctuations in the Rio Grande do Sul Continental Shelf, Brazil. *Quaternary of South America and Antarctic Peninsula* 4: 237-247. Rotterdam: A.A. Balkema Publishers.

Corrêa, I.C.S. & V.R. Ponzi 1978. Bioclastic Carbonate Deposits along Albardão and Mostardas in Rio Grande do Sul inner continental shelf. *Memorias:* 67-91, Seminario Sobre Ecología Bentónica y Sedimentación de la Plataforma Continental del Atlántico Sur, UNESCO. Montevideo.

Corrêa, I.C.S., R. Baitelli, R.N. Ayup-Zouain & E.E. Toldo Jr. 1991. Translation de la ligne de rivage sur le plataforme continentale du Rio Grande do Sul – Brésil, pendant l'Holocene. *Pesquisas* 18: 161-163. Brazil.

Dominguez, J.M.L., L. Martin & A.C.P.S. Bittencourt 1987. Sea-level history and Quaternary evolution of river mouth-associated beach-ridge plains along the east-southeast Brazilian coast: a summary. In: Nummedal, D., O.H. Pilkey & J.D. Howard (eds) *Sea-level fluctuations and coastal evolution,* Special Publication 41: 115-127. Society of Economic Paleontologists and Mineralogists.

Emiliani, C. 1955. Pleistocene temperatures. *Journal of Geology* 63: 538-575.

Figueiredo Jr., A.G. 1975. Geologia dos depósitos calcários biodetríticos da plataforma continental do Rio Grande do Sul. MSc. Thesis. Porto Alegre, Instituto de Geociências da Universidade Federal de Rio Grande do Sul. 72 pages.

Fontana, R.L. 1990. Investigações geofísicas preliminares sobre o Cone de Rio Grande, Bacia de Pelotas, Brasil. *Acta Geologica Leopoldensia* 13(30):161-170.

Godolphim, M.F., L. Artusi, B.A. Dehnhardt, J.A. Villwock, I.R. Forti-Esteves 1989. Novas evidências da transgressão holocênica na porção média da Planície Costeira do Rio Grande do Sul. *Acta Geologica Leopoldensia* 12(29): 23-36.

Imbrie, J., J. Hays, D. Martinson, A. McIntyre, A. Mix, J. Morley, N. Pisias, W. Prell & N. Shackleton 1984. The orbital theory of Pleistocene climate: support from a revised chronology of the marine $\delta^{18}O$ record. In: Berger, A. et al., (eds). *Milankovitch and Climate.* Part 1: 269-305. Riedel.

Kowsmann, R.O., M.P.A. Costa, M.A. Vicalvi, M.G.N. Coutinho & L.P. Gamboa 1977. Modelo de sedimentação holocênica na plataforma continental sul-brasileira. In: *Evolução sedimentar holocênica da plataforma continental e talude do sul do Brasil* V.2: 1-96. Petrobrás, DNPM, CPRM, DHN, CNPq. Rio de Janeiro.

Martin, L., K. Suguio, J.M. Flexor, A.C.S.P. Bittencourt. & G.S. Vilas-Boas 1980. Le Quaternaire marin bresilien (littoral pauliste, sud-fluminense et bahianais). *Cahier ORSTOM* (Série Géologie) 9(1): 96-124.

Martins, L.R., C.M. Urien & B.B. Eichler 1967. Distribuição dos sedimentos modernos da plataforma continental sul-brasileira e uruguaia. In: *Anais Congresso Brasileiro de Geologia* 21: 20-43.

Shackleton, N.J. & N.D. Opdyke 1973. Oxygen isotope and palaeomagnetic stratigraphy of Pacific Core V 28-238: Oxygen isotopic temperatures and ice volumes on a 10^5 and 10^6 year scale. *Quaternary Research* 3: 39-55.

Shakleton, N.J. & N.D. Opdyke 1977. Oxygen isotope and palaeomagnetic evidence for early Northern Hemisphere glaciation. *Nature* 261: 547-50.

Suguio, K., L. Martin, A.C.S.P. Bittencourt, J.M.L. Dominguez, J.M. Flexor & A.E.G. Azevedo 1985. Flutuações do nível relativo do mar durante o Quaternário Superior ao longo do litoral brasileiro e suas implicações na sedimentação costeira. *Revista Brasileira Geociências* 15(4): 273-286.

Tomazelli, L.J. 1993. O Regime dos Ventos e a Taxa de Migração das Dunas Eólicas Costeiras do Rio Grande do Sul, Brasil. *Pesquisas* 20(1): 18-26. Brazil.

Tomazelli, L.J. & J.A. Villwock 1989. Processos erosivos na costa do Rio Grande do Sul, Brasil: evidências de uma provável tendência contemporânea de elevação do nível relativo do mar. In: *Congresso da Associação Brasileira de Estudos do Quaternário*, 2, Resumos: 16. Rio de Janeiro.

Tomazelli, L.J. & J.A. Villwock 1992. Considerações sobre o ambiente praial e a deriva litorânea de sedimentos ao longo do Litoral Norte do Rio Grande do Sul, Brasil. *Pesquisas* 19(1):3-12. Brazil.

Villwock, J.A. 1984. Geology of the Coastal Province of Rio Grande do Sul, Southern Brazil. A Synthesis. *Pesquisas* 16: 5-49. Brazil.

Villwock, J.A. & L.J. Tomazelli 1995. Geologia Costeira do Rio Grande do Sul. *Notas Técnicas* 8: 1-45. Centro de Estudos de Geologia Costeira e Oceânica, Instituto de Geociências, Universidade Federal de Rio Grande do Sul, Porto Alegre. Brazil.

Villwock, J.A., L.J. Tomazelli, E.L. Loss, E.A. Dehnhardt, N.O. Horn, F.A. Bachi, & B.A. Dehnhardt 1986. Geology of the Rio Grande do Sul Coastal Province. *Quaternary of South America and Antarctic Peninsula* 4: 79-97. Rotterdam: A.A. Balkema Publishers.

Villwock, J.A., L.J. Tomazelli, M.F. Godolphin, B.A. Dehnhardt, L. Artusi, N.O. Horn, F.A. Bachi, E.L. Loss & E.A. Dehnhardt 1990. Advancements in geological mapping of Coastal Province of Rio Grande do Sul, Brazil. *International Symposium on Quaternary Shorelines: Evolution, Processes and Future Changes*, Abstracts: 68. INQUA, La Plata.

Holocene coastal evolution in Buenos Aires Province, Argentina

16

FEDERICO IGNACIO ISLA
CONICET-UNMDP, Centro de Geología de Costas y del Cuaternario,
Mar del Plata, Argentina.

ABSTRACT: The coast of Buenos Aires Province evolved under conditions that changed during the last 6000 years. The sea level raised until that age and has been dropping since then. Some authors believe that some radiocarbon dates from beach ridges mean short-term sea-level rises instead of the simple topographic expression of a storm deposit. In some areas, topographic differences have been explained by tectonic behaviour or different tidal ranges or wave set-up.

In low-gradient areas, there was a progradation of beach-ridge plains interfingered with former tidal flats and coastal lagoons. In these areas local vegetational changes took place, when marshes developed and barriers helped to expand the psammophytic assemblages.

In high-gradient regions, the Holocene coastal record is presented as beach ridges attached to palaeocliffs or infilled estuaries carved into Pleistocene continental deposits. These estuaries recorded the evolution of salinity indicators towards fresh-water assemblages.

Sea-surface temperature changes have been proposed to explain displacements of mollusk taxa bearing on beach ridges. Several reasons were suggested to explain mass-mortalities of infaunal mollusks in former tidal flats. Changes in the composition of littoral deposits have also been recorded.

Present barrier dunes (Eastern and Southern) were the consequence of sand availability, dominant wind directions and heavy littoral drifts. However, many beaches are today under critic erosive balances: some are the consequence of a very-modern sand scarcity or southeastern storm recurrence; other erosion problems are induced by improper beach managements or to the anachronistic policy in coastal defense.

RESUMEN: La costa de la Provincia de Buenos Aires evolucionó bajo condiciones que cambiaron en los últimos 6000 años. El nivel del mar aumentó hasta esa edad y desde entonces ha estado bajando. Algunos autores creen que las dataciones radiocarbónicas de dorsales de playa indican aumentos efectivos del nivel del mar, en lugar de la simple expresión topográfica del depósito de una tormenta. En algunas áreas, diferencias altimétricas han sido explicadas por efectos tectónicos, diferencias en el rango de mareas o en el 'setup' de las olas.

En áreas de baja pendiente, se produjo la progradación de una planicie de crestas de playa interdigitada con planicies mareales y lagunas costeras. En estas áreas, hubo cambios en la vegetación cuando se desarrollaron las marismas o cuando las barreras arenosas ayudaron a expandir las asociaciones psamofíticas.

En áreas de mayor pendiente regional, el registro costero holocénico se presenta como dorsales de playa adosadas contra paleoacantilados o estuarios colmatados emplazados entre depósitos continentales pleistocenos. Estos estuarios registraron la evolución de indicadores salinos hacia asociaciones de agua dulce.

Se han propuesto cambios en la temperatura superficial del agua para explicar desplazamientos de las especies de moluscos componentes de estas dorsales de playa. Varias explicaciones se han dado para explicar mortandades en masa de moluscos infaunales en antiguas planicies mareales. También se han registrado variaciones en la composición de los depósitos litorales.

Las actuales barreras arenosas (Oriental y Austral) fueron la consecuencia de una disponibilidad de arena, patrones de circulación de vientos e importantes derivas litorales. Sin embargo, muchas playas están actualmente bajo críticos balances erosivos: algunos son la consecuencia de una escasez de arena muy reciente o de una mayor recurrencia de tormentas sudestadas; otros problemas erosivos son inducidos por desafortunados manejos de la playa o por técnicas anacrónicas de defensa costera.

1 INTRODUCTION

The Holocene is, as a general rule, a climatically stable period, but characterized by significant variations in the landscape we like most to live on: the coast. Constructed by sea-level variations and storm behaviour, the Northern Hemisphere coasts are dominantly in recession whereas in the Southern Hemisphere much of the progradation is related to a sea-level fall with development of beaches and cheniers plains. In South America, much of the Holocene coast information is dispersed. Updated review papers are necessary to construct a framework from different sources to recognize the existing trends.

In the present paper, several papers of the Holocene coastal sequences of Buenos Aires (Argentina) are considered. Following the analysis of the data, sea-level trends, composition variations, morphodynamic changes, environmental conditions, tectonic behaviour, and progradation and infilling rates has been recognized and discussed.

2 STUDIED SITES

2.1 *Entre Ríos Province*

Holocene coastal deposits corresponding to the left margin of the former Río de la Plata embayment occur in Argentina (Fig. 1). They are found at the Province of Entre Ríos. By means of satellite images it is possible to recognize a regressive beach-ridge plain (Fig. 2). Guida & González (1984) and González et al. (1986) studied this area at the Irazusta quarry, close to Ibicuy. At this locality, estuarine deposits were sampled at 5 m giving ages between 6440 and 5610 years ^{14}C BP (González et al. 1986, Table 1).

2.2 *Paraná Delta – Río de la Plata*

The evolution of the delta of Paraná-Río de la Plata was reinterpreted by Parker & Marcolini (1992). Although there are no radiocarbon dates supporting it, they have interpreted a transgressive Basal Unit (subdelta) and a regressive Deltaic Unit (deltaic platform). The *Basal Unit* is comprising the basal sands (Atalaya Formation sensu Parker, 1980) and clayey facies of the Playa Honda Formation. On the other hand, the regressive phase, or *deltaic unit,* is characterized by the deposition of silts grading from 50 (interior) to 10 microns (inlet). These deposits occupied a restriction of the alluvial valley caused by a sea-level fall produced between 6000 and 4000 ^{14}C years BP (Table 1). Parker & Marcolini (1992) discriminated the fluvial and marine domain by the positions of the bars of Playa Honda and Barra del Indio. However, at the fluvial domain there are important tidal banks (Chico, Magdalena, Ortiz) and erosive basins. The top sets comprised marginal levees, swamps, small estuaries and deltas, and banks.

2.3 *Martín García Island*

In the middle of the Río de la Plata, at the Martín García Island, estuarine sediments were described at +5.2 m above mean sea level (MSL) yielding ages of 5800 ± 120 and 5740 ± 130 ^{14}C years BP (González et al. 1986, Table 1).

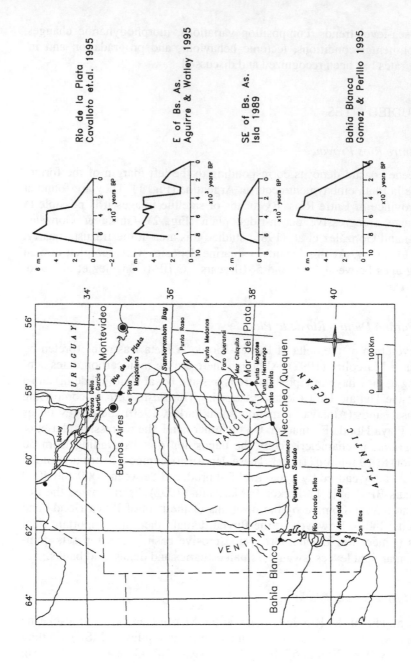

Figure 1. Location map of studied sites. Sea-level curves sensu Isla (1989), Aguirre & Whatley (1995), and Cavalotto et al. (1995); Bs.As.: Buenos Aires Province.

Figure 2. Beach plain
of Entre Ríos delta
complex (MSS image,
infrared band).

Table 1. Location and altitude of Holocene radiocarbon dates collected from the provinces of Buenos Aires and Entre Ríos.

Location	Lat.	Long.	^{14}C dates	Elevation (m)
Ibicuy (Entre Ríos)	33°25'	58°35'	5490 ± 110	4.5
	33°25'	58°35'	5530 ± 110	4.5
	33°25'	58°35'	5410 ± 110	4.5
	33°25'	58°35'	5280 ± 100	4.5
	33°16'	59°27'	5680 ± 110	5.0
	33°16'	59°23'	6440 ± 110	5.0
	33°15'	59°30'	5610 ± 110	5.0
	33°15'	59°30'	5760 ± 110	5.0
	33°14'	59°28'	5720 ± 110	4.5
	33°13'	59°31'	6030 ± 140	5.0
	33°13'	59°31'	5620 ± 110	5.0
	33°12'	59°37'	5680 ± 110	4.0
	33°19'	59°17'	5960 ± 110	5.0
Martín García Island	* 34°11'	58°15'	5800 ± 120	5.2
	* 34°11'		5740 ± 130	5.2
La Plata	* 34°52'	57°52'	6180 ± 150	6.5
	* 34°52'	57°52'	5150 ± 70	4.9
	* 34°52'	57°52'	5140 ± 140	4.0
	* 34°52'	57°52'	3990 ± 70	
	* 34°52'	57°52'	4250 ± 70	
	* 34°52'	57°52'	4760 ± 120	
	* 34°52'	57°52'	3820 ± 80	
Bahía Samborombón	36°07'	57°26'	4860 ± 110	4.5
	36°07'	57°26'	5580 ± 120	5.0
	36°14'	57°25'	4960 ± 110	5.0
	* 36°20'	57°22'	6980 ± 130	
	35°58'	57°27'	3320 ± 220	5.0
	35°58'	57°27'	4100 ± 110	4.0
	* 35°45'	57°22'	3000 ± 220	

Table 1. Continued.

Location	Lat.	Long.	^{14}C dates	Elevation (m)
Bahía Samborombón	35°46'	57°23'	2540 ± 80	2.0
Punta Rasa	* 36°18'	56°49'	1610 ± 100	
	36°36'	56°46'	1660 ± 110	2.0
	36°19'	56°45'	1720 ± 100	2.0
	36°24'	56°45'	1970 ± 100	2.0
	36°24'	56°49'	2770 ± 100	2.0
	36°27'	57°09'	3000 ± 110	2.5
	36°24'	56°46'	3370 ± 100	2.0
	36°25'	56°46'	5810 ± 100	3.0
Faro Querandí	* 37°30'	57°15'	5250 ± 200	
	* 37°30'	57°15'	4640 ± 120	
	* 37°30'	57°15'	4180 ± 120	
	* 37°30'	57°15'	5090 ± 200	
	* 37°30'	57°15'	3000 ± 100	
	* 37°30'	57°15'	3370 ± 100	
Mar Chiquita	37°27'14"	57°15'00"	3850 ± 60	1.7
	37°27'14"	57°15'00"	3620 ± 60	1.8
	37°27'53"	57°13'10"	3110 ± 80	1.9
	37°37'03"	57°18'50"	2880 ± 90	1.7
	37°37'03"	57°18'50"	2820 ± 80	1.7
	37°44'05"	57°26'24"	2920 ± 80	1.7
	37°44'05"	57°26'24"	2700 ± 50	1.7
	37°47'08"	57°27'20"	1340 ± 50	2.2
	37°44'53"	57°25'04"	540 ± 100	2.0
	37°25'55"	57°14'10"	4270 ± 70	
	37°42'41"	57°23'06"	2750 ± 160	2.3
Arroyo Las Brusquitas	38°14'40"	57°46'30"	6190 ± 160	2.2
	38°14'40"	57°46'30"	2380 ± 95	1.8
Punta Hermengo	38°17'10"	57°50'00"	6680 ± 160	1.1
	38°17'10"	57°50'00"	3325 ± 107	2.1
Costa Bonita	38°29'30"	58°36'	6000 ± 70	4.0
Río Quequén	38°34'00"	58°42'30"	7640 ± 90	0.8
	38°34'00"	58°42'30"	7140 ± 90	1.6
	38°34'00"	58°42'30"	7130 ± 90	2.2
	38°34'00"	58°42'30"	6230 ± 80	2.5
	38°34'00"	58°42'30"	5340 ± 80	2.6
Arroyo Claromecó	38°51'	60°05'	8430 ± 80	
	38°51'	60°05'	840 ± 70	
Pehuen Có	38°30'	61°34'	6260 ± 100	
	38°30'	61°34'	9400 ± 130	
	38°30'	61°34'	6080 ± 100	
	38°30'	61°34'	4320 ± 90	
	38°30'	61°34'	4060 ± 100	
	38°30'	61°34'	4450 ± 100	
Bahía Blanca	* 38°56'43"	62°00'44"	1890 ± 100	
	* 38°56'43"	62°00'44"	1740 ± 120	
	* 38°56'	62°03'	5980 ± 130	

302

Table 1. Continued.

Location	Lat.	Long.	^{14}C dates	Elevation (m)
	* 38°48'25"	62°09'	6490 ± 105	
Bahía Blanca	* 38°52'44"	62°12'	994 ± 190	
	* 38°47'	62°17'	3373 ± 205	
	* 38°47'24"	62°16'21"	4200 ± 190	
	* 38°46'14"	62°15'35"	4470 ± 95	
	* 38°46'06"	62°15'18"	4820 ± 120	
	* 38°45'48"	62°16'33"	4660 ± 95	
	* 38°44'15"	62°21'30"	6650 ± 100	
	* 38°41'15"	62°22'15"	5400 ± 120	
	* 38°59'04"	62°21'10"	5406 ± 227	
	* 38°59'04"	62°21'10"	7500 ± 120	
	* 38°59'04"	62°21'10"	3600 ± 95	
	* 38°59'04"	62°21'10"	5100 ± 100	
Río Colorado Delta	* 39°21'15"	62°07'	2170 ± 86	
	* 39°22'45"	62°16'30"	3920 ± 60	
	* 39°24'15"	62°12'45"	2850 ± 80	
	* 39°22'	62°09'30"	3580 ± 90	
	* 39°28'15"	62°09'30"	5750 ± 170	
	* 39°26'	62°04'	1640 ± 115	
	* 39°31'	62°06'	1240 ± 80	
	* 39°38'30"	62°14'15"	6930 ± 130	
	* 39°40'	62°15'30"	6000 ± 150	
	* 39°42'	62°19'30"	5310 ± 120	
	* 39°41'15"	62°11'	3060 ± 120	
	* 39°39'30"	62°05'30"	407 ± 100	
	* 39°41'	62°09'45"	2590 ± 110	
	* 39°50'	62°13'30"	3740 ± 90	
	* 39°49'15"	62°11'	2790 ± 90	
	* 39°50'30"	62°22'15"	4890 ± 110	
	* 39°51'	62°25'	5510 ± 110	
	* 39°51'45"	62°25'30"	5140 ± 110	
	* 39°52'45"	62°24'45"	5100 ± 110	
	* 39°53'30"	62°23'45"	3860 ± 95	
	* 39°59'30"	62°23'	9420 ± 150	
	* 39°40'45"	62°23'30"	5900 ± 100	
	* 39°42'	62°23'30"	4850 ± 90	
	* 39°42'	62°23'30"	6760 ± 100	
B. Anegada	* 39°53'	62°15'	5310 ± 110	
	* 39°53'	62°15'	5980 ± 90	
	* 39°53'	62°15'	5570 ± 120	
	* 39°53'	62°15'	5200 ± 110	
	* 39°53'	62°15'	5630 ± 170	
San Blas	* 40°35'	62°10'	4100 ± 95	7.0
	* 40°35'	62°10'	5370 ± 110	7.0
	* 40°35'	62°10'	2320 ± 80	2.5
	* 40°35'	62°10'	3450 ± 110	3.5

* approximate location. References in the text.

2.4 *La Plata area*

Close to La Plata, at terrains surrounding the Maldonado swamp, Cortele-zzi et al. (1992) recognized the Querandí and La Plata Formations. Three shell ridges (I, II and III) were dated (*Adelomedon brasiliana* remains) at ages between 6160 and 3820 ^{14}C years BP (Table 1).

2.5 *Bahía Samborombón*

At the counties of Castelli, Chascomús and Magdalena, Fidalgo et al. (1972) recognized two formations of coastal origin. The Destacamento Río Salado Formation constitutes tidal-flat and tidal-channel deposits. Above it, the Las Escobas Formation comprises beach deposits (Cerro La Gloria Member) whereas the Canal 18 Member has finer sediments (greenish silty sands; Fig. 3). These authors stated clearly that at that area the transgressive sea was very shallow with many islands composed of Pleistocene deposits of the 'Pampiano Formation' (Fidalgo et al. 1972).

Along Samborombón Bay coastline, Codignotto & Aguirre (1993) dated samples belonging to Holocene shelly beaches and spanning be-

Figure 3. A palaeobeach exceeding 4 m altitude is cut by the Río Salado very close to the route 11. Below the beach, there are tidal-flat deposits.

tween 5580 ± 120 and 2540 ± 80 ^{14}C years BP (Table 1). In the sense of these authors, Punta Rasa evolved as a spit towards the north during that same interval. Eight radiocarbon samples indicate an evolution between 5810 ± 100 and 1610 ± 100 years BP. Codignotto & Aguirre (1993) believed that a barrier island began to grow 6000 years ago and that at 3500 years BP there was a slight fall of 1.5 m above MSL.

2.6 Punta Médanos

At Punta Médanos, Parker (1980) described three coastal formations of Holocene age. Pozo N° 8 Formation corresponds to the transgressive phase with two facies (Pinamar and La Victoria). An age of 5200 ± 130 ^{14}C years BP was obtained. The Pozo N°17 Formation was deposited during the regressive phase with ages younger than 3000 years. Two facies (La Ernestina and Mar de Ajó) were identified in this unit. The Banco Punta Médanos Exterior Formation is the name given to the transgressive sands of the inner platform facies. Ages of $11,610 \pm 140$ and $10,380 \pm 180$ [14]C years BP were obtained (Table 1).

2.7 Faro Querandí

Violante & Parker (1993), at the region of Faro Querandí, distinguished two Holocene formations related to the sea-level oscillations. The Medaland Formation (with 3 facies) is the transgressive phase (barrier facies), giving a radiocarbon (*Tagelus plebeius*) age of 5250 ± 200 years BP. The regressive phase (4 facies) is the Mar Chiquita Formation (defined by Schnack et al. 1982), with ages spanning from 5090 to 3000 ^{14}C years BP (Table 1).

2.8 Mar Chiquita

In the marginal plain surrounding Mar Chiquita coastal lagoon, Schnack et al. (1982) sampled shells contained in ancient tidal flats (today supratidal channelized marshes or swamps). Living-position shells of *Mactra isabelleana* and *Tagelus plebeius* gave radiocarbon ages between 3850 ± 60 and 1340 ± 50 years BP at altitudes of 2.2 to 1.7 m above MSL (Table 1). They recognized that these deposits belong to the regressive phase (Fig. 4). Diatoms from these sediments indicate salinity conditions expected for the evolution suggested by Schnack et al. (1982) (in Espinosa 1994).

Figure 4. At the coast of
Mar Chiquita coastal la-
goon (San Gabriel), *Tagelus
plebeius* can be sampled in
living position.

2.9 *Punta Mogotes*

Close to Mar del Plata, at Punta Mogotes, Isla & Selivanov (1993) sampled *Glycimeris longior* shells from a shelly beach elevated up to 6 m
height (Fig. 5). Although these samples gave ages of 35,000 ± 3000 and
27,350 ± 1450 ^{14}C years BP, they assigned a Holocene age for this high-
energy deposit, bearing shells that should have been removilized.

2.10 *Arroyo Las Brusquitas*

Along the Southern Barrier, Holocene coastal facies occur as infilled small
estuaries. At Las Brusquitas Creek, Espinosa et al. (1984) dated a basal
layer of 6190 ± 160 ^{14}C years BP, with *Tagelus plebeius* in living position.
Another level was dated using *Ostrea spreta* shells in 2380 ± 95 years ^{14}C
BP (Isla & Selivanov 1993, Table 1). The sequence is characterized by
progressive fresh-water conditions indicated by diatoms, forams and ostracods assemblages (Isla et al. 1986).

Figure 5. At Punta Mogotes, close to Balneario La Reserva, a beach deposit of more than 3 m thick is dipping seawards.

2.11 *Punta Hermengo*

Same infilling estuarine conditions with different salt-water proportions are present in Punta Hermengo, very close to Miramar city. Fidalgo & Tonni (1983) reported radiocarbon dates of 6680 ± 160 years BP (*Littoridina parchappei* shells) at the bottom and 3325 ± 107 years BP (carbonate) at the top (Table 1). The microfossil content indicate oligohaline conditions although *Brachidontes rodriguezi* shells suggest a marine connection (Espinosa et al. 1984, Isla et al. 1986).

2.12 *La Ballenera*

Frenguelli (1928) recognized the profile outcropping at the inlet of La Ballenera Creek. This estuarine sequence is characterized by a coarsening-upwards grain-size record. It is interpreted that maximum turbidity dominates during the infilling of the bottom of the channel, and that bedload transport becomes dominant at the final infilling stages of the estuary (Isla et al. 1996). A radiocarbon date of 6120 ± 80 was performed on *Littoridina sp* remains, and another of 4120 ± 60 ^{14}C years was obtained from *Littoridina* sp and *Planorbis* sp remains (Table 1).

2.13 *Costa Bonita*

Very close to the present beach, a Holocene palaeobeach was discovered

Figure 6. East of Costa Bonita, a trench discovered a beach deposit (sands with boulders, more than 4 m thick) very close to the present beach, but at a higher altitude.

east of Costa Bonita (Fig. 6). It is composed of very coarse sand with caliche boulders on top of what was described as storm berms (Isla et al. 1996). Shells of *Tegula patagonica* gave an age of 6000 ± 70 ^{14}C years BP (Table 1).

2.14 *Río Quequén*

The 2 m thick estuarine sequence at Rio Quequén inlet was interpreted as transgressive (Fasano et al. 1987). The maximum age provided by broken shells at the base is 7640 ± 90 ^{14}C years BP. The top of the sequence is also composed of broken shells of *Tagelus plebeius* of 5340 ± 80 ^{14}C years BP. *Tagelus plebeius* shells in living position with an age of 7140 and 7130 ^{14}C years BP (Table 1) confirmed low-energy conditions and a rapid infilling rate. In opposition to Las Brusquitas regressive sequence, at Río Quequén polyhalobous diatom assemblages increase towards the top confirming its transgressive character (Isla et al. 1986). On the other hand, this sequence has halophobous taxa, something expected at a partially stratified river.

308

2.15 Claromecó

One km inlands of the inlet of Claromecó Creek, Frenguelli (1928) described an estuarine sequence composed of muds increasing the fine-sand fraction to the top (Fig. 7). Ash deposits became also important to the top of the sequence (Isla et al. 1996). A peat layer was dated in 2430 ± 80 ^{14}C years BP towards the bottom whereas another horizon rich in organic matter was dated in 840 ± 70 ^{14}C years BP (Table 1).

2.16 Río Quequén Salado

In the Quequén Salado Holocene section, close to the inlet, Farinati & Zavala (1995) sampled *Tagelus plebeius* valves in living position and dated them at 7720 ± 1000 ^{14}C years BP (Table 1). The deposits were interpreted as mixed tidal flats and are overlying beach deposits (foreshore facies) composed of broken shell up to 5 cm in length.

2.17 Pehuén Co

González (1996) described 3 generations of *Callianasa* burrows affecting the last of the 3 marine stages that occurred at Pehuén-Co beach since Up-

Figure 7. At Claromecó, the estuarine infilling sequence is composed of dark muds at the bottom and light sands to the top. Arrow is pointing the erosion surface of a palaeochannel.

per Pleistocene. This last mid Holocene transgression has 2 transgressive pulses. The first pulse is constituted by a conglomerate composed of quartzitic clasts in a coarse sandy matrix. Remains of *Adelomedon* and *Ostrea* gave a ^{14}C age of 6260 ± 100 yr BP while shells of *Amiantis purpurata* gave ages significantly older (9400 ± 130 ^{14}C yrs BP). The second pulse is a grayish sandstone corresponding to foreshore and tidal-inlet facies. Different remains yielded ages spanning between 6080 and 4060 ^{14}C years BP (Table 1). In the sense of González (1996), these pulses should be assigned to a probable age of 6020 and 4060 years, respectively.

2.18 *Bahía Blanca*

Holocene deposits were sampled from the tidal flats surrounding Bahía Blanca. Farinati (1984) and González (1989) determined that these shells have ages spanning between 7500 and 994 ^{14}C years BP (Table 1). From cores obtained at the Principal Channel of Bahía Blanca harbour, Aliotta et al. (1996) assigned sandy sediments (estuarine marine facies), with different proportions of gravels and shell fragments, to the Holocene transgressive cycle.

2.19 *Río Colorado Delta*

Weiler (1983) sampled several mollusks specimens from the gravel beach ridges related to the delta of the Río Colorado. The regression of the sea spanned between 6930 ± 130 and 407 ± 100 ^{14}C years BP (Table 1). The delta was reported to be cuspate, assymmetric and subject to a northward drift, storms and floods (Codignotto & Marcomini 1993).

2.20 *Bahía Anegada*

At Bahía Anegada, Weiler (1993) discriminated barrier islands and coastal lagoons as the two ancient environments related to the postglacial transgression. At heights between 4 and 2.5 m above MSL, living-position valves were dated between 5980 and 5200 ^{14}C years BP (Table 1). In regard that the maximum transgression occurred 6000 years ago and reaching a maximum altitude of 5 m above MSL, the deposits of Bahía Anegada would correspond to high-energy regressive events (Weiler 1993).

2.21 *Bahía San Blas*

The San Blas area is a complex of gravel barriers, and salt marshes and tidal flats. Some of these barriers are beaches of Late Pleistocene (Sangamon) age. Only the stages IV and V are of Holocene age (Witte 1916).

The stage III is at 10 m height and assigned to the 'Querandinense' in the sense of F. Ameghino. Unfortunately, there were no radiocarbon dates verifying the ages of this detailed study. More recently, Trebino (1987) dated some shells from his Level III giving ages from 5370 ± 110 to 2320 ± 80 ^{14}C years BP (Table 1).

3 SEA-LEVEL AND CLIMATIC TRENDS

The first sea-level curve was presented by Auer (1952) for the Río Negro inlet. He recognized sea level variations and related them to fluvial terraces.

Isla (1989) compiled the information of southeastern Buenos Aires Province with other published curves of the Southern Hemisphere.

The sea-level fluctuation at Faro Querandí was correlated to a humid and warm climate (Violante & Parker 1992).

At the area of Bahía Anegada, maximum transgression is thought to have reached the 5 m level 6000 ^{14}C years ago (Weiler 1993).

Codignotto & Aguirre (1993) believed that the sea-level maximum reached an altitude of 10 m above present MSL before 7000 ^{14}C years BP. After a regression, another transgression established a barrier between Punta Piedras and General Conesa about 6890 ^{14}C years BP.

Aguirre et al. (1995) suggested a sea-level fall between 4500 and 4000 years ago, in coincidence with a lowering in sea surface temperature suggested by a decrease of warm-water mollusks. However, the curve drawn by these authors indicates a sea-level fall (not below present MSL) between 3700 and 2500 ^{14}C years BP (Aguirre & Whatley 1995, Fig. 1).

Cavalotto et al. (1995), for the outer Río de la Plata, proposed a maximum sea-level height of 6.5 m at 6000 ^{14}C years BP. They believed in a stepwise sea-level fall: from 6000 to 5000 ^{14}C years BP it fell from 6.5 to 5 m above present MSL; from 5000 to 3500 ^{14}C years BP it remained stable; between 3500 and 2900 ^{14}C years BP there was a rapid fall to +2.5 m above present MSL; from 2900 ^{14}C years BP to present it continued falling (see Fig. 1).

Gómez & Perillo (1995) proposed a sea-level curve for Bahía Blanca based on datings performed by González (1989).

4 SEDIMENT COMPOSITION AND AVAILABILITY

Shelly coastal deposits are related either to beach ridges corresponding to the maximum transgression, or to cheniers originated by the storm reworking of tidal flats and marshes during the regressive phase (Samborombón Bay). Usually these shelly coastal deposits characterized the region between Río de la Plata and Mar Chiquita.

On the other hand, present beaches of this region have less than 20% of bioclastics; they are composed of quartz, feldspar, volcanic fragments and heavy minerals (magnetite, piroxenes, amphiboles; see Mazzoni 1977, Isla 1991).

Between Mar del Plata and Bahía Blanca, the fossil beaches related to the maximum transgression (Punta Mogotes, Costa Bonita, Claromecó) are composed of terrigenous coarse sand with volcanic gravels and broken shells (Isla et al. 1996).

South of Bahía Blanca, fossil beaches corresponding to this maximum transgression are dominantly composed of volcanic gravels in a sandy matrix (Weiler 1993).

The different composition between transgressive and present beaches led to relate them to geomorphological evidences suggesting drift reversals during the Holocene (Schnack et al. 1982, Violante & Parker 1992, Isla & Espinosa 1994). However, new evidence from Faro Querandí indicates that this differentiation is not as sharp: non-operating washover deposits are composed of volcanic ('Tehuelche Gravels') and orthoquartzitic (Balcarce Formation) gravels. Present washovers do not contain these clasts; they are composed mostly of beachrock and shells. This is indicating that the northward drift (induced by storms) was larger than present, and transported clasts that today could not arrive north of the Mar Chiquita inlet; volcanic gravels could only dominate at beaches as north as Chapadmalal or Mar del Sud.

On the other hand, and based on geomorphological evidences from the Eastern and Southern barriers, Isla & Espinosa (1995) recognized that there was significant availability of sand at the initial stages of the postglacial transgression, and that sediment became scarce due to a logic consumption during a sea level position which was stable for approximately 6000 years.

5 MORPHODYNAMIC CHANGES

In many descriptions, it has been stated the major morphological changes: former tidal flats and coastal lagoons were overlain by barrier islands (Schnack et al. 1982, Codignotto & Aguirre 1993, Violante & Parker 1992, Bértola 1995). Other papers, specifically described the morphodynamic differences between reflective shelly beaches belonging to the maximum transgression, and those beaches (intermediate or dissipative) of the present clastic shoreline of the same areas (Isla & Espinosa 1995, Isla et al. 1996).

Mostly, in the urbanized sites, there is an increase in erosive processes.

In other sites, present erosion is caused by very poor coastal defences. Groynes block the littoral drift to the north and cause erosion of foredunes (Fig. 8). Storms produce severe problems where management of the beach sand resources is improper (Fig. 9). Much of this transgressive problem is man-made (Fig. 10) as tidal-gauges records are not suggesting an increase in MSL.

Figure 8. South of Cariló (Pinamar), scarps of foredunes are indicating an intense erosive process.

Figure 9. The storms of 1993 (mainly July, 8) and 1994 (June 24) destroyed many of the resort facilities.

Figure 10. These storms are episodically affecting constructions, streets and parking places. However, city authorities insist on paving the coastal avenue.

6 ENVIRONMENTAL CONDITIONS

Environmental conditions along the coast of Buenos Aires during the Holocene are not conclusive. Mollusks, pollen and diatoms records gave different approximations.

At the distal portions of the Chico or de las Gallinas Creek (Mar Chiquita Lagoon), Espinosa (1994) studied the evolution of water salinity based on diatom assemblages. Before 3000 years BP, freshwater conditions dominate, with mesohalobous excursions. After 3110 ^{14}C years BP, polyhalobous (marine) conditions became dominant while the large number of epiphytic diatoms indicate a salt marsh. The top of the sequence is again dominated by freshwater diatoms that have been related to the restriction of the tidal excursion within the coastal lagoon (Espinosa 1994).

Aguirre & Whatley (1995) assigned an age of 7030 ^{14}C years BP for the 'Querandinense' of Samborombón Bay; and correlated it with the 11,000 years date from the shells collected at 115 m depth, at the platform shelf (Richards & Craig 1963). They did not stated clearly whether the 'Querandinense' corresponds to marshy or platform-shelf deposits or whether it deposited during the Hypsithermal, the Late Glacial or the Younger Dryas cool event (Aguirre & Whatley 1995).

A sea-surface temperature higher than present was proposed for a climatic optimum spanning between 8000 and 4500 ^{14}C years; the Hypsithermal peak corresponds to beach ridges deposited between 8000 and 6000 ^{14}C years (Aguirre & Whatley 1995). However, this hypsithermal was proposed according to different sources and varied ages: atmospheric patterns (8000-3000 years; Iriondo & García 1993), glacier advances at Southern Andes (6000-4500 years; Rabassa 1987), or shifts of oceanic water masses in South Africa (7000-6500 years; Partridge 1993).

Aguirre & Whatley (1995) thought that a post-Hypsithermal climate deterioration correlates with a falling sea level after 4500 ^{14}C years BP, and a northwards displacing of mollusk taxa.

Prieto (1996) stated the climatic changes and vegetational evolution of the grassland pampa, based on the pollen record in sediments. Prior to 10,500 ^{14}C years BP, there was an herbaceous psammophytic steppe. Between 10,500 and 8000 ^{14}C years, moisture was similar to present, and the region was characterized by ponds, swamps and floodplains. Grasses became dominant between 8000 and 7000 ^{14}C years BP. Closer to the coast, the sea-level maximum about 6200 ^{14}C years brought local halophytic communities. The marine regression was dominated by fresh-water pollen assemblages in coincidence with diatoms remains. He finally stated that changes in the Atlantic and Pacific anticyclones controlled precipitation, while sea-level variations conditioned continentality. Both, precipitation and distance to the coast, affect vegetational changes (Prieto 1996).

In southwestern Buenos Aires Province (Monte Hermoso-Bahía Blanca), Quattrocchio et al. (1995) studied Holocene vegetational changes based on the palynological content in sediments. The development of palaeosols during the Late Pleistocene-Early Holocene suggests an episode of stability. Temperate and humid climate is inferred for the Middle Holocene related to the grass-dominated steppe recorded in a section of Sauce Grande river. On the other hand, the Late Holocene was dominated by arid and semiarid conditions characterized by a psammophytic herbaceous steppe. These arid conditions prevailed until our days, although in the Sauce Grande river, an increase in the halophytic steppe was recorded (Quattrocchio et al. 1995). In the Napostá estuary, the dissapearance of cysts of dinoflagellates during the last 3000 years should be indicating a dominance of continental inputs (Grill & Quatrocchio 1996).

Many authors have been interested on the present reduction in abundance of certain mollusks (*Tagelus plebeius, Mactra isabelleana*) from very recent to present tidal flats. Some of them have considered the probable cause of assumed mass mortalities:

a) Schnack et al. (1982) pointed to a rapid sea-level drop or an inlet obstruction at the Mar Chiquita Lagoon.

b) Farinati et al. (1992), at the Naposta Grande Creek, considered a catastrophic burial by fluvial sands (flood).

c) Isla & Espinosa (1995) analyzed the regional character of these dissapearances with the diatom-based salinity records of estuaries of the Southeastern Buenos Aires. They pointed that taphonomic processes could have caused a replacement of infaunal communities with epifaunal communities (Isla & Rivero D'Andrea 1993) at the final stages of a sea-level fluctuation of less than 3 m.

d) Iribarne et al. (1997), based on the comparison between present and ancient valves, proposed that *T. plebeius* would have preferently predated by birds *(Haematopus pallictus)*. In this sense, the specimens could remain in living position while having different ages.

7 TECTONIC BEHAVIOUR

Isla et al. (1986) suggested that tectonics could explain slight differences in the altitude of estuarine sequences of the southeastern Buenos Aires. In their sense, the altitude of the fossil beach of Punta Mogotes should only be explained by a tectonic effect.

Codignotto et al. (1992) related the elevations of Holocene beaches and their radiocarbon dates to recognize the tectonic behaviour of different sectors of the Argentine coast. Although they did not consider the tidal effects on berm altitudes, they compared the uplift affecting different systems and even the sedimentary basins (Codignotto et al. 1992).

In Faro Querandí, Violante & Parker (1992) believed that subsidence affected the area during middle and upper Pleistocene. This subsidence would have finished by the end of the Pleistocene, when the area began to rise, reactivating the relief.

Cavalotto et al. (1995) accepted the tectonic effect on relative sea-level fluctuations, as a secondary process of the triggering climatic changes.

8 PROGRADATION AND INFILLING RATES

The extension of more than 100 km of the Paraná River emerged delta is indicating the huge progradation at the headlands of the Río de la Plata.

Bértola (1995) estimated the progradation at different positions of the Samborombón Bay. In Punta Piedras (northern end), he calculated an erosion rate of 0.8 m/yr. The bay prograded in the order of 1 m/yr for the last 6500 years. In Punta Rasa (southern end), the progradation is of 10 m/yr (Bértola 1995). In General Conesa, the coast of Samborombón Bay prograded 60 km across shore (to the SE; Codignotto & Aguirre 1993).

From Villa Gesell to San Clemente del Tuyú, the Eastern Barrier prograded 100 km alongshore (northwards; sensu Codignotto & Aguirre 1993).

Isla & Espinosa (1995) compared the infilling rate of some estuarine environments (Mar Chiquita, Las Brusquitas, Punta Hermengo and Quequén Grande) of southeastern Buenos Aires Province.

The Colorado delta prograded 12 km in the last 6000 years (Codignotto & Marcomini 1993).

9 DISCUSSION

A significant part of the discrepancies in different proposed models about sea-level behaviour during the Holocene (Fig. 1) seems to have been produced by a lack of topographic control and correct (morphological) interpretation of palaeoenvironments in relation to sea level:

1. The contour intervals of topographic charts (1/50.000) is usually of 5 or 2.5 m. Thus, they do not represent morphologic variations of 1 or 2 m, and should not be therefore used to interpolate the altitude of sampling sites.

2. Beach ridges are ancient storm or tidal berms deposited above MSL. In Mar del Plata, present berms are ca. 2.2 m and 1 m (storm and tidal berms, respectively) above the MSL; intertidal sand bars or ridges (ridge and runnel systems) are at sea level ± 0.5 m; and subtidal bars could be at heights of −1 m to −2 m depth. Cheniers are at the limit of the meteorological-forced tides (storms), more than 2.4 m above MSL. Tidal flats occupy the tidal excursion (tidal range) although some of them could be controlled by meteorological tides. Marshes are the vegetated portions of the supratidal flats; however, care should be taken to recognize if they are operating today. In prograding plains (Samborombón, Mar Chiquita), typical marshy vegetation is not controlled by the maximum height of the meteorological tide but by a salty substrate related to former tidal flats. Coastal-lagoon bottoms are today not more than 1 m below MSL (Mar Chiquita); when sea level was higher (less infilled) they could be deeper, sandier and with high-energy conditions (waves, tidal currents). The blocking of estuaries (banks, spits, barriers or dune migration) is another process to consider in palaeoenvironmental interpretations.

Much of the palaeoclimatic interpretations came from different techniques (mollusks, pollen and diatoms) and from different environments. Beach ridges were usually sampled to characterize the coastal evolution although they are composed of allochthonous assemblages related to the effects of storm deposition. This problem led to several authors to use minimum datings obtained from these environments (see Weiler 1993). Continuous sequences (estuaries, coastal lagoons, tidal flats) are better sites for sea-level studies than discrete sequences (beach ridges, cheniers).

In the construction of sea-level curves, care should be taken not to confuse sea-level indicators from different palaeoenvironments:

1. High-energy beach ridges usually correspond to the maximum reaches of the transgression,

2. Cheniers are related to storm reworking of tidal flats with infaunal remains, and

3. Coastal lagoons and tidal flats are usually constructed during the regressive phase.

1 and 2 correspond to discrete and episodic phenomena where mixing assemblages are expected, and up to 2 m above MSL. Coastal lagoons and tidal flats, instead, are low-energy environments where continuous records are expected, but they do not represent more than the maximum reaches of the storms (below the altitude of 2 m over the MSL).

Interpretations of mollusks assemblages should be related to taphonomic conditions (see Farinati & Zavala 1995) so as not to confuse allochthonous and autochthonous assemblages.

Coastal pollen records could also be pointing either regional conditions during the transgression, or local conditions (marsh, beach ridge, or dune associations) during the regressive phase.

10 CONCLUSIONS

1. *Sea level* has been dropping for the last 6000 years. However, different sea-level curves have been interpreted based on different palaeotopographical criteria.

2. Littoral *sediment composition* changed during the last thousands of years. Some of the present beaches are indicating that sand has become relatively scarce with time.

3. There were *morphodynamic changes* that are confirming variations in the sediment availability.

4. Several *environmental changes* were proposed for the last 6000 years based on diatoms, pollen and mollusk assemblages. These changes were triggered by climate, wind patterns, beach drift and estuarine infilling rate.

5. Although it was hard to believe at first, some evidences are pointing towards *tectonic movements* affecting the Holocene records of Buenos Aires Province coastal sites.

6. Holocene *progradation and infilling rates* are good indicators to predict coastal evolutionary trends of the Buenos Aires coastline.

ACKNOWLEDGEMENTS

Financial support was provided by the Buenos Aires (Comisión de Investigaciones Científicas, CIC) and Argentina (Consejo Nacional de Investigaciones Científicas y Técnicas, CONICET) Research Council, and the University of Mar del Plata (UNMDP). Figures were drafted by M. Farenga.

REFERENCES

Aguirre, M.L. & R.C. Whatley 1995. Late Quaternary marginal marine deposits and palaeoenvironments from northeastern Buenos Aires Province, Argentina: a review. *Quaternary Science Reviews* 14: 223-254.

Aguirre, M.L., D.Q. Bowen, G.A. Sykes & R.C. Whatley 1995. A provisional aminostratigraphical framework for Late Quaternary marine deposits in Buenos Aires province, Argentina. *Marine Geology* 125: 85-104.

Aliotta, S., G.O. Lizasoain, W.O. Lizasoain & S.S. Ginsberg 1996. Late Quaternary sedimentary sequence in the Bahía Blanca Estuary, Argentina. *Journal of Coastal Research* 12(4): 875-882.

Auer, V. 1952. Evolución postglacial del valle inferior del Río Negro y variaciones cuaternarias de la línea costanera. *Revista Investigaciones Agrarias* 5. Buenos Aires.

Bértola, G.R. 1995. Geomorfología y sedimentología de los ambientes mareales de la Bahía Samborombón (Buenos Aires, Argentina). Ph.D.Dissertation, Facultad de Ciencias Naturales, Universidad Nacional de La Plata. La Plata. 88 pages. Unpublished.

Cavalotto, J.L., G. Parker & R.A. Violante 1995. Relative sea level changes in the Río de la Plata during the Holocene. In L.Ortlieb (ed.) *Late Quaternary coastal records of rapid change: Application to present and future conditions.* IInd. Annual Meeting, Abstracts: 19-20. Antofagasta, Chile, 12-28 November.

Codignotto, J.O. & M.L. Aguirre 1993. Coastal evolution in sea level and molluscan fauna in northeastern Argentina during the Late Quaternary. *Marine Geology* 110: 163-175.

Codignotto, J.O., R.R. Kokot & S.C. Marcomini 1992. Neotectonism and sea-level changes in the coastal zone of Argentina. *Journal of Coastal Research* 8(1): 125-133.

Codignotto, J.O. & S.C. Marcomini 1993. Argentine deltas morphology. In Kay R. (ed.) *Deltas of the world. Coastlines of the world:* 323-336. Coastal Zone '93, ASCE, New Orleans.

Cortelezzi, C.R., R.E. Pavlicevic, C.A. Pitori & A.V. Parodi 1992. Variaciones del nivel del mar en el Holoceno en los alrededores de La Plata y Berisso. *IV Reunión Argentina de Sedimentología,* Actas II: 131-138. La Plata.

Espinosa, M.A. 1994. Diatoms paleoecology of Mar Chiquita lagoon delta, Argentina. *Journal of Paleolimnology,* 10: 17-23.

Espinosa, M.A., J.L. Fasano, L. Ferrero, F.I. Isla, A. Mujica & E.J. Schnack 1984. Micropaleontología y microestratigrafía de los sedimentos holocenos aflorantes en la desembocadura Arroyo Las Brusquitas (Partido de General Pueyrredón) y en Punta Hermengo (Partido de General Alvarado), Provincia de Buenos Aires. *IX Congreso Geológico Argentino,* Actas III: 520-537. San Carlos de Bariloche.

Farinati, E.A. 1984. Dataciones radiocarbónicas en depósitos holocenos en los alrededores de Bahía Blanca, Provincia de Buenos Aires. In E.J. Schnack (ed.). *Simposio Oscilaciones del Nivel del Mar y evolución costera en el Holoceno tardío:* 27-31. Mar del Plata.

Farinati, E.A., S. Aliotta & S.S. Ginsberg 1992. Mass mortality of a Holocene *Tagelus plebeius* (Mollusca, Bivalvia) population in the Bahía Blanca estuary, Argentina. *Marine Geology,* 106: 301-308.

Farinati, E. & C. Zavala 1995. Análisis tafonómico de moluscos y análisis de facies en la serie holocena del Río Quequén Salado, Provincia de Buenos Aires, Argentina. *VI Congreso Argentino de Paleontología y Bioestratigrafía,* Actas, 117-122. Trelew.

319

Fasano, J.L., F.I. Isla, W.G. Mook & O. Van de Plassche 1987. Máximo transgresivo postglacial de 7.000 años en Quequén, Provincia de Buenos Aires. *Revista Asociación Geológica Argentina*, 42(3-4): 475-477. Buenos Aires.

Fidalgo, F., U.R. Colado & F.O. De Francesco 1972. Sobre ingresiones marinas cuaternarias en los partidos de Castelli, Chascomús y Magdalena (Provincia de Buenos Aires). *V Congreso Geológico Argentino*, Actas III: 227-240. Buenos Aires.

Fidalgo, F. & E.P. Tonni 1983. Geología y paleontología de los sedimentos encauzados del Pleistoceno tardío y Holoceno en Punta Hermengo y Arroyo Las Brusquitas (Partidos de General Alvarado y General Pueyrredón, Provincia de Buenos Aires). *Ameghiniana* 20(3-4): 281-296. Buenos Aires

Gómez, E.A. & G.M.E. Perillo 1995. Submarine outcrops underneath shoreface-connected sand ridges, outer Bahía Blanca Estuary, Argentina. *Quaternary of South America & Antarctic Peninsula* 9: 23-37. Rotterdam: Balkema Publishers.

González, M.A. 1989. Holocene levels in the Bahía Blanca Estuary, Argentina Republic. *Journal of Coastal Research* 5: 65-77.

González, M.A. 1996. Mid-Holocene littoral deposits at Pehuen-có Beach (38°30'S, 61°34'W, Buenos Aires Province). Geochronology and correlations. *Proceedings of the Bahía Blanca International Coastal Symposium:* 265-271. October 7-11, 1996. Universidad Nacional del Sur. Bahía Blanca.

González, M.A., N.E. Weiler & N.G. Guida 1986. Late Pleistocene transgressive deposits from 33° S.L. to 40° S.L., Republic of Argentina. *Journal of Coastal Research*, Special Issue 1: 39-47.

Guida, N.G. & M.A. González 1984. Evidencias paleoestuáricas en el sudeste de Entre Ríos. Su evolución con niveles marinos relativamente elevados del Pleistoceno superior y Holoceno. *IX Congreso Geológico Argentino*, Actas III: 577-594. San Carlos de Bariloche.

Grill, S.C. & M.E. Quatrocchio 1995. Fluctuaciones eustáticas durante el Holoceno a partir del registro del paleomicroplancton; Arroyo Napostá Grande, sur de la Provincia de Buenos Aires. *Ameghiniana* 33(4): 26-12. Buenos Aires.

Iribarne, O., J. Valero, M. Martínez & L. Lucifora 1997. Predación por aves playeras explicaría el origen de los depósitos cuaternarios de almejas navaja (*Tagelus plebeius*). *XVIII Reunión Argentina de Ecología*: 65. Asociación Argentina de Ecología. Buenos Aires.

Iriondo, M. & N. García 1993. Climatic variations in the Argentine Plains during the last 18,000 years. *Palaeogeography, Palaeoclimatology and Palaeoecology* 101: 209-220.

Isla, F.I. 1989. The Southern Hemisphere sea-level fluctuation. *Quaternary Science Reviews* 8: 359-368.

Isla, F.I. 1991. Spatial and temporal distribution of beach heavy minerals: Mar Chiquita, Argentina. *Ocean & Shoreline Management* 16, 161-173.

Isla, F.I. & I. R. D'Andrea 1993. Procesos retro-tafonómicos en secuencias costeras poco profundas. *Pesquisas* 20(2): 90-95. Universidade Federal do Rio Grande do Sul. Brazil.

Isla, F.I. & A. Selivanov 1993. Radiocarbon contributions to the Quaternary eustatism of Buenos Aires, Chubut and Tierra del Fuego. *Taller Internacional 'El Cuaternario de Chile'*, Resúmenes: 47, Santiago, 1-9 Nov. 1993.Chile.

Isla, F.I. & M.A. Espinosa 1995. Environmental changes associated to the Holocene sea-level fluctuation: Southeastern Buenos Aires, Argentina. *Quaternary International* 26: 55-60.

Isla, F.I., L.C. Cortizo & E.J. Schnack 1996. Pleistocene and Holocene beaches and estuaries along the southern barrier of Buenos Aires. *Quaternary Science Reviews* 15(8-9): 833-841.

Isla, F.I., L. Ferrero, J.L. Fasano, M.A. Espinosa & E.J. Schnack 1986. Late Quaternary marine-estuarine sequences of the Southeastern coast of the Buenos Aires Province, Argentina. *Quaternary of South America & Antarctic Peninsula* 4: 137-157. Rotterdam: A.A. Balkema Publishers.

Mazzoni, M.M. 1977. Características composicionales de la fracción pesados de arenas de playa frontal del litoral atlántico bonaerense. *Revista Asociación Argentina de Mineralogía, Petrología y Sedimentología* VIII(3-4): 73-91. Buenos Aires.

Parker, G. 1980. Estratigrafía y evolución morfológica durante el Holoceno en Punta Médanos (planicie costera y plataforma interior), Provincia de Buenos Aires. In: *Simp. Problemas Geológicos del Litoral Atlántico Bonaerense*. Resúmenes: 205-224. CIC, Mar del Plata.

Parker, G. & S. Marcolini 1992. Geomorfología del delta del Paraná y su extensión hacia el Río de la Plata. *Revista Asociación Geológica Argentina* 47(2): 243-249. Buenos Aires.

Partridge, T.C. 1993. Warming phases in Southern Africa during the last 150,000 years: an overview. *Palaeogeography, Palaeoclimatology and Palaeoecology* 101 (3-4): 237-244.

Prieto, A.R. 1996. Late Quaternary vegetational and climatic changes in the Pampa grassland of Argentina. *Quaternary Research* 45: 73-88.

Quattrocchio, M.E., A.M. Borromei & S. Grill 1995. Cambios vegetacionales y fluctuaciones paleoclimáticas durante el Pleistoceno Tardío-Holoceno en el sudoeste de la Provincia de Buenos Aires (Argentina). *VI Congreso Argentino de Paleontología y Bioestratigrafía*, Actas: 221-229. Trelew.

Rabassa, J. 1987. The Holocene of Argentina: a review. *Quaternary of South America and Antarctic Peninsula* 5: 269-290. Rotterdam: A.A. Balkema Publishers.

Richards, H. & J. Craig 1963. Pleistocene mollusks from the Continental Shelf off Argentina. *Proceedings of the Academy of Natural Sciences of Philadelphia* 115: 127-152.

Schnack, E.J., J.L. Fasano & F.I. Isla 1982. The evolution of Mar Chiquita lagoon, Province of Buenos Aires, Argentina. En D.J. Colquhoun (ed.) *Holocene Sea-Level Fluctuations: Magnitudes and Causes:* 143-155. IGCP 61, University of South Carolina, Columbia, SC.

Trebino, L. G. 1987. Geomorfología y evolución de la costa en los alrededores de San Blas, Provincia de Buenos Aires. *Revista Asociación Geológica Argentina* 42(1-2): 9-22. Buenos Aires.

Violante, R.A. & G. Parker 1992. Estratigrafía y rasgos evolutivos del Pleistoceno medio a superior-Holoceno en la llanura costera de la región de Faro Querandí (Provincia de Buenos Aires). *Revista Asociación Geológica Argentina* 47(2): 215-227. Buenos Aires.

Weiler, N.E. 1983. Rasgos morfológicos evolutivos del sector costanero comprendido entre Bahía Verde e Isla Gaviota, Provincia de Buenos Aires. *Revista Asociación Geológica Argentina* 38(3-4): 392-404. Buenos Aires.

Weiler, N.E. 1993. Niveles marinos del Pleistoceno tardío y Holoceno en Bahía Anegada, Provincia de Buenos Aires. *Revista Asociación Geológica Argentina* 48 (3-4): 207-216. Buenos Aires.

Witte, L. 1916. Estudios geológicos de la región de San Blas. *Revista Museo de La Plata*, XXIV, 99 pp. La Plata.